Progress in Physical Chemistry Volume 2

Materials Dominated by their Interfaces

by
Rolf Hempelmann (Ed.)

Series Editor: Helmut Baumgärtel

Oldenbourg Verlag München Wien

Preface to the Series

Like all natural sciences Physical Chemistry also is strongly affected by a development which leads to an inflation of information details which cannot be thoroughly covered in its depth by the regular journal world. Therefore the editors and the publisher of Zeitschrift für Physikalische Chemie have decided to provide a platform for scientists to present their research and results on a broader basis. Thus the book series “Progress in Physical Chemistry” has been created.

The first volume of the series is devoted to “Different Aspects of Intermolecular Interaction”. It collects most important reviews on the topic published in Zeitschrift für Physicalische Chemie between 2004 and 2006.

Volume 2 covers the results of the Collaborative Research Center (SFB) 277 of the German Research Foundation (DFG)

Helmut Baumgärtel
(Series Editor)

Bibliografische Information der Deutschen Nationalbibliothek

Die Deutsche Nationalbibliothek verzeichnet diese Publikation in der Deutschen Nationalbibliografie; detaillierte bibliografische Daten sind im Internet über <http://dnb.d-nb.de> abrufbar.

Rosenheimer Straße 145, D-81671 München
Telefon: (089) 45051-0
oldenbourg.de

Lektorat: Angelika Sperlich
Herstellung: Tina Bonertz
Coverentwurf: Kochan & Partner, München
Gedruckt auf säure- und chlorfreiem Papier
Gesamtherstellung: Memminger MedienCentrum, Memmingen

ISBN 978-3-486-58629-9

Preface

Volume 2 of "Progress in Physical Chemistry" comprises 13 invited mini-review articles elucidating the progress achieved in the chemistry, physics and materials science of interface-dominated condensed matter. All of these papers were written by previous members of the Collaborative Research Centre (SFB) 277 *"Grenzflächenbestimmte Materialien: Synthese, Charakterisierung, Physikalische Eigenschaften, Modelle"* (Materials Dominated by their Interfaces: Synthesis, Characterization, Physical Properties, Models), which was funded by the German Research Foundation (Deutsche Forschungsgemeinschaft) during the period from 1994 to 2006. The contributions have also been published in the special topical issue of "Zeitschrift für Physikalische Chemie" (2008/2–3).

Interfaces or grain boundaries play an important role for special properties (mechanical, magnetic, transport) even in coarse-grained solids, although the fraction of atoms in or adjacent to grain boundaries is vanishingly small. The scientific interest of SFB 277, however, was directed to materials with a large fraction of those atoms, *i.e.*, to nanostructured materials and to nanoparticles.

Materials of this kind, although not under this label, have been used by mankind since ever: we mention the pigments in stone-age cave paintings, the old-Egyptian ink and the medieval gold-ruby church windows. Colloidal chemistry, comprising nanoparticles, was prospering in the 1920s, summarized in Wolfgang Ostwald's book "Die Welt der vernachlässigten Dimensionen". The interest of the physicists was roused and a renaissance of nanoscience was brought about by Richard Feynman 1959 in his famous lecture "There is plenty of room at the bottom". For nano-analytics 1984 is the key year: Ernst Ruska, Gerd Karl Binnig and Heinrich Rohrer received the Physics Noble Price for the electron microscope and the scanning tunnelling microscope, respectively.

Nowadays, *nano* is a word that has left the science reservation and entered the public consciousness. The field of nanoscience with its promise of amazing nanotechnologies is one of today's most challenging, exciting, multi-disciplinary and competitive fields. Prerequisite for all achievements and promises is a synthetic access to nanomaterials, a fundamental understanding of the underlying atomic mechanisms of nanocrystal formation and a basic knowledge of chemical, physical and engineering effects. Actually, these were the main objectives and the working field of SFB 277 as will be outlined in the following.

The **synthesis** by inert gas condensation (IGC), originally developed in Saarbrücken by Gleiter and coworkers, has meanwhile been replaced by a technique with streaming reactive gas in an aerosol reactor (chemical vapour synthesis) with substantially higher preparation rate than IGC. Magnetic-field-assisted self-assembly of iron nanoparticles in the streaming gas phase results in the formation of nanorods. Via MO-CVD (metal organic chemical vapour deposition), starting from heterometallic alkoxide precursor molecules ("single-source-precursor") ternary and even quarternary metal oxide layers with precise stoichiometry as well as metal/metal oxide composite layers have been prepared. A speciality of SFB 277 is the electrochemical route to nanomaterials with narrow grain/particle size distribution. With this approach, called pulsed electrodeposition, nanostructured metals with substantially improved mechanical properties as well as noble metal nanoparticles with enhanced electrocatalytical activity have been made available.

For the **structural characterization** preferentially electron microscopy and scattering methods (X-ray and neutron diffraction and small angle scattering) have been applied, as also is the case for the less-common hyperfine interaction technique perturbed angular correlation which yields valuable local structural information on an atomic scale. By means of high resolution photoemission spectroscopy clean and rare gas covered noble metal surfaces, a special form of interfaces, have been investigated.

Fundamentally novel and unexpected **physical effects** emerge when the sample size and/or some characteristic length scale of the microstructure become comparable or even smaller than a property-dependent characteristic length scale such as the exchange length (magnetism) or the distance of distortion bands (mechanics). This holds also true for cryoliquids condensed in mesoporous hosts with respect to the thermodynamics and structure of spatially mesoscale confined systems. For mechanical studies on this length scale, *i.e.*, on the scale of the microstructure of the nanomaterials, scanning probe methods are ideally suited. Atomic force acoustic microscopy as a newly developed technique has been validated by experiments on nanocrystalline metals. Nanoindentation combined with imaging within an atomic force microscope elucidated deformation mechanisms specific to the nanometre scale. A combinatorial workflow has been developed to produce mixed oxided thin films and screen them for piezoelectric properties by atomic force microscopy in the ultrasonic piezo-mode.

Theoretical modelling has been used in shedding light on the structure and energetics of metallic and semiconductor nanoparticles (by various density-functional methods) and on the scaling behaviour of the energy barriers in spatially confined systems (via Monte-Carlo simulations). A more extended review is given on the spatio-temporal equilibrium and non-equilibrium behaviour of ferrofluids, covering experimental results and results from various theoretical methods ranging from purely analytical calculations to full numerical approaches.

The transfer of this and other accumulated knowledge from academia into industrial applications, which in a few cases is already taking place, remains one of the great challenges for the near future.

At this place, we would like to thank the numerous reviewers for their valuable comments and the editorial staff of *Zeitschrift für Physikalische Chemie* for their extremely helpful support. The funding of the German Research Foundation over the full project time and the continued monitoring and advice by its representatives Dr. Dieter Funk and Dr. Barbara Jörg are gratefully acknowledged by all members and co-workers of this Collaborative Research Centre.

Saarbrücken, December 2007

Rolf Hempelmann

Magnetic Nanorods: Genesis, Self-Organization and Applications

By Rainer Birringer*, Helmut Wolf, Christian Lang, Andreas Tschöpe, and Andreas Michels

Universität des Saarlandes, Technische Physik, Postfach 151150, D-66041 Saarbrücken, Germany

Magnetic Nanostructures / Self-Organization / Magnetoviscosity / Nanorods

Magnetic-field-assisted self-assembly of magnetic dipole moment carrying iron nanoparticles is shown to result in the formation of magnetic and mechanically stiff nanoscale rods. The cooperative behavior of an ensemble of such rods and bundles thereof exhibits self-organized pattern formation on different length scales. Pattern formation on large length scales reveals great similarity with physical systems undergoing spinodal decomposition. Possible applications for dipolar magnetic nanorods in the field of perpendicular storage media are highlighted. We discuss an aerosol-synthesis-route allowing to prepare ferrofluids (FF) with shape-anisotropic particles constituting the magnetic phase immersed in the nonmagnetic carrier fluid. These so-called nanorod FF unveil a two orders of magnitude increase of viscosity enforced by an applied field of 10 mT even at shear rates larger than 10^{-2} s. This raises prospects for applications in microfluidics and MEMS.

1. Introduction

Recent progress on magnetism and magnetic materials have made nanostructures a particularly interesting class of materials for both scientific and nanostructured technological explorations. Studies on subjects such as interlayer coupling, giant magnetoresistance, colossal and tunnelling magnetoresistance, exchange bias, half-metallic ferromagnets, spin injection and current-induced switching have eventually led to the exciting possibility of utilizing electron spin for information processing or spintronics [1, 2]. The materials used for either putting the ideas discussed above into practice but also for assembly of prototypical devices merely belong to the class of layered materials-thin film

* Corresponding author. E-mail: birringer@nano.uni-saarland.de

Z. Phys. Chem. **222** (2008) 229–255

and multilayer systems. However, not only nanoscale layered materials manifest fascinating and novel properties, a story of scientific and technological success is evidenced by the large scale application of nanostructured soft and hard magnetic materials [3], which belong to the category of nanostructured bulk materials. Nanoscale particulate composites represent a scenario that covers a broad diversity of materials from biomaterials to superspin glasses, and granular magnetic materials serving as model systems for the study of ageing, rejuvenation and memory phenomena [4]. The fabrication of ordered nanostrutures – essentially utilizing dot-like or rod-like nanometer-sized objects as building units – encompasses a variety of preparation techniques such as different lithography and nanoimprint techniques, copolymer nanolithography, as well as self-assembled and nanotemplate-assisted growth of nanostructures [5].

Fundamentally, novel and unexpected physical effects will emerge when the sample size and/or some characteristic length scale of microstructure becomes comparable or even smaller than a property-dependent characteristic length scale such as the carrier mean free path, various magnetic exchange lengths, or the spin diffusion length [6]. This rationale has been the driving force for the development of increasingly sophisticated materials, as discussed above. Dimensional analysis of the well-known micromagnetic free energy yields two fundamental length scales. The wall-width parameter $\delta_0 = (A/K)^{1/2}$ (A being the exchange stiffness, and K the magnetocrystalline anisotropy) that is varying from a few nm in extremely hard materials to more than 100 nm in very soft materials. It determines the thickness of magnetic domain walls and the spatial response of the magnetization to local perturbations. Secondly, the exchange length $l_{\text{ex}} = \sqrt{A/(\mu_0 M_s^2)}$ (M_s: saturation magnetization) which describes the competition between interatomic exchange and magnetostatic self-interactions; experimentally found exchange lengths are on the order of 10 nm [7]. There are other relevant length scales such as the critical single-domain size [6] $R_{\text{SD}} = 36\sqrt{AK}/(\mu_0 M_s^2)$ of a sphere ($R_{\text{SD}}(\text{Fe}) \cong 6$ nm, $R_{\text{SD}}(\text{Co}) \cong 34$ nm, $R_{\text{SD}}(\text{SmCo}_5) \cong 764$ nm) and domain sizes specifying the real space scale above which inhomogeneous magnetization states extend throughout the material. In magnets with random magnetic anisotropy, the magnetic correlation length $L \propto D^{-3}$ yields a measure of how many grains of size D exhibit strong intergranular correlations [8]. A length scale, particularly relevant to storage media, specifying the size below which the material appears no longer ferromagnetic is the superparamagnetic limit characterized by $KV \approx k_B T$, where V denotes the particle volume; we note that the observation of superparamagnetic behaviour depends on the time frame of measurement. In the superparamagnetic state, thermal fluctuations are sufficiently strong to overcome an effective energy barrier, basically set by the magnetocrystalline- and shape anisotropy. Finally, there are two limiting length scales in magnetism, the Fermi wavelength reflecting intra-atomic electrostatic interactions, and the correlation length describing the physical behaviour near critical points *e.g.* at the ferro- to paramagnetic phase transition [6].

A recent upsurge in research on magnetic nanostructures puts much emphasis on understanding of ordered arrangements or patterned arrays of building units which are confined in two (rod-like) or three dimensions (dot-like) [5]. Advanced sample growth and patterning techniques allows one to control and modify shape and size of building blocks and likewise the geometry of patterned arrays of such entities. As a result, highly ordered arrangements of magnetic building units serve now as candidate systems for studying and understanding mesoscopic effects generated by the simultaneous manifestation of geometric confinement, physical proximity and presence of structural order. Altogether, they govern the interplay between the relevant physical length scales (as discussed above) and the scales and measures related to size, shape and degrees of imperfections of the building blocks as well as the geometry/topology of their spatial arrangements.

In studying arrays of nanometer-sized building units, it has been found that in particular *proximity* in many cases interferes with the physical length scale associated with long-ranged interaction phenomena, in this context represented by the dipolar interaction which decays $\propto 1/r^3$. In a simplistic approach, assuming building blocks which have single domain character and therefore may be approximated by magnetic dipoles, the dipolar field generated by an individual building unit and its interaction with nearest neighbor dipoles may yield a rather intuitive picture of the magnetic configuration and some of the properties of the overall array. However, whenever the distance between the building units becomes comparable to their size, the simple dipolar approach fails. A more realistic estimate of the effects of dipolar fields in arrays of nanostructures requires a minimization of the magnetostatic self-energy, which in turn yields the stable magnetic configuration of the array. This is a formidable task, since the local values of the magnetization $\mathbf{M}(\mathbf{r})$ and the local demagnetizing field $\mathbf{H}_\mathrm{d}(\mathbf{r})$ are to be integrated over all sample volume. Nevertheless, many new phenomena have been discovered and are essentially related to proximity effects in patterned arrays of nanostructures, such as enhancement or decrease of coercivity, changes of the switching field width, shape- and interaction-induced anisotropies as well as shifts of the spin-wave excitation frequencies to name just the most prominent effects [5].

In summary, it seems fair to conclude that the dipolar interaction manifests a constitutive force in determining and controlling the properties of ordered arrays of nanostructures, artificially grown or prepared by utilizing *e.g.* lithography-, nanoimprinting- or template-assisted growth techniques. In what follows, we are going to discuss a complementary phenomenon focussing on the *self-organized* pattern formation of magnetic nanoparticles which serve as building units. Again, the dipolar interaction has been identified as the force of creation, thus controlling and dominating the morphogenesis. As discussed above, we will treat the individual nanoparticles in a simplistic manner as magnetic dipoles. In chapter two, their single-particle behavior is discussed in terms of the Stoner–Wohlfahrt model, moreover, the limiting cases of two interacting dipoles and an ensemble of otherwise unconfined dipoles are addressed. Chapter three deals with the understanding and modelling of the evolution of regular patterns observed during morphogenesis. Potential ramifications to nanofabrication in the area of high-density perpendicular storage media are highlighted. In chapter four, we discuss how the magnetoviscous properties of ferrofluids can be enhanced by tailoring the shape-anisotropy of the magnetic phase dispersed in a nonmagnetic carrier fluid.

2. Magnetic properties of nanoparticles

2.1 The Stoner–Wohlfarth model

The Stoner–Wohlfarth (SW) model [9] is one of the simplest approaches for describing hysteresis effects in magnetism. It can be used to describe the magnetization curve of a collection of noninteracting identical single domain particles. In the simplest case, the magnetic energy E of an individual particle has contributions from uniaxial magnetic anisotropy and from the interaction of the particles' magnetization vector $\mathbf{M}$ with an applied magnetic field $\mathbf{H}$,

$$E = KV\sin^2\theta - \mu_0 V \mathbf{H}\cdot\mathbf{M}\,. \tag{1}$$

Here, K denotes the effective uniaxial anisotropy constant, V is the volume of the particle, θ represents the angle between the anisotropy axis and $\mathbf{M}$, and μ_0 is the permeability of free space. Within the SW model, the particles are assumed to be sufficiently small so that strong exchange forces hold all the atomic spins parallel and, consequently, any space dependence of the magnetization can be neglected. On the other hand, the particles should be large enough to avoid temperature effects (see below).

For given materials parameters and for a given orientation of the anisotropy axis with respect to the applied field, the magnetization curve of a single particle can be obtained by minimizing Eq. (1) with respect to the angle between $\mathbf{M}$ and $\mathbf{H}$. When the uniaxial anisotropy axis is parallel to $\mathbf{H}$, the hysteresis loop has the well-known rectangular shape, with a remanence M_r which is equal to the saturation magnetization M_s and with a coercivity (switching field) $H_c^0 = 2K/(\mu_0 M_s)$. On the other hand, no hysteretic behavior results for a perpendicular orientation between the anisotropy axis and $\mathbf{H}$. For a sample of randomly oriented noninteracting uniaxial particles, $M_r = \frac{1}{2} M_s$ and $H_c \cong 0.479\, H_c^0$ [9].

2.2 Magnetic anisotropy

The origin of the assumed uniaxial magnetic anisotropy in Eq. (1) may be due to the dipole–dipole interaction (shape anisotropy) and/or due to the combined effect of the crystal-field and spin–orbit interaction (magnetocrystalline anisotropy). Other types of anisotropy which may be of relevance in the context of nanoparticle magnetism such as surface anisotropy or exchange anisotropy will be ignored in the following discussion. The general expression for the magnetostatic shape-anisotropy term E_M of a homogeneously magnetized ellipsoid is [10]

$$E_M = \frac{1}{2}\mu_0 V \left(N_x M_x^2 + N_y M_y^2 + N_z M_z^2\right), \tag{2}$$

where N_x, N_y, and N_z denote the demagnetizing factors along the principal (*e.g.*, Cartesian) axes of the ellipsoid, and M_x, M_y, and M_z are the respective Cartesian components of the magnetization. For the particular case of an elongated cigar-shaped prolate spheroid (with two equal axes), Eq. (2) simplifies to

$$E_M = \frac{1}{2}\mu_0 V(N_x - N_z) M_s^2 \sin^2\theta + \text{const.}, \tag{3}$$

where we have assumed $N_x = N_y$ and $M_z = M_s \cos\theta$. Equation (3) reveals the same functional form as the first term on the right hand side of Eq. (1), with the effective shape-anisotropy constant $K = K_s = \frac{1}{2}\mu_0 (N_x - N_z) M_s^2$.

The magnetocrystalline anisotropy energy E_K expresses the dependency of the magnetic energy of a particle on the orientation of its magnetization **M** relative to the crystal axes. For a particle with uniaxial magnetocrystalline anisotropy, E_K is generally expanded in terms of the angle θ between **M** and the anisotropy axis, which is *e.g.* the *c*-axis in hcp Co,

$$E_K = K_1 V \sin^2\theta + K_2 V \sin^4\theta + \dots, \tag{4}$$

where K_1 and K_2 are functions of temperature. The use of cubic magnetocrystalline anisotropy in Eq. (1) instead of or in addition to uniaxial anisotropy as well as the incorporation of higher-order (K_2) terms is straightforward.

It is of interest to compare the magnitude of magnetocrystalline anisotropy with the strength of shape anisotropy. At room temperature, the magnetocrystalline anisotropy constants of single crystals of Ni, Fe, and Co are, respectively, $-5.7 \times 10^3\,\text{J/m}^3$, $4.7 \times 10^4\,\text{J/m}^3$, and $4.5 \times 10^5\,\text{J/m}^3$ [11]. By using the saturation-magnetization values $M_s(\text{Ni}) = 500\,\text{kA/m}$, $M_s(\text{Fe}) = 1700\,\text{kA/m}$, and $M_s(\text{Co}) = 1400\,\text{kA/m}$, we see that for an elongated particle (with $N_x = 1/2$ and $N_z = 0$), the corresponding values for the shape-anisotropy constants, $7.9 \times 10^4\,\text{J/m}^3$, $9.1 \times 10^5\,\text{J/m}^3$, and $6.2 \times 10^5\,\text{J/m}^3$, are (at least for Ni and Fe) much larger than their magnetocrystalline counterparts.

2.3 Pole avoidance

The dipole–dipole interaction is usually the most complicated energy term, since it requires evaluating the interaction of each atomic magnetic moment

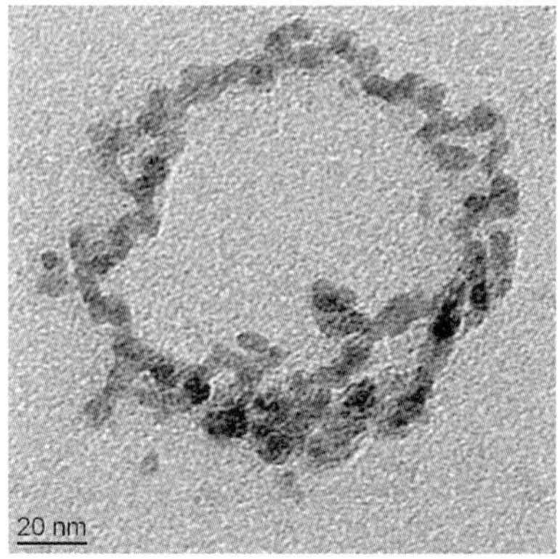

Fig. 1. Bright-field TEM image of an ensemble of Fe nanoparticles. The pole avoidance principle supports the formation of flux-closure patterns (ring structures).

with the magnetic field which is created by all the other dipoles. The complexity of the problem can easily be grasped by considering the well-known textbook case of two isolated dipoles. Equation (5) describes the mutual interaction energy $E_{1,2}$ between two magnetic moments $\mathbf{m}_1$ and $\mathbf{m}_2$ separated by a distance $\mathbf{r}_{1,2}$ ($r_{1,2} = |\mathbf{r}_{1,2}|$),

$$E_{1,2} = \frac{\mu_0}{4\pi}\left[\frac{\mathbf{m}_1 \cdot \mathbf{m}_2}{r_{1,2}^3} - \frac{3\,(\mathbf{m}_1 \cdot \mathbf{r}_{1,2})\,(\mathbf{m}_{1,2} \cdot \mathbf{r}_{1,2})}{r_{1,2}^5}\right]. \tag{5}$$

When the dipoles are forced to align in parallel with the moment direction along the separation distance, then the parallel "head-to-tail" configuration is the one with the lowest energy. On the other hand, when the collinear dipoles are constrained to have their moments oriented perpendicularly to the separation distance, antiparallel alignment is energetically favoured. This relatively simple example suggests that in realistic experimental situations, where usually a large number of particles, wires, *etc.* are involved, the problem of finding the ground state of the magnetostatic energy is rather complicated, and it is difficult to predict whether the effect of the dipolar interaction is *e.g.* to prefer parallel or antiparallel alignment of magnetic moments.

Nevertheless, based on the energetics, some general statements on the qualitative behaviour of a large number of building blocks can be made [10]. The magnetostatic energy is positive definite, $E_M \geq 0$, and prefers magnetization structures with no so-called surface and volume poles, *i.e.*, configurations where both terms $\mathbf{n} \cdot \mathbf{M}$ (surface charges) and $-\nabla \cdot \mathbf{M}$ (volume charges) vanish; here, $\mathbf{n}$ denotes a unit vector which is normal to the surface of the magnetic specimen. The above statement is known as the pole-avoidance principle. The most prominent manisfestation of this rationale is the formation of so-called flux-closure patterns which are characterized by the tendency to minimize E_M. As an example, Fig. 1 shows the realization of pole avoidance on the nanoscale: eventually, a chain of sufficiently large nanoparticles will bend and form a closed ring structure, once the number of particles reaches a certain critical value.

2.4 Magnetic relaxation

As indicated in Sect. 1, the SW model is valid only at zero temperature. At nonzero temperatures, the thermal energy $k_B T$, where k_B denotes the Boltzmann constant, may induce transitions between local energy minima. When the associated relaxation time is comparable to the characteristic time of the measurement, which in the case of dc magnetization is typically a few seconds, time-dependent effects become important, since the magnetic properties change during measurement.

Within the Néel–Brown theory of superparamagnetism, the magnetic relaxation time τ_N of a particle exhibiting uniaxial magnetic anisotropy is given by [10]

$$\tau_N = \tau_0 \exp\left(\frac{KV}{k_B T}\right) , \tag{6}$$

where $\tau_0 \cong 10^{-9} - 10^{-10}$ s, and KV denotes the anisotropy-energy barrier. At zero applied magnetic field, Eq. (6) describes the average time it takes for the magnetic moment to jump between the two equivalent energy minima at $\theta = 0$ and $\theta = \pi$ (compare Eq. (1)).

When the particles are dispersed in a fluid of viscosity η, an additional relaxation mechanism, the so-called Brownian rotational-diffusion mechanism comes into play. Brownian relaxation is characterized by a switching time

$$\tau_B = \frac{3\,V_H\,\eta}{k_B\,T} , \tag{7}$$

where V_H denotes the hydrodynamic volume of a single particle. In general, both mechanisms Néel and Brown relaxation take place, resulting in an effective relaxation time [12]

$$\tau_{eff} = \frac{\tau_N\,\tau_B}{\tau_N + \tau_B} . \tag{8}$$

3. Self-organized pattern formation of magnetic nanoparticles

3.1 Self-organization

Ordinary states of matter like a layer of fluid, a mixture of chemicals, granular or genetic material can exhibit under appropriate conditions a multitude

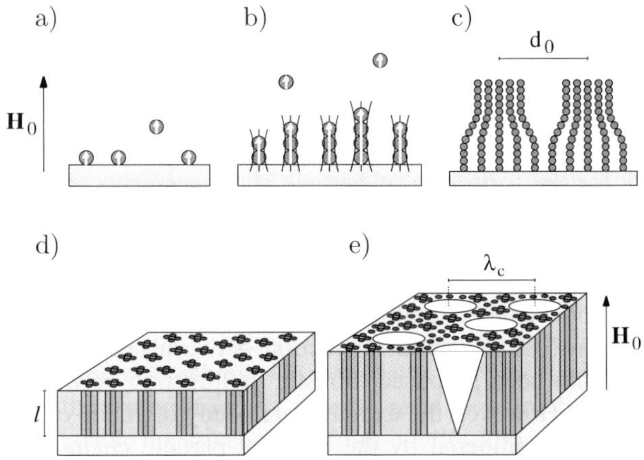

Fig. 2. Schematic representation of the evolution of form and patterns during magnetic field assisted deposition of iron nanoparticles on a non-magnetic substrate. (a) The particles' magnetic moment (white arrows) becomes aligned parallel to the external field H_0 during deposition. (b) Magnetic field assisted self-assembly of individual dipolar rods (for more details see Sect. 3.3). The build-up of field gradients is symbolized by flux-lines. (c) Further growth of individual rods is accompanied by a change in growth mode resulting in the formation of bundles of rods. Each bundle having a typical diameter of 1 μm constitutes a network of a few hundred rods. The center-to-center distance d_0 between nearest neighbor bundles is about 5 μm. (d) The system of bundles carrying the magnetic flux gives now the appearance of a carpet with homogeneous bundle density $n_0 \propto 1/d_0^2$, and bundle length ℓ. The consequences of bundle–bundle interactions are summarized in (e). (e) When the carpet grows beyond a critical length of $\ell_c \approx 40\,\mu\text{m}$, the bundles start bending so that funnel-like pores of zero-bundle-density with a typical distance of $\lambda_c \approx 300\,\mu\text{m}$ develop. Since the number of bundles is conserved, pore formation is associated with an enhancement of bundle density in the space around pores.

ual bundles gives rise to an elastic restoring force, which limits the amplitude of fluctuations to infinitesimally small values. With increasing bundle length ℓ, the system of homogeneous bundle density is driven into non-equilibrium and becomes sensitive to infinitesimal density fluctuations. Eventually, symmetry becomes broken by spontaneously developing pores. It appears that individual pores emerge by tilting bundles radially outward relative to the pore-center axis thereby generating bundle density gradients in the space between pores. We describe the geometry of a pore by introducing the tilt angle $a(\mathbf{r}, t)$ that measures the obliquity of bundles relative to its pore-center axis. The lateral arrangement of pores across the substrate exhibits hexagonal symmetry implying that the relevant constituents of the system must be able to communicate in a highly cooperative manner over distances which are comparable to the system size of tens of millimeters.

The fundamental observation, also manifesting a mode of collective behavior, concerns the tilt angle a: tilt angles of individual pore forming bundles

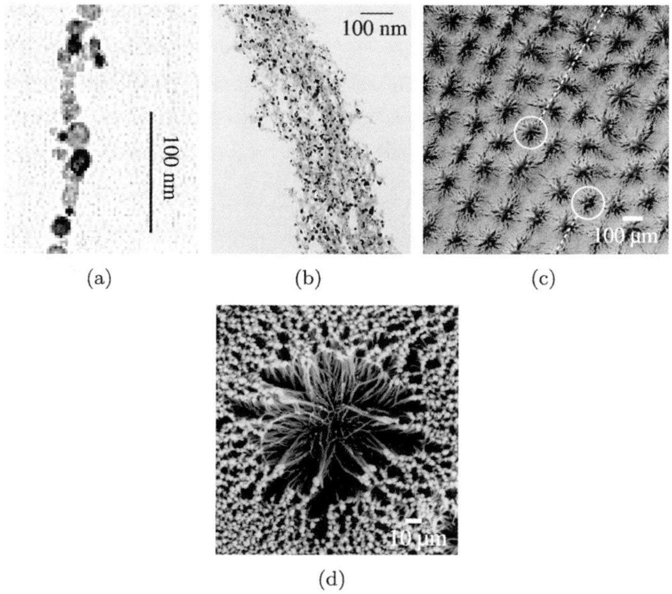

Fig. 3. Pattern formation in a 40 µm thick layer of iron aerosol particles deposited in a homogeneous magnetic field. (a) TEM bright-field micrograph of an isolated Fe-particle chain, obtained by ultrasonic disintegration of a Fe-chain bundle. (b) TEM bright-field micrograph of a single Fe-chain bundle, grown on a TEM grid at a low deposition rate, illustrating entanglement of individual particle chains at large ℓ. (c) Top-view SEM micrograph of a 40 µm thick layer of Fe-chain bundles manifesting self-organized density modulations in the form of blossom-like pores arranged in a pattern exhibiting (local) hexagonal symmetry. The core regions of edge dislocations in the ordered arrangement of pores are marked by circles. (d) Magnified single blossom-like pore of (c). All SEM and TEM viewgraphs were taken *ex situ*. Exposure of the as prepared structures and patterns to ambient atmosphere resulted in thin oxide layer protecting the Fe-core structures [24].

cancel out when averaged over the total substrate area A so causing $\langle a \rangle_A = 0$. In other words, the tilt angle is a globally conserved quantity and, therefore, must satisfy a continuity equation

$$\langle a \rangle_A = 0 \quad \Rightarrow \quad \frac{\partial}{\partial t} a = -\nabla \cdot \mathbf{J}\,, \tag{9}$$

where $\mathbf{J}$ is interpreted as a tilt-angle current in full analogy to diffusive flux of matter in a system of conserved particle number. In such systems the current is proportional to the local chemical potential gradient and the chemical potential itself is given by the change of free energy density with respect to particle density. Following this guideline, we assume a linear response ansatz for $\mathbf{J}$

$$\mathbf{J} = -M\nabla\left(\frac{\delta w}{\delta a}\right), \tag{10}$$

here M describes the mobility of the bundles' top ends when bending occurs, and w is an appropriate energy functional density.

The experimental evidence $\langle a \rangle_A = 0$ together with the fundamental Eqs. (9) and (10) are the basic ingredients we will use to model and understand pore-pattern formation. Combining Eqs. (9) and (10) yields the nonlinear partial differential equation of the problem $\partial a/\partial t = +M\nabla^2(\delta w[a]/\delta a)$, the solution of which is a formidable task. We prefer simplifying this equation based on assumptions motivated by experimental observations. As we observe bundles being always tilted towards increasing bundle density and never orthogonal to the density gradients, likewise, any chiral character of bundle arrangement is missing, it seems feasible to describe structure formation by a one dimensional approach. Beyond the critical length where the symmetry of the homogeneous system becomes broken, it appears appropriate to orient our coordinate axis to be parallel to the substrate and parallel to the wave vector k representing the observed density modulation of the pattern. The remaining task is to construct an expression for the total energy density w.

We consider three relevant energy terms entering w. Modelling individual bundles by magnetic dipolar rods, the magnetic dipolar interaction energy density w_{m} is an obvious contribution to w. Taking only nearest-neighbor interactions into account, w_{m} is obtained by a Taylor expansion about the homogeneous bundle configuration $a \equiv 0$, and constitutes a nonlinear functional of $a(x)$ and its gradients. The energy associated with bending of bundles with solid-like rigidity is modelled by the bending energy density of a single solid beam, w_{el}, multiplied by the local bundle density $n(x) = n_0 + \delta n(x) = n_0 - n_0\ell\,(\partial_x a)$ with n_0 being the bundle density of the homogeneous carpet; overall, we obtain $w_{\mathrm{el}} \propto a^2/\ell$. Similarly, the interaction energy density w_{h} with the external field H_0 is modelled by that of a single dipole with H_0 multiplied by the local bundle density $n(x)$; $w_{\mathrm{h}} = -\mu_0\, n(x)\, m H_0\, \cos[a(x)]$ where m is approximated by the saturation magnetization M_{s} of a single rod ($m \propto M_{\mathrm{s}}\ell$). Finally, we obtain for the total energy density

$$w[a(x)] = c_0 + c_2 a^2 + c_4 a^4 + \eta(\partial_x a) + \kappa(\partial_x a)^2 + \cdots\,, \tag{11}$$

where c_0, c_2, c_4, η, κ are coefficients representing the individual energy densities or combinations thereof.

The one dimensional version of the combined Eqs. (9) and (10) inserted with the expression for w reads after linearization

$$\frac{\partial}{\partial t} a(x,t) = (c_2\, M)\,\frac{d^2 a}{dx^2} - (2\kappa M)\,\frac{d^4 a}{dx^4}\,. \tag{12}$$

Equation (12) reflects full analogy to the linearized Cahn–Hilliard equation (C–H eq.) of spinodal decomposition [25–27]. The coefficient

$$c_2 = \frac{ED^4}{40\ell d_0^2} - \frac{3\mu_0 m^2}{d_0^5} + \frac{\mu_0 m H_0}{2d_0^2} \tag{13}$$

is positive for short rod length, thus, being dominated in the early growth stage by mechanical stiffness of the rods (w_{el}). However, finally c_2 becomes negative due to the power law increase of the dipolar interaction term ($m^2 \propto \ell^2$) with increasing bundle length. The coefficient κ remains positive at all length scales. As a result, our analysis clearly identifies ℓ as the external control parameter, and the critical length ℓ_c marking the onset of instability is determined by $c_2(\ell_c) = 0$. The finding that the external field contribution becomes eventually dominated by the dipolar interaction term justifies our notion of magnetic-dipole-interaction-driven spinodal decomposition as a plausible mechanism which triggers spontaneous growth of long-wavelength perturbations in a spatially uniform state. In its essence, the emergence of pores is caused by repulsive dipolar forces; attractive dipolar interactions are effectively suppressed by the missing lateral and horizontal degrees of freedom of movement of individual bundles relative to the substrate plane. Nevertheless, the externally applied field is a necessary prerequisite for pattern formation since, first, H_0 forces the magnetization of the incoming particles to orient parallel to it, and second, the interaction of already deposited particles with H_0 generates the field gradients which stimulate the growth of linear chains and bundles.

We conclude by discussing the central quantitative results obtained from the C–H eq. (12) which is solved by

$$a(x,t) = a_k \exp\{\mathcal{A}(k)\,t\} \exp(ikx)\,, \qquad (14)$$

with $\mathcal{A}(k) = -M(c_2 k^2 + 2\kappa k^4)$, and k denotes the wave number of a bundle-density-modulation wave. The solution consists of a spatially periodic term and a term describing the time evolution of the amplitude. If $\mathcal{A}(k) < 0\,\forall k$, all modes k will decay. For $\ell > \ell_c$, modes exist for which $\mathcal{A}(k) > 0$, consequently, the amplitudes grow exponentially. We assume that the fastest growing mode k_c, determined by $\mathrm{d}\mathcal{A}/\mathrm{d}k = 0$, selects the characteristic length $\lambda_c = 2\pi/k_c$ of the evolving pattern. We find $\lambda_c \approx 1000\,\mu\mathrm{m}$ and $\ell_c \approx 47\,\mu\mathrm{m}$ which agree, despite the restrictive assumptions, fairly good with the experimentally found values: $\lambda_c^{ex} \approx 300\,\mu\mathrm{m}$ and $\ell_c^{ex} \approx 40\,\mu\mathrm{m}$. Strong evidence for the validity of our simplistic analysis of pore-pattern formation has been obtained by experimentally studying the external field dependence of λ_c. Based on the condition $\mathrm{d}\mathcal{A}/\mathrm{d}k = 0$, theory predicts $k_c^2 = |c_2|/(4\kappa)$ resulting in $\lambda_c \propto H_0^{-1/2}$– if H_0 is not too small. This power law dependence of λ_c as a function of the externally applied magnetic field H_0 has been reasonably well verified by experiment, see Fig. 4. The observed influence of the field confirms that the pattern formation occurs during deposition and not during removal of the applied field at the end of the deposition process. Finally, we notice that the observed non-equilibrium pore-pattern formation belongs to a subclass of dynamical systems, namely, gradient or potential systems characterized by $\partial a/\partial t = \delta F[a]/\delta a$ [14]. For such sys-

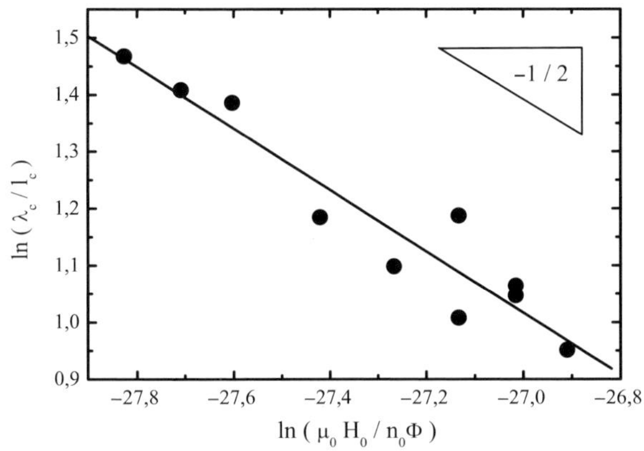

Fig. 4. Scaling behavior of critical density modulation wavelength λ_c. Normalized average nearest neighbor distance λ_c/ℓ_c of blossom-like pores plotted as a function of the normalized applied magnetic field $\mu_0 H_0/n_0\Phi$ at fixed bundle length ℓ_c on a log–log-scale (n_0: homogeneous bundle density; μ_0: permeability constant; Φ: magnetic flux through a single bundle). The solid line represents a least-squares fit to the data points. From the slope, we determine the power law $\lambda_c \propto H_0^{-0.54\pm0.06}$.

tems the dynamics consists of relaxation towards the minimum in the energy functional F in agreement with the behavior predicted by the C–H equation.

3.4 Possible applications

Apart from pattern formation, magnetic field directed self-assembly may have potential ramifications to nanofabrication, especially in the area of high-density perpendicular magnetic storage media [28–30]. We suggest a manufacturing process that combines lithographically defined generation of ordered structures, which can be created over large areas, with magnetic field directed self-assembly of nanometer-sized static magnetic moment carrying aerosol particles [31]. In fact, our idea relies upon the manufacturing of a lithographically defined master substrate encoding the desired width of and distance between magnetic tracks, as well as enabling to adapt rotationally symmetric arrangement of tracks – the standard read and write geometry. As shown in the previous paragraphs, magnetic field assisted deposition of magnetic nanoparticles on such a substrate would result in rod-like particle arrangements decorating with high preference the lithographically defined magnetic tracks of the master substrate. Diameter of rods, and shape anisotropy – which may become much larger than the intrinsic anisotropies particularly in nanostrucutres –, as well as material selection would allow to tailor the key parameter of magnetic recording media [5, 32, 33]. Figure 5 illustrates the two-step process suggested to prepare nanoscale perpendicular magnetic storage media.

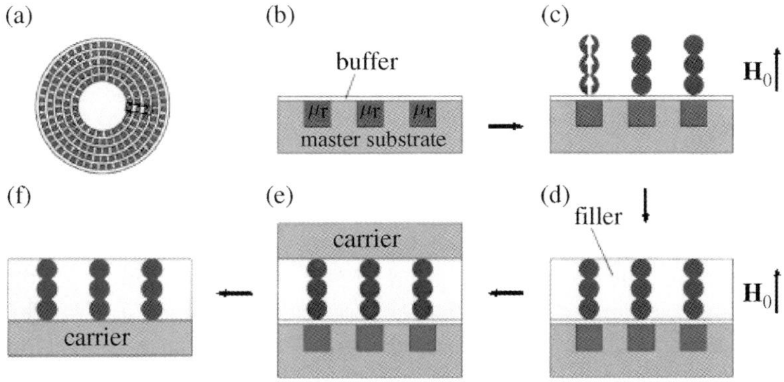

Fig. 5. Manufacturing process of a perpendicular magnetic storage medium by magnetic field directed self-assembly. (a) Schematic topview of a perpendicular magnetic storage medium master substrate which is a prerequisite for high throughput mass production. A cross-sectional view of the marked rectangular frame in (a) is shown in (b). (b) Lithographically generated high permeability (high μ_r) tracks. The buffer layer enables to lift off the deposit from the master substrate. In what follows, we describe the steps necessary to build-up the magnetic storage medium: Step 1 (c), magnetic field assisted deposition of magnetic aerosol particles forming nanorods by sintering. Step 2 (d), mechanical stabilisation and chemical passivation by a filler material. Step 3 (e), attachment of a carrier substrate on the magnetic storage layer. Step 4 (f), perpendicular magnetic storage medium lifted off from the master substrate.

4. Ferrofluids (FF)

Interestingly, the impact of tailor-made magnetic nanoparticles has not only influenced the development of novel solid state nanostructures and devices but also the realm of complex fluids has benefitted from the availability of an ever growing variety of magnetic nanoparticles. Ferrofluids (FF) [34] manifest a scientifically and commercially relevant example of a smart complex fluid, consisting of magnetic nanoparticles, typically spherically-shaped, with diameters on the order of 10 nm and with adsorbed dispersant layers of approximately 1–2 nm thickness, enabling to stably immerse such particles in nonmagnetic carrier fluids. They decorate magnetic field gradients, which gave raise to a number of commercially viable applications of FF such as rotary seals, stepper motor dampers or heat transfer fluids in audio speakers. Owing to the possibilities for designing and modifying the physical and chemical properties of both the suspending fluid and the particles as well as the adsorbed dispersant layer, FF are gaining popularity in several novel applications, functioning as magneto-responsive colloidal extractants, targeted magnetic carriers, *in vivo* imaging agents and in magnetic cell-sorting schemes [35]. Those nanoparticles represent either permanent magnets, or behave as superparamagnetic particles. Suspended in carrier fluids, they undergo rotational and translational Brownian motion. The free rotation of magnetically hard particles in a carrier liquid exhibits a magnetization versus field behavior resembling that of superparamagnetic particles, yet is of different origin. It is the fascinating interplay between magnetic and rheological properties that has generated much cross-disciplinary interest in the nature and behavior of FF [36].

A central issue of fundamental research on FF is magnetoviscosity, discovered by McTague and Rosensweig *et al.* [37, 38], delineating that in the presence of a homogeneous magnetic field the magnetic torque on magnetically hard particles hampers or even prevents them from free rotation in a shear flow as dictated by the fluid dynamics of the carrier liquid. By taking into account the magnetic and mechanical torques as well as the thermally induced Brownian motion, Shliomis [39] delivered the first theoretical description of the magnetoviscous effect for a FF in the dilute regime of magnetically non-interacting, spherical particles. As a result, the viscosity rises with increasing magnetic field H and the relative viscosity $\Delta\eta/\eta_0$ reaches a plateau-value of $1.5\,\phi$, where $\eta_0 = \eta(H = 0)$, and ϕ is the nanoparticle volume fraction. Recently, Odenbach and coworkers [40] investigated the magnetoviscous behavior of highly concentrated ($\phi \gtrsim 0.1$), commercially available FF and found changes of viscosity more than a hundred times larger than the values expected from theory [41]. It has been verified that this dramatic increase of viscosity is due to dipolar-interaction-driven formation of chain-like aggregates of individual particles. A second observation manifests that with increasing shear rates, $\dot{\gamma} \gtrsim 10\,\mathrm{s}^{-1}$, the giant viscosity enhancement disappears, which has been attributed to the fragile nature of the dipolar coupled chain-like aggregates; in fact, it has been suggested that they become disintegrated by shear forces supplied by the carrier fluid [42].

Irrespective of fundamental research issues, the envisioned novel applications or the diversity of commercially viable applications of FF, the currently used FF manifest a common characteristic: namely, the *spherical/equiaxed* shape of the constituent magnetic particles. Regarding their magnetic properties, the effect of strength of the magnetocrystalline anisotropy energy on the magnetoviscosity has been theoretically investigated quite recently [43]. We propose to explore a plausible, yet heretofore scarcely noticed degree of freedom: the *shape-anisotropy* of the constituent particles, which can be tailored to enhance various FF-properties. In particular, we will prepare and characterize rod-like magnetic nanoparticles, which behave as Brownian particles when immersed in a carrier fluid, therefore, named: nanorod-FF. The rod-like instead of sphere-like geometry of the magnetic phase, first, suppresses superparamagnetism as well as typically overcomes the crystalline anisotropy of the particles due to the large shape anisotropy energy and, secondly, anisometric permanent magnets with solid-like rigidity are expected to resist shear flow and hence contribute to magnetoviscosity far more effectively than equiaxed particles.

4.1 Nanorod ferrofluids: basic ideas

Based on a Fokker–Planck equation, Martsenyuk, Raikher and Shliomis [44] derived an expression for the magnetic-field-dependent change of viscosity, $\Delta\eta$, of a dispersion of non-interacting nanoparticles exposed to homogeneous shear flow, given as:

$$\Delta\eta = \frac{1}{4}\phi\,\tau_{\perp} M(B)\,B\,. \tag{15}$$

The applied magnetic field is labeled by B, the transversal relaxation time $\tau_{\perp}$ translates into the Brownian relaxation time τ_{B} *via* $\tau_{\perp} = 2\tau_{\mathrm{B}}/(2+\xi L(\xi))$ where L denotes the Langevin function $L(\xi) = \coth(\xi) - 1/\xi$ with its argument $\xi = mB/k_{\mathrm{B}}T$, and m stands for the magnetic moment of an individual particle. $M(B)$ indicates the field-dependent magnetization of the FF which for small shear flow according to [44] may be approximated by $M(B) = M_{\mathrm{s}}\,L(\xi)$ and $M_{\mathrm{s}} = m^{\mathrm{tot}}/V_{\mathrm{mag}}$ is the saturation magnetization of the sample. The overall number of particles making up the magnetic volume V_{mag} and occupying the volume fraction ϕ of the FF have the overall magnetic moment m^{tot}. It seems instructive to consider the limiting case of saturation, where a sufficiently large applied field, B_{sat}, assures that the Langevin function approaches a quasi-plateau-value characterized by the saturation magnetization M_{s}. It is straightforward to derive that by approaching saturation at fixed temperature T, say room temperature, Eq. (15) simplifies to

$$\Delta\eta(B_{\mathrm{sat}}) \propto \phi\,\tau_{\mathrm{B}}\,. \tag{16}$$

The particle geometry is hidden in $\tau_{\mathrm{B}} = (2D_n)^{-1}$, with D_n being the coefficient of rotational diffusion. As shown by Brenner [45, 46], D_n of solids of revolution – here we consider long slender axisymmetric bodies, *e.g.* cylinders – can be written as:

$$\frac{1}{D_n} = \frac{K(n)}{D_{\mathrm{s}}}\,, \tag{17}$$

where $K(n)$ is a shape function, n specifies the aspect ratio of bodies and $D_{\mathrm{s}} = k_{\mathrm{B}}\,T/6\,\eta\,V^{\mathrm{s}}_{\mathrm{hyd}}$ is the Stokes rotational diffusion coefficient of a sphere with equal hydrodynamic volume, $V^{\mathrm{s}}_{\mathrm{hyd}} \equiv V^{\mathrm{cyl}}_{\mathrm{hyd}}$. Mathematical expressions for shape functions are given in references [45–49]. For the sake of argument, it seems sufficient to display the functional form of $K(n)$ as shown in Fig. (6) for prolate ellipsoids and circular cylinders [45, 46], where

$$K(n) = \frac{2}{9}\left[\frac{n^2}{\ln n} + \frac{n^2}{(\ln n)^2}(\ln 2 - 1) + \frac{3\,L}{8\pi}\right] \tag{18}$$

denotes the shape function of a blunt-ended cylinder with aspect ratio n. The numerical constant L depends critically on the precise shape of the blunt ends

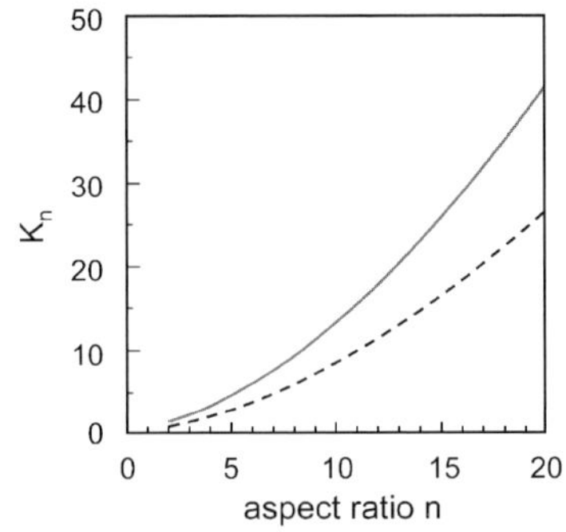

Fig. 6. Shape function $K(n)$. Functional behavior of $K(n)$ *versus* aspect ratio n for cylinders (dashed line) and prolate ellipsoids (full line).

of the body; $L \approx 5.45$ has been determined for a circular cylinder. Requiring $V^{s}_{hyd} \equiv V^{cyl}_{hyd}(n)$ and substituting Eq. (18) into Eq. (17) yields an expression for the rotational diffusion coefficient and, hence, also for τ_B which depends apart from temperature on the material parameters η, M_s, ϕ and n. Let us consider spheres ($n = 1$) as reference system, and further assume that all material parameters except n are fixed at constant temperature. Then, increasing n at conserved V^{s}_{hyd} consequently contributes with a nonlinear enhancement of τ_B and thus $\Delta\eta$ due to the effect of shape anisotropy on D_n imparted by $K(n)$. We note that $\tau_B = (1/2\pi)\, f_B^{-1}$ is inversely proportional to the Brownian relaxation frequency f_B, which is quantitatively accessible by means of ac-susceptibility measurements. This allows one to deduce an overall aspect-ratio of anisometric objects, probed in the state of being immersed in a carrier fluid, based on the relation

$$\frac{1}{f_B(n)} = \frac{\pi K(n)}{D_s}\,, \tag{19}$$

provided that all other material parameters are prescribed. Finally, since $f_B(n)$ renders $\tau_\perp$ becoming n-dependent, the magnetic-field-dependent viscosity of nanorod FF [Eq. (15)] is suggested to crucially rely upon n, which, when exceeding 10, may give rise to a *giant magnetoviscous effect*. We will scrutinize this expectation by performing field and frequency dependent viscosity measurements on a nanorod FF in a squeeze flow rheometer.

4.2 Synthesis and characterization

Iron nanorods were generated in a modified aerosol flow condenser [50] as shown in Fig. 7. Before evaporation, the reactor was evacuated to 10^{-6} mbar. Then the high vacuum pump was disconnected and replaced by a rotary vane

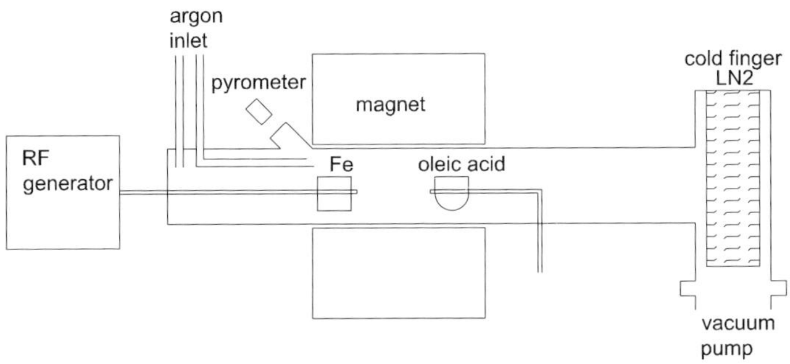

Fig. 7. Magnetic-field-assisted aerosol synthesis of Fe-nanorods. Iron particles are grown in an Ar gas flux and become linearly aligned in the presence of a homogeneous magnetic field. Sintering promotes rigid nanorod formation. Oleic acid is evaporated as a surfactant to avoid agglomeration between as-grown nanorods. Further downstream the oleic-acid-coated nanorods are collected on a cold finger held at 77 K.

pump in order to establish a continuous Argon gas flow of 40 sccm reading a pressure of 20 mbar. Elemental iron was heated and evaporated from a tungsten crucible into the Argon gas flux using an induction coil operated at a frequency of 700 kHz and a maximum power output of 20 kW. Supersaturation of Fe in the Ar gas leads to nucleation of Fe clusters. Subsequent growth by condensation and coagulation results in Fe-nanoparticles which further downstream tend to form, depending on their local density in the gas phase, particle aggregates – dimers, trimers, *etc.* Since the evaporator was located in the homogeneous field regime at the entrance of a Halbach-type permanent magnet, linear aggregation of individual nanoparticles becomes strongly favored. First, due to the magnetic torque experienced by all nanoparticles, which make their magnetic moments rotate parallel to the field direction. Whenever the magnetic torque couples sufficiently strong to an easy axis and/or the shape of the particles, it impels a rotation of these particles themselves. Secondly, owing to the dipolar interaction which favors a head-to-tail arrangement of particulate dipoles among all other possible configurations, the emergence of linear particle aggregates seems highly reasonable. The enhanced temperature extending along the Halbach magnet enables sintering [18] of these anisometric particle aggregates along their way downstream. It is the reduction of the particles free surface energy that yields the driving force for sintering, which causes neck formation between neighboring particles resulting in rod-like magnetic objects with solid-like rigidity, concisely called nanorods.

In order to prevent agglomeration of already grown nanorods in the Ar gas, we coated individual nanorods by evaporation of oleic acid using a thermally heated quartz glass crucible placed at the exit of the Halbach magnet. Variation of the distance between the rf coil and the surfactant source allowed us to roughly control the rod length: a shorter rod length appeared when decreasing this distance. The aerosol jet exiting the Halbach magnet is directed toward a LN_2 cooled cold finger where the oleic acid coated nanorods become deposited due to thermophoresis. After turning off the evaporation sources, the Ar gas flux was still maintained until the cold finger temperature approached room temperature. Then the Ar gas pressure was increased to ambient pressure and *via* a leak valve Ar gas was slowly exchanged by air in order to passivate the surface of the oleic acid coated nanorods by a thin oxide layer thus avoiding full oxidation of available iron and rendering the samples appropriate for handling under atmospheric conditions. The soot-like material collected on the cold finger was rinsed off using heptane and accumulated in a container. Adding a polar solvent (acetone) makes the oleic-acid-coated particles to agglomerate, which allows then to separate them from the solvent by using a permanent magnet. Repeated redispersing of the coated particles in heptane and separation eventually leads to a stable FF. The volume concentration of the magnetic phase immersed in the carrier fluid was determined by magnetometry yielding the magnetic moment which is directly proportional to the volume of magnetic objects whenever they become saturated through an applied field. We preferred to prepare FF with a volume concentration of magnetic phase of 1–2 vol.-‰ to assure that such FF agree with the limiting case of a dilute, non-interacting system. The structure and morphology of nanorods has been investigated by means of TEM (Jeol JEM 2010).

As discussed in the theory section, the Brownian relaxation frequency f_B, which enables access to the aspect ratio, is related to the complex ac-susceptibility $\chi(f)$. In fact, f_B coincides with the frequency f correlated to the maximum value of the imaginary part $\chi''(f)$ of $\chi(f) = \chi'(f) + i\chi''(f)$. Using a Quantum Design PPMS (Model 6000), we measured $\chi(f, T)$ and also obtained $\chi''(f, T)$ by means of sinusoidally varying the applied magnetic field between ± 10 Oe at frequencies covering the range between 10–10^4 Hz.

The magnetoviscous behavior of nanorod FF has been investigated by utilizing a PAV (piezo-axial-vibrator) rheometer [51]. The specimen chamber has a disc-shaped geometry of 20 mm diameter, a height of 5 μm and can be operated under closed-system conditions. The bottom part of the chamber is designed to act as a piezo-driven membrane vibrating at frequency f and an amplitude of 5 nm in this way forcing the nanorod FF to sustain dynamical squeeze-flow. The frequency-dependent viscosity is deduced from the complex spring constant of the vibrating system [52]. Using a solenoid oriented with its axis perpendicular to the driving membrane, a magnetic dc-field of variable strength can be applied to the specimen chamber.

4.3 Results and discussion

In Fig. 8 we show a representative bright field micrograph of an assemblage of nanorods prepared by the magnetic-field-assisted aerosol process discussed in the previous paragraph. A statistical evaluation of the geometry of objects was, in general, hampered by the circumstance that they exhibited a strong tendency

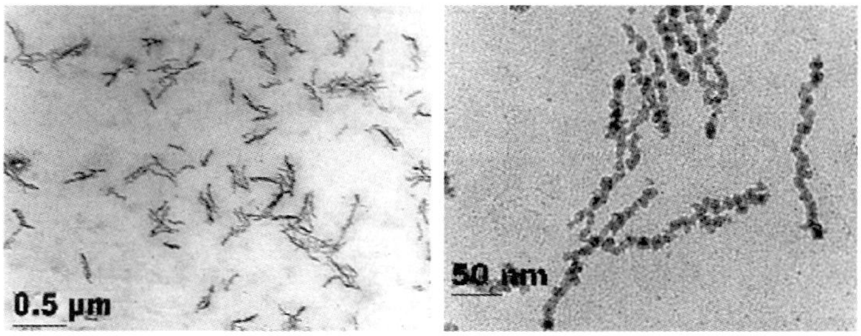

(a) (b)

Fig. 8. Morphology of nanorods. Representative TEM bright field micrographs of aerosol-grown nanorods at low (a) and high (b) magnifications.

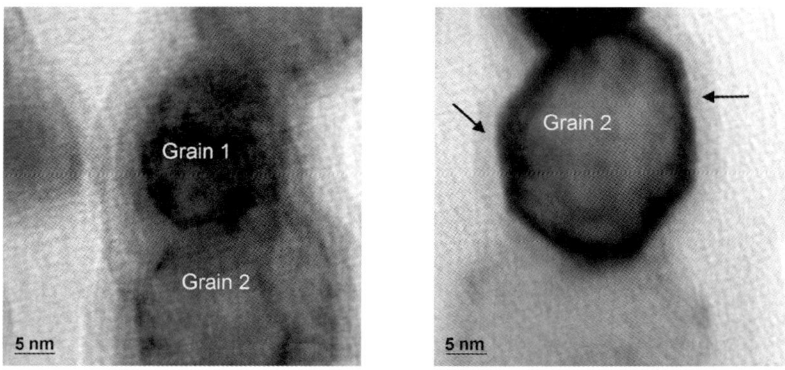

(a) (b)

Fig. 9. Neck formation and iron-oxide core shell structure. (a) TEM dark field image displaying intergrain neck formation. (b) TEM dark field micrograph revealing the existence of a thin oxide layer, marked by arrows, surrounding the Fe-core of grain 2.

to agglomerate during preparation of the TEM specimen, so causing projection artefacts which in turn gave rise to high uncertainty of the values determined for their individual length. This problem is due to attractive magnetic dipolar forces between individual rods which tend to impel agglomeration as the solvent evaporates. As a consequence, we restrict our analysis to estimating the overall diameter of nanorods; an effective aspect ratio of nanorods will be obtained from the ac-susceptibility as shown in the next paragraph. TEM micrographs displayed in Fig. 9 reveal clear evidence for neck formation between individual particles – so providing the nanorods with rigidity – and the formation of a thin oxide layer protecting the core volume of particles/nanorods from further oxidation. The ac-susceptibility measurements allowed to deduce the imaginary part $\chi''(f)$ of the complex susceptibility of a nanorod FF. The

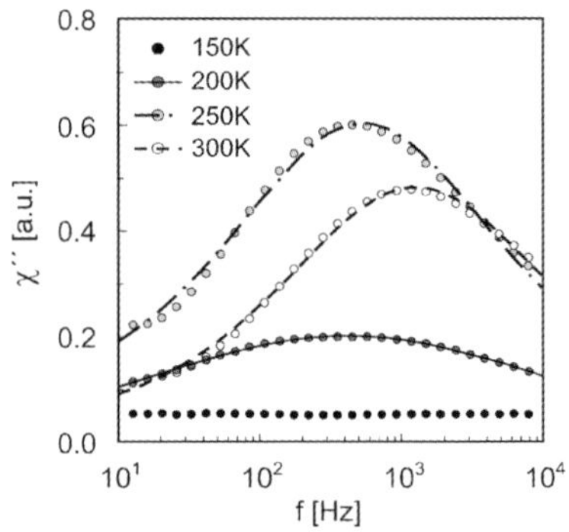

Fig. 10. Ac-susceptibility of a nanorod FF. Imaginary part of the susceptibility of a nanorod FF at 4 different temperatures and $H_{dc} = 0$. The different lines represent fits to the data points using Eq. (20). The data set taken at $T = 150\,\text{K}$ displays the signal of a frozen fluid, indicating that Brownian relaxation is missing. Brownian relaxation frequencies can be deduced from the maxima of the upper three curves.

experimental data shown in Fig. 10 were analyzed in terms of the Cole–Cole model [53]

$$\chi''(2\pi f) = \frac{\chi_0 - \chi_\infty}{2}\left(\frac{\sin\left(\frac{\pi\beta}{2}\right)}{\cosh\left(\beta \ln\left(2\pi f \tau_{CC}\right)\right) + \cos\left(\frac{\pi\beta}{2}\right)}\right), \tag{20}$$

which could fit all data points and yielded the frequencies f_B associated with the maximum values of $\chi''(f)$. The Cole–Cole model represents an empirical modification of the Debye-relaxation. The parameter β has been introduced as a measure of the width of a size distribution of individual objects exposed to the ac-field and τ_{CC} denotes the average relaxation time. According to Eq. (19), determination of the effective aspect ratio, n_{eff}, based on experimental data for $f_B(n)$ relies upon a reasonable approximation to the geometry of nanorods. For the sake of simplicity, we assume a cylinder geometry so implying $V_{hyd}^{nrod}(n) \approx V_{hyd}^{cyl}(n)$. The hydrodynamic nanorod volume is modeled by

$$V_{hyd}^{nrod} \approx (\pi(r+t)^2)(n2r+2t)\,, \tag{21}$$

where t is a measure of the thickness of the surfactant layer (oleic acid) bonded to the solid core-structure of the nanorods, r is the rod diameter and n is the aspect ratio representing the number of spheres of diameter $2r$ that would fit into the hollow cylinder with an inner radius of r and an outer radius of $(r + t)$. Concerning further data analysis, we fixed $r \simeq 5\,\text{nm}$ based on TEM results, and supposed t is reasonably approximated by a monolayer of oleic acid having a thickness of 1–1.5 nm. We note that the data shown in Fig. 10 suggest that the onset of Brownian relaxation should correlate with the melting transition of the carrier fluid; below the melting point, Brownian relaxation

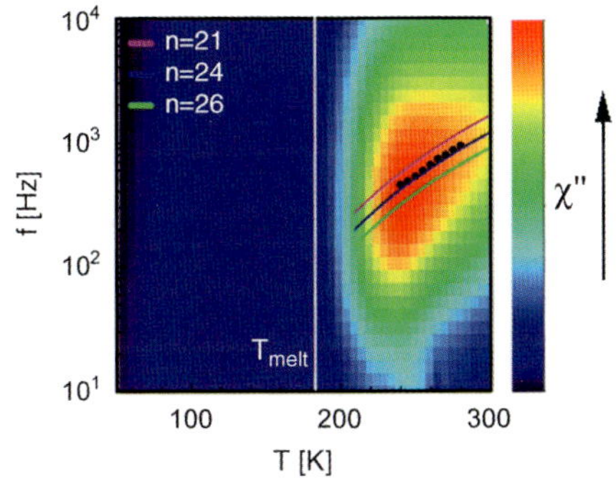

Fig. 11. Aspect ratio deduced from the temperature-dependent ac-susceptibility. Pseudo-3d chart of the imaginary susceptibility $\chi''(f, T)$ of a nanorod-FF at $H_{dc} = 0$ [see Fig. 10]. Black dots (•) indicate maximum values of χ'' characteristic of Brownian relaxation frequencies. Full lines show computed relaxation frequencies for different aspect ratios n and a nanorod diameter $d = 11$ nm, representing an average diameter obtained from TEM for the investigated nanorod-FF. The vertical line at $T = 183$ K indicates the melting point of the carrier fluid heptane.

should cease. Any signal in this regime would originate from superparamagnetic (Neelian) relaxation [54]. In order to verify this conjecture, we measured frequency- *and* temperature-dependent ac-susceptibilities to obtain $\chi''(f, T)$ as shown in Fig. 11. It is straightforward now to analyze the maximum of $\chi''(f, T)$ in terms of $f_B(n, T)$ where n is treated as a parameter, implying that V_{hyd}^{s} in D_s is substituted by $V_{hyd}^{nrod}(n)$. The temperature dependence of f_B reveals the temperature dependence of the rotational diffusion coefficient. The Brownian relaxation frequencies derived from the Cole–Cole analysis are depicted as black dots. The full lines are fits to the data points using Eq. (19). Superparamagnetic relaxation below T_M seems absent. For integer values of n, it is shown that $n = 24$ is a fairly good approximation to the data points. In this context n is characteristic of an effective aspect ratio, n_{eff}, ensemble-averaged over an *a priori* unknown distribution of geometrical/morphological degrees of freedom of nanorods probed by $\chi''(f, T)$. We found that n does not depend on temperature and frequency. What is more, we also verified that n is independent of the magnitude of a superimposed magnetic dc-field. Dc-bias-fields were successively applied up to a magnitude of 20 mT and, as suggested by Waldron *et al.* [55], we observed for all fields applied only a shift of the relaxation frequency, however, no concomitant change of n. The fairly large value of n seems to indicate that among all others the largest nanorods may control the overall behavior of the complex fluid. We conclude by pointing out that all observations made so far imply that nanorods manifest a pronounced aspect ratio in conjunction with solid-like rigidity.

The effect of shape anisotropy on the magnetoviscous behavior of nanorod-FF has been investigated by utilizing a PAV (piezo-axial-vibrator) rheome-

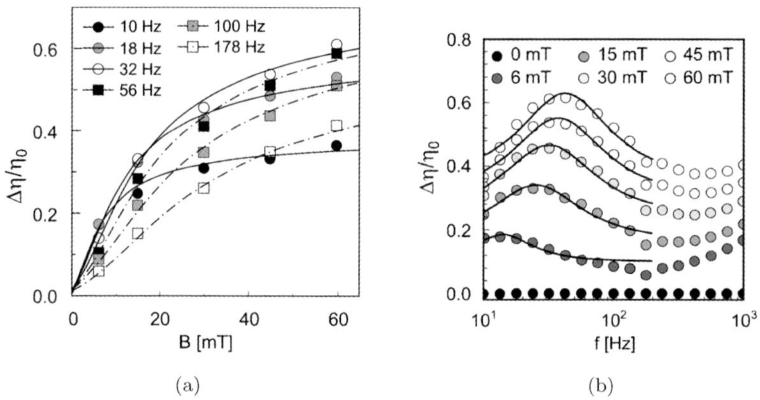

Fig. 12. Magnetoviscous effect of a nanorod FF. (a) Magnetic field dependent viscosity change as obtained from a PAV viscosimeter at different frequencies. Lines represent least squares fits to the experimental data based on Eq. (15). The volume concentration of the magnetic phase corresponds to $\phi = 0.9$ vol.-‰. (b) Frequency-dependent viscosity change for different magnetic fields. Full lines reflect least squares fits based on Eq. (20). The shift of maxima to higher frequencies with increasing field strength indicates a deviation from pure orientational relaxation behavior.

ter [51]. In Fig. 12 we display the relative change of viscosity of a nanorod-FF as a function of applied field and frequency, respectively. With the volume fraction ϕ of the magnetic phase determined by magnetometry to be 0.9 vol.-‰ – assuming that all magnetic material is made up of Fe – we analyzed the data points shown in Fig. 12a in terms of Eq. (15). The full lines depict a least squares fit with $\tau_\perp$ being the central fit parameter. Fig. 12b reveals a magnetic field-dependent shift of maxima of $\Delta\eta/\eta_0$ to higher frequencies with increasing field strength; the fit to the data points is based on the Cole–Cole formula, Eq. (20). This shift supports the idea that, apart from mere orientational relaxation behavior of the nanorods, resonance behavior becomes superimposed with the applied homogeneous magnetic field acting as the restoring torque.

Finally, we compare the magnetoviscous effect of nanorod FF with conventional commercially available FF. In order to make this comparison convincing, we normalize the relative change of viscosity to the volume fraction ϕ of the magnetic phase immersed in the carrier fluid. In Fig. 13 the quantity $\Delta\eta/\phi\eta_0$ for a conventional FF, investigated by Odenbach and coworkers, is contrasted with the nanorod FF. A direct comparison of the two sets of data would require a relation between the shear-rate- and frequency-dependent viscosity. Although there is no general theory describing the relation between $\eta(\dot{\gamma})$ and $\eta(f)$, for simple liquids the Cox–Merz theorem postulates in its rigorous version $\eta(\dot{\gamma}) = |\eta^*(f)|$, where $\eta^*(f)$ is the complex viscosity; for complex fluids this identity brakes down. Recently, Chae *et al.* [56] investigated the validity of the Cox–Merz rule for magnetic dispersions. They could demonstrate

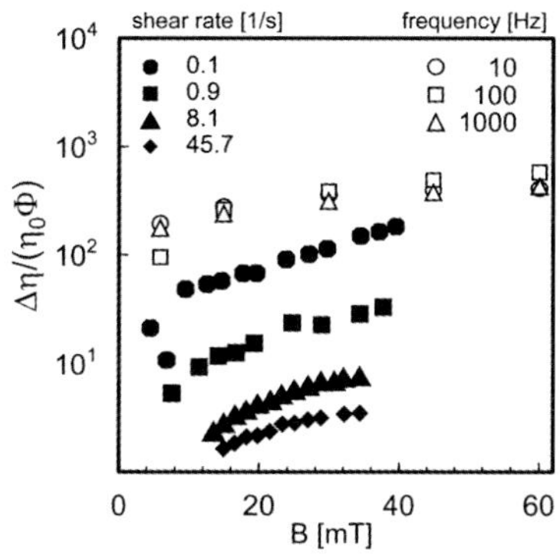

Fig. 13. Shear-rate- and frequency-dependent magnetoviscous effect of a conventional and a nanorod-FF. Full black symbols denote the conventional, highly concentrated FF and open symbols characterize the nanorod-FF. The relative viscosity change is normalized to the volume fraction ϕ of magnetic phase. Frequency values can be translated into shear rates through division by 10 for direct comparison.

that the slopes of the viscosity and the complex viscosity as a function of $\dot{\gamma}$ or f, respectively, are nearly identical. When they shifted the frequency by a slightly concentration dependent but otherwise constant factor $a \approx 0.1$, they found the relation $\eta(\dot{\gamma}) \approx |\eta^*(a\,f)|$ reasonably good obeyed. Therefore, by dividing the frequencies given in Fig. 13 by a factor of 10, the viscosity data of the nanorod FF and the conventional FF are expected to be comparable on a one to one basis. A certainly striking feature is that $\Delta\eta/\phi\eta_0$ of conventional FF exhibits a decrease over two orders of magnitude with increasing shear rate, whereas, only little change was detected for nanorod FF. In detail, at a shear rate of $10^{-1}\,\mathrm{s}^{-1}$ and raising field strength, $\Delta\eta/\phi\eta_0$ of conventional FF reveals a restrained increase (concave-up), in contrary, the nanorod FF is characterized by a pronounced enhancement up to virtually saturation behavior within less than 10 mT of applied field. Such a behavior is in agreement with the idea that in conventional FF, with the given high volume fraction $\phi \gtrsim 0.1$ of magnetic phase, raising field strength triggers chain formation of dipolar interacting nanoparticles, and growing chain length goes along with an increase in $\Delta\eta/\phi\eta_0$ [40]. On the other hand, the solid-like rigidity of nanorods, which are present in the carrier fluid also at zero field, makes this complex fluid susceptible to small applied fields ($B \lesssim 10\,\mathrm{mT}$) and as a consequence results in a concomitant severe change of $\Delta\eta/\phi\eta_0$. In the shear rate or frequency regime which is of technical relevance ($\dot{\gamma}$, $f \gtrsim 10\,\mathrm{s}^{-1}$), $\Delta\eta/\phi\,\eta_0$ of the conventional FF drops by two orders of magnitude, in contrast to the nanorod-FF which yields evidence for, on that scale, a nearly frequency-independent behavior. As conjectured, the solid-like rigidity of nanorods renders them resistant against mechanical forces/torque mediated by the fluid phase. By contrast, the applied-field-enhanced build-up of dipolar coupled chain-like aggregates in conventional FF results in shear unstable objects, which become basically disintegrated into their individual building blocks whenever the shear rate exceeds $10\,s^{-1}$.

In summary, we succeeded in preparing a nanorod-FF, a complex fluid made up of shape-anisotropic magnetic particles – called nanorods – immersed in a nonmagnetic carrier fluid. These anisometric permanent dipoles have solid-like rigidity and behave as Brownian particles. Nanorod-FF manifest a giant magnetoviscous effect, particularly, in the low field regime ($\lesssim 10\,mT$) and at measured frequencies/shear rates up to $10^3/10^2\,s^{-1}$. Therefore, nanorod-FF recommend themselves as a novel class of complex fluids with a major potential for technical applications. Since the magnetoviscous effect can be "switched on" at low magnetic fields as well as in the regime of technical relevant shear rates, applications in the areas of microfluidics and MEMS and/or NEMS appear plausible.

Acknowledgement

We are grateful to P. Pechold (Ulm University) for having supported us with the PAV measurements. We thank D. Junk for having made TEM results available to us.

References

1. D. Sellmyer and R. Skomski (Eds.), *Advanced Magnetic Nanostructures*. Springer, New York (2006).
2. S. D. Bader, Rev. Mod. Phys. **78** (2006) 1.
3. R. C. O'Handley, *Modern Magnetic Materials: Principles and Applications*. Wiley, New York (2000).
4. S. Sahoo, C. Binek, and W. Kleemann, Phys. Rev. B **68** (2003) 174431.
5. J. I. Martin, J. Nogues, K. Liu, J. L. Vicent, and I. K. Schuller, J. Magn. Magn. Mater. **256** (2003) 449.
6. R. Skomski, J. Phys. C Condens. Matter **15** (2003) R841.
7. C. Kittel, Rev. Mod. Phys. **21** (1949) 541.
8. G. Herzer, in *Handbook of Magnetic Materials*. K. H. J. Buschow (Ed.), Elsevier, Amsterdam (1997) vol. 10, pp. 415–62.
9. E. C. Stoner and E. P. Wohlfarth, Phil. Trans. R. Soc. London A **240** (1948) 599.
10. A. Aharoni, *Introduction to the Theory of Ferromagnetism*. Clarendon Press, Oxford, 2nd edn. (1996).
11. S. Chikazumi, *Physics of Ferromagnetism*. Clarendon Press, Oxford (1997) chap. 12.
12. P. C. Fannin, A. Slawska-Waniewska, P. Didukh, A. T. Giannitsis, and S. W. Charles, Eur. Phys. J. Appl. Phys. **17** (2002) 3.
13. P. Ball, *The Self-Made Tapestry: Pattern Formation in Nature*. Oxford University, Oxford (1999).
14. M. Cross and P. Hohenberg, Rev. Mod. Phys. **65** (1993) 851.
15. H. Haken, *Advanced Synergetics: Instability Hierarchies of Self-Organizing Systems and Devices*. Springer, Berlin (1987).
16. G. Gaeta, Phys. Rep. **189** (1990) 1.

17. S. Fauve, in *Hydrodynamics and Nonlinear Instabilities*. C. Godreche and P. Manneville (Eds.), Cambridge University, Cambridge (1998) p. 387.
18. H. Wolf, H. M. Sauer, and R. Birringer, Europhys. Lett. **60** (2002) 573.
19. K. Butter, P. Bomans, P. Frederik, G. Vroege, and A. Philipse, Nat. Mater. **2** (2003) 88.
20. P. Teixeira, J. Tavares, and M. Telo da Gama, J. Phys. C Condens. Matter **12** (2000) R411.
21. P. Jordan, Mol. Phys. **25** (1973) 961.
22. K. I. Morozov and M. I. Shliomis, J. Phys. C Condens. Matter **16** (2004) 3807.
23. W. Kingery, H. Bowen, and D. Uhlmann, *Introduction to Ceramics*. Wiley, New York (1975) chap. 10.
24. Y. Kwok, X. Zhang, B. Qin, and K. Fung, Appl. Phys. Lett. **77** (2000) 3971.
25. J. Cahn and J. Hilliard, J. Chem. Phys. **28** (1958) 258.
26. J. Cahn, Acta Metall. **9** (1961) 795.
27. K. Binder and P. Fratzl, in G. Kostorz (Ed.), *Phase Transformations in Materials*. Wiley-VCH, New York (2001) p. 470.
28. M. Plumer, J. van Eck, and D. Weller, *The Physics of Ultra-High-Density Magnetic Recording*. Wiley, New York (1975) chap. 10.
29. A. Fert and L. Piraux, J. Magn. Magn. Mater. **200** (1999) 338.
30. M. Turnbull, T. Sugimoto, and L. Thompson, *Molecule-Based Magnetic Materials*. Oxford University, Oxford (1999).
31. B. Yellen, G. Friedman, and A. Feinerman, J. Appl. Phys. **91** (2002) 8552.
32. D. J. Sellmyer, M. Zheng, and R. Skomski, J. Phys. C Condens. Matter **13** (2001) R433.
33. R. Skomski, H. Zeng, M. Zheng, and D. J. Sellmyer, Phys. Rev. B **62** (2000) 3900.
34. R. E. Rosensweig, *Ferrohydrodynamics*. Dover, New York (1985).
35. In S. Odenbach (Ed.), *Ferrofluids: Magnetically Controllable Fluids and Their Applications, Lecture Notes in Physics*. Springer, Berlin (2002).
36. B. Huke and M. Lücke, Rep. Prog. Phys. **67** (2004) 1731.
37. J. McTague, J. Colloid Interf. Sci. **51** (1969) 133.
38. R. E. Rosensweig, R. Kaiser, and G. Miskolczy, J. Chem. Phys. **29** (1969) 680.
39. M. I. Shliomis, Sov. Phys. JETP **34** (1972) 1291.
40. S. Odenbach, J. Phys. C Condens. Matter **16** (2004) R1135.
41. M. I. Shliomis, in *Ferrofluids: Magnetically Controllable Fluids and Their Applications, Lecture Notes in Physics*. S. Odenbach (Ed.), Springer, Berlin (2002) p. 85.
42. A. Y. Zubarev, J. Fleischer, and S. Odenbach, Physica A **358** (2005) 475.
43. K. Morozov, M. I. Shliomis, and M. Zahn, Phys. Rev. E **73** (2006) 066 312.
44. M. A. Martsenyuk, Y. L. Raikher, and M. I. Shliomis, Sov. Phys. JETP **38** (1974) 413.
45. H. Brenner, Int. J. Multiphase Flow **1** (1974) 195.
46. H. Brenner, J. Coll. Interf. Sci. **23** (1967) 407.
47. A. Y. Zubarev and L. Y. Iskakova, Phys. Rev. E **61** (2000) 5415.
48. P. Illg and M. Kröger, Phys. Rev. E **66** (2002) 021 501.
49. P. Illg and M. Kröger, Phys. Rev. E **67** (2003) 049 901.
50. V. Haas, R. Birringer, H. Gleiter, and S. E. Pratsinis, J. Aerosol Sci. **28** (1997) 1443.
51. L. Kirschenmann and W. Pechhold, Rheol. Acta **41** (2002) 362.
52. J. J. Crassous, R. Régisser, and M. Ballauffa, J. Rheol. **49** (2005) 851.
53. K. Cole and R. Cole, J. Chem. Phys. **9** (1941) 341.
54. L. Néel, Rev. Mod. Phys. **25** (1953) 293.
55. J. T. Waldron, Y. P. Kalmykov, and W. T. Coffey, Phys. Rev. E **49** (1994) 3976.
56. B. S. Chae, A. M. Lane, and J. M. Wiest, Rheol. Acta **40** (2001) 599.

Thermodynamic and Structural Investigations of Condensates of Small Molecules in Mesopores

By Klaus Knorr*, Patrick Huber, and Dirk Wallacher**

Technische Physik, P.O. Box 151150, D-66041 Saarbrücken, Germany

Mesopores / Phase Transition / Capillary Condensation / Capillary Sublimation

Liquids and solids consisting of small, mainly van-der-Waals interacting building blocks, such as Ar, Kr, N_2, O_2, and CO, are among the most simple systems of condensed matter imaginable. As we shall demonstrate in this microreview on our work sponsored within the Sonderforschungsbereich 277, these cryoliquids condensed in mesoporous hosts with typical mean pore diameters of 7 to 10 nm are also particularly suitable for the investigation of fundamental questions regarding the thermodynamics and structure of spatially mesoscale confined systems. An exploration of phase transitions like the vapour–liquid (capillary condensation), the vapour–solid (capillary sublimation), the liquid–solid (freezing and melting) and some solid–solid transformations of such pore condensates reveals a remarkably rich, sometimes perplexing phenomenology. We will show, however, that by experiments combining sorption isotherm, X-ray and neutron diffraction, calorimetric and optical transmission measurements, and by referring to concepts, intermediate between surface and bulk physics, a deeper understanding of the mesoscale mechanisms ultimatively responsible for this complex behaviour can indeed be accomplished, both on a qualitative and a quantitative level.

1. Introduction

Matter subject to spatial confinement has properties different from bulk matter [1–4]. To some part this is just the consequence of the geometric constraint which *e.g.* leads to the discreteness of wave vectors, but for the systems studied the interaction of the confined material with the confining interfaces is of more importance. The differences to the bulk state are most conspicuous at first

* Corresponding author. E-mail: knorr@mx.uni-saarland.de

** *Present address:* Hahn-Meitner Institute, Berlin (Germany).

Z. Phys. Chem. **222** (2008) 257–285

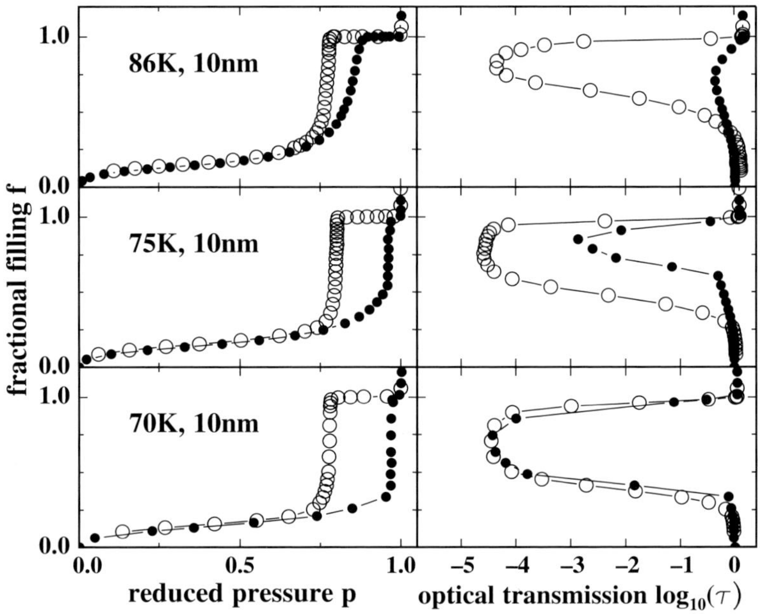

Fig. 1. Sorption isotherms (fractional filling *vs.* reduced vapour pressure) of Ar in Vycor glass. Solid symbols refer to filling/adsorption, open symbols to emptying/desorption. Also shown is the optical transmission as function of the filling fraction. At 86 K the "mobile" part of the pore filling is liquid, at 75 K and 70 K the capillary condensed part of the pore filling is solid. From [5].

order phase transitions. The sorption isotherm of Ar in a mesoporous glass at 86 K (Fig. 1, upper left panel) may serve as a first example [5]. We anticipate that the increase of the filling fraction f at a reduced vapour pressure p of about 0.8 is due to capillary condensation. This process can be regarded as the legitimate analogue of the vapour–liquid transition of the bulk system, even though strictly speaking the free energy of a system enclosed in a tubular pore does not show singularities. Note that the transition does not occur at the saturated but at a significantly reduced vapour pressure. Analogous shifts in the thermodynamic variable relative to the bulk state also occur for other transitions.

The present article deals not only with this, but also with the vapour–solid and the liquid–solid transition and some solid–solid transitions of small atoms and molecules such as Ar, Kr, N_2, O_2 or CO in mesoporous varieties of silica and silicon [5–25]. Furthermore we will investigate the structure of the solidified pore fillings. The mean pore diameters are of the order of 7 to 10 nm which corresponds to roughly 20 molecular diameters. The intermolecular and the interaction of the molecules with the substrates are of the van-der-Waals type which is weak compared to the binding energy within the substrate. Hence confinement affects the filling but the substrate can be considered inert.

2. Experimental

Vycor glass [26], the xerogels known as "controlled pore glasses", and the mesoporous substrate SBA-15, produced by means of a template of an ordered array of micelles [27], are practically pure SiO_2 with terminal Si–OH groups on the pore walls. The pores of SBA-15 are also linear and parallel with respect to one another and even form a self-assembled, hexagonal array with a high degree of 2D translational order.

The pores of Vycor and of the xerogel form an irregular network with a lot of pore junctions, the variation of the pore diameter is large, of the order of 15%. Porous Si has been produced by the electrochemical etching of (100)Si-wafers [7, 28]. Here the pores are linear and aligned perpendicular to the face of the wafer. The pore walls carry Si–H groups. The variation of the pore diameter is of the order of 10%.

The pore condensates have always been characterized by means of adsorption/desorption isotherms. The samples have been kept in a closed cell, held at a constant temperature T, and small volumetrically controlled portions of gas have been admitted sequentially to, for adsorption, or removed from the sample cell, for desorption. Thereby fractional fillings f could be prepared in a controlled and reproducible way, with the state of the pore condensate specified by three thermodynamic variables T, f, and its reduced vapour pressure p. p can be converted into the difference $\Delta\mu$ of the chemical potential with respect to the bulk reference state, $\Delta\mu = k_B T \ln(p)$ (assuming that the vapour can be treated as an ideal gas which is a good approximation for the cases studied).

The crystallographic structure of the solidified state of some samples has been examined by means of conventional wide angle X-ray powder diffractometry, working with monochromatized Cu K_α radiation emanating from a rotating anode [8]. For Kr in SBA-15 the diffraction pattern of the hexagonal pore lattice has been investigated by means of small angle X-ray diffraction [9]. Here the Bragg intensities give information on the radial partition of the condensate in the pores.

Our calorimetric set up allows both adiabatic and scanning calorimetry [10]. The merit of the scanning mode is that it can be used not only in heating but also in cooling runs and that it avoids problems of the adiabatic heat pulse technique that usually appear at hysteretic phase transitions. The cooling/heating rate was about two orders of magnitude smaller than in routine DSC runs.

For the optical transmission measurements the quantity of interest is simply the ratio τ of the transmitted and the incoming intensity of the light of a He-Ne laser at perpendicular incidence [11].

Depending on the type of experiment the dependence of the signal as function of f at constant T (isotherms) and/or as function of T at constant f (isosters) has been recorded.

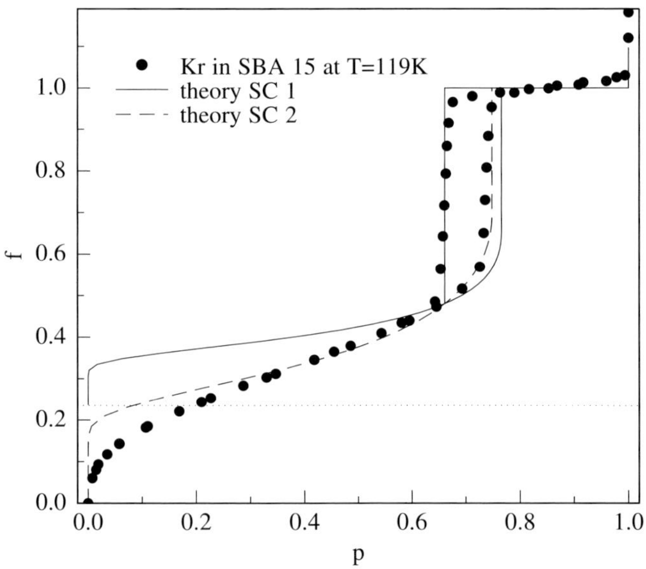

Fig. 2. Sorption isotherm of Kr in SBA-15 at 119 K. The curves shown as solid and dashed line are two attempts to fit the SC-model to the experimental data. From Ref. [9].

3. Sorption isotherms of the liquid regime, vapour–liquid transition

In the reversible low vapour pressure section of sorption isotherms (see Ar in Vycor at 86 K, Fig. 1, and Kr in SBA-15 at 119 K, Fig. 2), a film adsorbs on the pore walls and f increases gradually with p. Thinking of a homogeneous film of thickness t in a cylindrical pore of radius R, f can be translated into t *via* the geometric relation $f = 1 - r^2/R^2$ with $r + t = R$, r being the radius of the cylindrical liquid–vapour boundary. As long as t is small compared to R, the $f(p)$ isotherm of this film regime is little different from what is observed on planar substrates. At some critical value t_c and hence f_c and p_c, the cylindrical boundary becomes instable (due to a divergent amplitude of long wavelength undulation type capillary excitations of the boundary) and bridges of capillary condensate form terminated by concave menisci. See the mean field theory of Saam and Cole (SC) [29]. This transition is discontinuous, of first order, since it involves a qualitative change of the shape of the liquid–vapour boundary. The further filling process of the pores up to $f = 1$ then proceeds at constant p by condensation of vapour onto these menisci which then advance along the pore. The fact that there is hysteresis, that the filling pressure p_{ads} is higher than the emptying pressure p_{des}, is due to the fact that a metastable film can be grown to a thickness beyond t_c. According to SC, the emptying pressure p_{des} is identified with p_c, since pore emptying starts with the evaporation of the pore liquid at the pore mouths such that the menisci just retreat into the interior at a constant vapour pressure.

Sorption isotherms, often of N_2 at 77 K, are considered a standard tool for the determination of pore sizes in porous or granular media [30]. The determination is usually based on the Kelvin equation $\Delta\mu = kT \ln(p_i) = -\sigma v_m/R'$ for the vapour pressure above a curved meniscus with a radius of curvature R'. R' is related to the pore radius R *via* $R = R' \cos\theta$. v_m is the specific volume *per* molecule in the liquid, σ the surface tension of the liquid, θ the contact angle. Since the equilibrium pressure is not accessible in the experiment, it is standard practice to refer either to p_{ads} or to p_{des}. For our systems the contact angle θ is zero and in a first approximation the pore radius R can be obtained from the radius of curvature R' of the meniscus, if the thickness of the preadsorbed film at the onset of capillary condensation is properly accounted for. This t-correction is important since for pores in the 10 nm-range about one third up to one half of the pore volume is filled by film adsorption on the pore walls.

The predictions of the more advanced SC-model have been tested by X-ray small angle diffraction [9] on the isothermal adsorption/desorption of Kr in SBA-15, the matrix that is believed to be the closest realization of an ensemble of independent linear pores. The experiment supplies the intensities of the first five Bragg reflections of the hexagonal pore lattice. A slab model of the radial dependence of electron density has been fitted to the data. The primary parameter of the model is the radius r_I, which is just the radius of the cylindrical boundary between the adsorbed film and the vapour in the pore centre at low f and the boundary between the film part of the pore filling and the part in the pore centre where the capillary condensate still coexists with the vapour for $f_c < f < 1$. The f-dependence of r_I is shown in Fig. 3. The fact that r_I decreases with f below f_c (f_c is about 0.5, see Fig. 2), is constant beyond f_c, somewhat lower for adsorption than for adsorption is in qualitative agreement with the SC-theory. The unusually high value of f_c and most importantly the fact that r_I does not obey the geometric relation $f = 1 - r^2/R^2$ at low f are in disagreement with the model. The model assumes smooth walls, but the analysis of the diffraction data rather suggests that the pore walls contain niches that have to be filled first before a cylindrical liquid–vapour front can propagate inwards, towards the pore centre. The onset of capillary condensation is therefore delayed, since the instability of the surface modes requires a free surface which is only established after all such niches are filled. It is therefore by no means surprising that the model cannot fit the experimental sorption isotherm (see Fig. 2). Nevertheless the SC-model appears to grasp the essentials of adsorption and capillary condensation.

The hysteresis showing up between condensation and evaporation is the main problem in understanding sorption isotherms. Extending the reasoning of SC, the hysteresis should be absent in a pore with one open and one blind end, since not only pore emptying but also pore filling can then be achieved at vapour–liquid equilibrium, namely by the advance of the concave meniscus

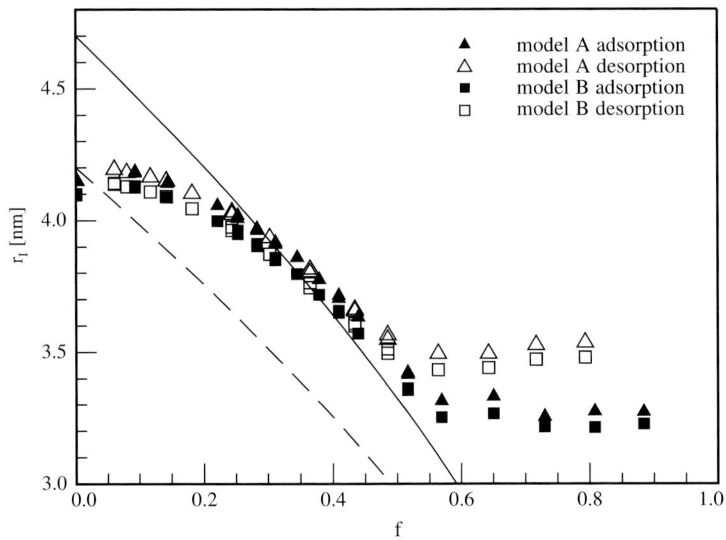

Fig. 3. The radial parameter r_I of two versions of a shell model that has been fitted to the intensities of the Bragg peaks of the pore lattice of SBA-15, partially filled with Kr at 119 K. See Fig. 2 for the corresponding sorption isotherm. r_I is the radius of the core in the pore centre that is vapour filled in the film regime of the sorption isotherm at lower fractional fillings f and that contains a coexistence of vapour and liquid in the regime of capillary condensation at higher f. The solid and the dashed lines are examples of the geometric relation between f and r_I for film growth in a cylindrical pore with smooth walls. From [9].

that already exists at low f-values in the blind end. The pore network of Vycor and the xerogels contains a lot of blind ends, nevertheless there is hysteresis (Fig. 1, left upper panel).

The approach of f and p to stationary values can be painstakingly slow in the hysteresis region [12, 31, 32], but on the other hand the asymptotic p-values are highly reproducible. One concludes that equilibrium values on f and p are not accessible in this region. Any attempt to derive exact values of the pore radius R or even pore size distributions $P(R)$ by referring to relations that assume equilibrium thermodynamics is a questionable effort. This is particularly obvious for Vycor glass where the slopes of the adsorption and the desorption branch are largely different (Fig. 1), hence a distribution $P(R)$ derived from the desorption branch is much narrower than one based on the adsorption branch, with the additional problem that for Vycor glass, the xerogel and porous Si there is no indication where the film grows ends and capillary condensation starts upon adsorption.

The different slopes in Vycor glass and similar substrates have been interpreted in terms of the concept of "invasive percolation" occurring on desorption but not on adsorption [33]. In the pore network of Vycor glass and the xerogels, the onset of pore emptying is indeed very sharp and is characterized by a very steep slope of the desorption branch for $1 > f > f_c$ (Fig. 1, $T = 86$ K). This suggests a phase-transition-like onset of desorption. The idea is that the vapour, as a non-wetting fluid, invades the pores and displaces the wetting liquid [34]. At bottle necks of the pore network this process comes to a halt and only continues after p has been lowered to a value that corresponds to the radius of the bottle neck, the relation between p and R being given by some monotonically increasing function $p = F(R)$ such as the Kelvin equation or varieties thereof. This means that on desorption there are always filled regions of the pore network that include pore segments of larger radius which would have already evaporated if they had free access to the vapour phase outside and were not blocked by liquid in the bottle necks. This situation is termed "pore blocking". For adsorption on the other hand it is argued that for any chosen p all segments with radii smaller than $R = F^{-1}(p)$ have been filled by capillary condensation whereas wider segments are still empty. Increasing p means that further condensation takes place such that not only the existing menisci move forward along the pores but that also new parcels of liquid are formed such that all menisci are at sites with the same local pore radius. In a complex pore network the approach of this state usually requires mass transport between filled regions across empty sections, by distillation processes or alternatively diffusion along the pore walls [31, 32, 35].

Figure 1 also shows the optical transmission τ recorded simultaneously with the sorption isotherm. Such data have been presented first by Page *et al.* on hexane in Vycor [36]. The present example refers to Ar in Vycor at 86 K. The glass matrix and the pore filling do not absorb visible light. The finite transmission is mainly due to scattering. τ is relatively high for the empty ($f = 0$), the completely filled as well as for the substrate just with an adsorbed film on the pore walls ($f < f_c$). There is some reduction of τ for capillary condensation along the adsorption branch ($f_c < f < 1$), but this is almost negligible compared to what is encountered in the same f-range upon desorption. Here τ is reduced by several orders of magnitude, indeed a drastic difference between adsorption and desorption.

We will argue in the following that the optical transmission data strongly supports the pore blocking concept presented in the last paragraph. Light scattering results from the mismatch of the refractive index of different regions of the probed sample volume and also depends on the size of these regions. The largest difference is clearly that between the glass matrix and the empty pores but obviously this variation occurs over distances (about some tens of nm, as given by the pore–pore distance which in turn is comparable to the pore diameter) that are much smaller than the wavelength of light. Hence there is little light scattering for $f = 0$. The same reasoning explains the relatively high transmission for $f = 1$ and even for the entire adsorption branch. For $f_c < f < 1$ regions which are completely filled coexist with "empty" regions

4. Solid state

Ar and similar molecules have reasonably large vapour pressures down to temperatures well below the triple point. Thus sorption isotherms can be easily measured even in a regime in which the pore filling is already solid [8, 13]. Two questions arise right away: Is the pore solid stable with respect to the bulk solid and if so, what is the structure of the pore solid? Sorption isotherms answer the first question. Indeed the pore solids formed by small van-der-Waals molecules are stable, they are formed at reduced vapour pressures, $p < 1$, but the hysteresis loops are shifted to higher p-values compared to the liquid regime. See Fig. 1 for Ar in Vycor glass at 70 K. This means the pore solid still profits from the attractive interaction with the substrate but has to pay an additional price for the matching to the pore geometry (see also further below). It is also for this reason that freezing/melting occurs at temperatures lower than in the bulk state (see next section).

There are a number of further differences with respect to the liquid state. The adsorption branch now shows a clear change of slope signalling the termination of the film growth on the pore walls and the onset of capillary condensation or more strictly spoken capillary sublimation (Fig. 1, 70 K-isotherm). The film on the walls is thinner than in the liquid state, f_c corresponds to the equivalent to about three monolayers, compared to about five in the liquid regime. The optical transmission is low for $f_c < f < 1$ not only for the desorption but also for the adsorption cycle, the hysteresis loop is wider with respect to both f and p (Fig. 1, 70 K), and finally – in contrast to the liquid regime – pore blocking does occur in the ink bottle pore, as is documented in the two bottom frames of Fig. 4.

Starting with the last observation, one concludes based on the discussion from above that the density fluctuations are significantly reduced, as one would expect for a solid. The low and practically identical τ values of the adsorption and the desorption branch suggest that filling and emptying proceed in an analogous fashion, namely by advancing resp. receding solid–vapour fronts resulting in a coarse partition into filled and empty regions. The nucleation of a large number of small aggregates does not occur any more along the adsorption branch. Obviously the equilibration of the vapour pressure between all empty pore segments independent of whether or not they are connected to outside can be no longer approached let alone established within the duration of the experiment. The effects of freezing/melting on the adsorbed film will be subject of the next section.

Figure 5 shows a series of X-ray diffraction patterns as obtained in parallel with an adsorption isotherm of Ar in a xerogel at 65 K. Thus the samples have been prepared by the condensation of the vapour at this temperature, by sublimation right into the solid state. At higher f-values the pattern shows the complete set of Bragg reflections expected for an fcc solid within the studied range of the scattering angle. Thus a part of the solid pore filling crys-

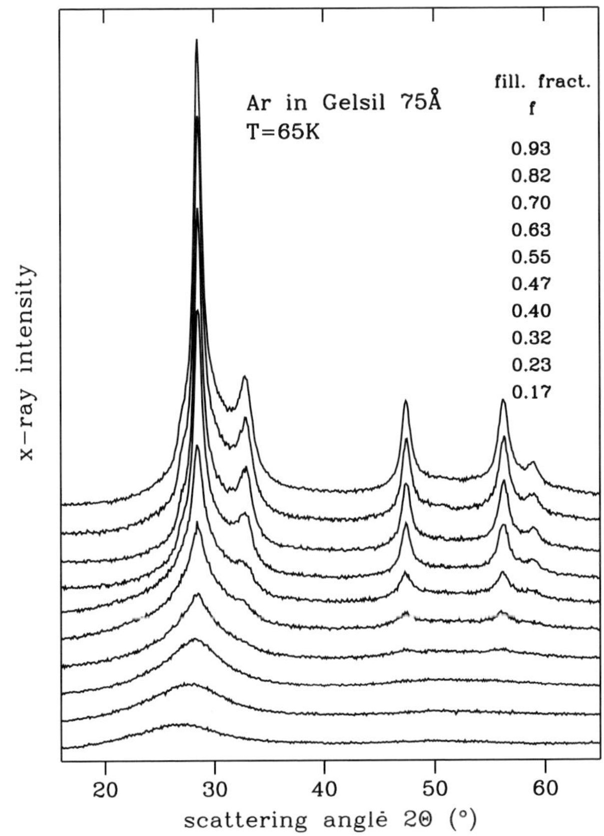

Fig. 5. X-ray diffraction pattern of various filling fractions of Ar in xerogel at 65 K. The background intensity of the empty substrate has been subtracted. The samples have been prepared right at 65 K by condensation of the vapour into the pores – capillary sublimation. (Cu K_α radiation, $\lambda = 1.542$ Å). From Ref. [8].

tallizes in the structure known from the bulk solid. Within the experimental error the lattice parameter is equal to that of the bulk solid. The Bragg intensities are roughly proportional to $f - f_c$, with $f_c = 0.40$, *i.e.* to that part of the filling that is added by capillary sublimation. The fcc diffraction profile of the capillary condensed component has been compared to a model that considers finite crystal size and stacking faults. The stacking fault probability, which is the probability that a stack of the three triangular (111) fcc net planes obeys the ABA rather than the ABC sequence of the fcc lattice, is of the order of 5% and more, if the pore solid is not prepared by sublimation but by cooling down pores initially filled with liquid [8]. Obviously this type of lattice defect (and other defects that are not visible in a powder pattern) is necessary to match the solid to the pore geometry. Values of the coherence length of up to 100 nm have been extracted. This means that the solid can grow coherently along the pores over distances that exceed the pore diameter by far.

For low values of f, $f < f_c$, within the film regime, the diffraction pattern is of the amorphous type, practically identical to the structure factor of the bulk liquid [39]. f_c corresponds to the equivalent of about three monolayers adsorbed on the pore walls. As we will show further below the first two monolayers are "dead" in the sense that they do not participate in the freezing/melting process. Hence one thinks of molecules immobilized in deep adsorption sites next to the rough pore walls. Alternatively one may think of these monolayers in terms of a disordered version of the 2D triangular net planes known from Ar monolayers adsorbed on planar substrates [40]. The principal peak of the pattern at low-f is then identified as the fundamental (10) reflection of the triangular lattice which then gradually develops into the (111) reflection of the fcc lattice as more and more such layers are on top of each other. This process involves a shift of the peak and a decrease of its width that are in good agreement with the experimental result.

Evidence for a partition of the pore filling into the film part and the quasi-bulk capillary sublimate and/or into a dead and a mobile part not only comes from sorption isotherms, diffraction and calorimetric (see next section) data, but also from dielectric results on Ar and CO in Vycor [19] and in case of N_2 from an infrared study of the stretching vibration [6].

Along the desorption branch the fcc peaks of the capillary condensate vanish not at $f_c = 0.40$ but at a lower value that corresponds just to the two dead layers without restoring the third layer. This reduction of the limiting thickness of the film on desorption is analogous to what is observed in the liquid regime and is in agreement with the interpretation of the sorption isotherm.

There is another observation that will be of importance in the next section. As f is increased it is not only that the Bragg peaks of the crystalline component grow, but also that the amorphous component representing the film next to the walls decreases. The decrease corresponds to about one monolayer equivalent. Obviously the "mobile" third layer on top of the two dead layers is incorporated into the crystalline fraction of the pore filling.

MD simulations on Ar in a cylindrical pore have shown that the crystallographic structure of the solid pore filling depends on the strength of the substrate potential [41]. In a strong potential the solid consists of a series of concentric monolayer shells, each of which is close to a 2D triangular lattice rolled up into a cylinder. For weaker potentials this layer wise organisation is lost and the 3D fcc structure is approached, with some intrinsic disorder due to hcp like local configurations. The diffraction results are in agreement with the latter case, apart from the dead layers next to the pore walls the solidified pore filling of Ar (and also N_2, CO, O_2, [8, 13, 14] crystallizes in the structures known from the bulk state [15]. There are no indications for a layered organization of the crystalline part of the pore filling.

5. Liquid–solid transition

It has been known for a long time that the melting transition of most pore fillings is shifted downward with respect to the bulk state and that there is thermal hysteresis between freezing and melting, both temperatures T_{f} and T_{m} being lower than the bulk triple point T_3 [2–4]. Hence the hysteresis is not simply the result of supercooling. For $T_{\mathrm{f/m}}$ in pores one usually refers to the Gibbs–Thomson equation, the analogue of the Kelvin equation where σ is now the liquid–solid interfacial energy. Thus the shift $\Delta T = T_3 - T_{\mathrm{f/m}}$ should scale with the inverse of a characteristic length R' which in turn has some connection to the pore radius R, thereby suggesting the possibility of pore size spectroscopy by investigations of the pore freezing/melting [42]. Note however that pore sizes obtained along these lines cannot really be considered independent from those obtained from the characteristic pressure of sorption isotherms, since both approaches are based on equivalent thermodynamic relations. Furthermore, any model that is based on the competition of volume and surface free energy produces a $1/R'$ scaling and therefore tells little about the shape of the interfaces involved, whether it is *e.g.* of hemispherical or of cylindrical shape.

Avoiding any reference to a particular geometry, ΔT can be related to experimental data on the pressure of the vapour coexisting with the pore solid and the pore liquid, taken from sorption isotherms or alternatively from isosteric heating and cooling runs. Resulting chemical potential μ-temperature T phase diagrams on Ar in Vycor and porous Si can be found in Refs. [8, 10, 16]. Such diagrams supply the chemical potential differences, $\Delta\mu_{\mathrm{liq}}$ and $\Delta\mu_{\mathrm{sol}}$ of the pore state with respect to the bulk state, both in the liquid and solid regime, and ΔT is then related to the $\Delta\mu$'s *via* $\Delta T = (\Delta\mu_{\mathrm{liq}} - \Delta\mu_{\mathrm{sol}})/\Delta S$ where ΔS is the entropy of fusion which is assumed to be equal in the pore and the bulk state, in agreement with the experimental results presented just below. The problem of this and any other thermodynamic relation is that equilibrium states are not accessible experimentally because of hysteresis.

The most detailed information on pore freezing and melting comes from calorimetric experiments [10]. We refer to them first and add supplementary information from diffraction and optical transmission later.

Data on Ar in Vycor, $R = 10\,\mathrm{nm}$, collected in both the adiabatic and the scanning mode (Fig. 6) of operation on the melting transition show sawtooth-shaped melting anomalies for higher fractional fillings f, $f < 0.28$. For lower fillings the anomaly is absent, supporting the idea of dead layers on the pore walls. For $f > 0.28$, *i.e.* for the capillary condensed component of the pore filling, the integrated area of the anomaly increases with f in a linear fashion. The derived entropy of fusion S_0 is within the error margins of the experiment identical to that of the bulk state. Lower values cited by previous authors were based on data on completely filled pores only, and therefore have not been corrected for the part of the pore filling that does not participate in the melting process [43].

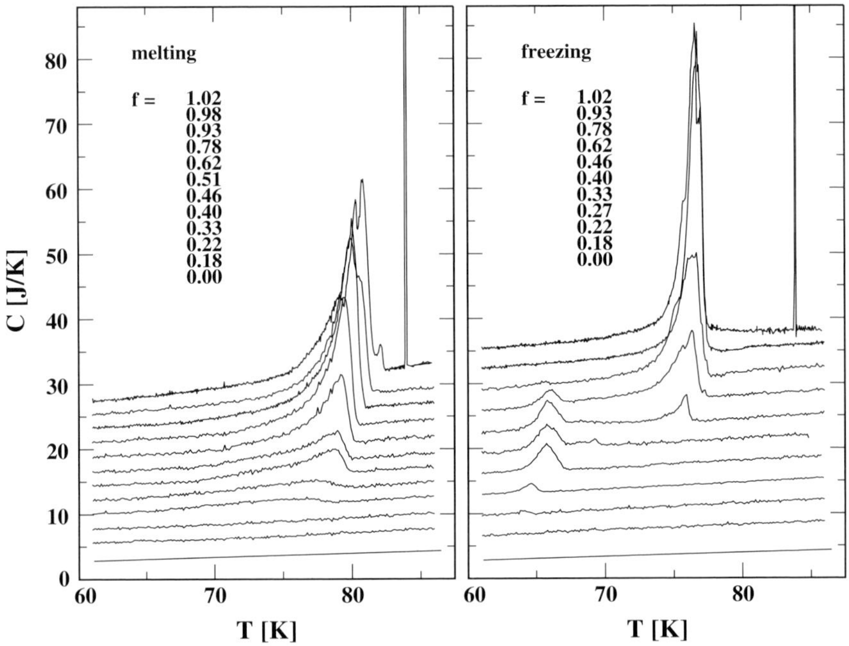

Fig. 6. The heat capacity of various filling fractions of Ar in Vycor glass as obtained with a scanning technique on heating (frame at the left) and cooling (panel at the right). The spike for $f = 1.02$ at about 84 K is due to the freezing/melting of the material excess outside the pores. The fillings have been prepared by condensation of the vapour at 86 K, the results shown are from the second cooling/heating cycle. From Ref. [10].

Scanning calorimetry also gives access to the freezing process (Fig. 6). The freezing anomaly for $f = 1$ looks similar to the melting anomaly, again with a sharp onset at high T, occurs at a lower temperature, is somewhat narrower than the melting peak such that the low-T wings of the freezing and the melting anomaly practically coincide.

Based on some kind of $\Delta T - 1/R'$ relation, *i.e.* on the idea that the large pores are the first to freeze on cooling and the last to melt on heating, that pore segments of different diameter freeze and melt independently, the anomalies can be translated into pore size distribution, with the problem that different distributions are obtained from the freezing and the melting data. Furthermore the special shape of the anomalies calls for rather peculiar distributions.

We therefore presented a simple thermodynamic model that assumes that in a completely filled pore the liquid–solid interface moves in radial direction with temperature rather than along the pore [10]. Melting starts at the pore walls or more correctly at the surface of the dead layers, a concept reminiscent of surface melting known to occur at free, solid surfaces [44]. As T is increased the thickness of the liquid shell increases and when a critical radius r_c of the

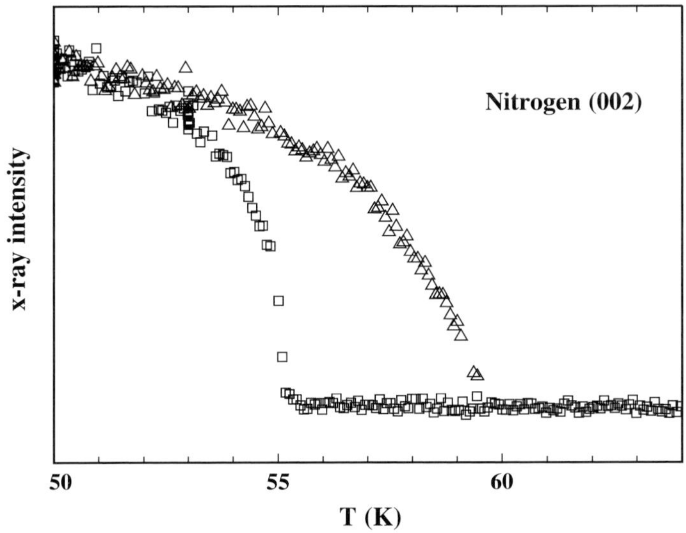

Fig. 7. The temperature dependence of the X-ray intensity of the (002) Bragg peak of the hcp component of N_2 in mesoporous Si, upon heating (triangles) and cooling (squares) through the melting region. The pores have been filled completely by condensation in the liquid regime. From Ref. [16].

solid core remaining in the pore centre is reached, this core melts all at once. In so far the model has some similarity to the SC model for sorption isotherms, the solid phase playing the role of the vapour phase in a sorption experiment. The model produces metastable states that account for the observed hysteresis. The calorimetric signature of this model are δ-like anomalies at the freezing and melting of the core at the spinodal temperatures T^- and T^+ with wings on their low-T side. Thus the model supplies a natural explanation not only for the hysteresis and the shift ΔT, but also for the sharp high-T cutoff of the anomalies.

Diffraction experiments show that Bragg intensities of the solid state develop only gradually with T (Fig. 7). This observation supports the idea that the liquid and solid phase coexist over some T-interval, the width of which compares favourably with the width of the heat capacity anomalies, both for freezing and melting. However the diffraction experiment integrates over the pore filling and therefore cannot tell whether the phases coexist in separate pore segments (with different diameters) or whether there is radial coexistence in every segment.

The complex freezing behaviour of partially filled pores (Fig. 6) that will be described in the following came to us as a surprise. The samples have been cycled twice between 86 K in the liquid regime, where they have been prepared by condensing the vapour into the pores and 60 K, a temperature deep in the solid regime. We refer to the results of the second cooling run which are somewhat simpler than those of the first cooling run. The melting anomaly starts showing up at $f = 0.22$ and has the usual shape, its high-T cutoff shifts slightly to higher T with increasing f, due to the progressive filling of larger pore segments. The freezing anomaly is however split, into an asymmetric peak at higher temperature that is identical with the temperature where the Bragg peaks set in, when cooling down an $f = 1$ sample and a second peak at about 66 K, roughly 10 K below the first peak.

The full entropy of fusion is only recovered when both peaks are considered, thus there is no doubt that the peak at 66 K is due to no other phase transition but solidification. The peak intensities vary with f in a peculiar way. The 66 K-peak is strongest at f-values of 0.33 and 0.40 where the high-T freezing signal has not yet appeared. These f-values still lie in the reversible film section of the sorption isotherm. One concludes that the high-T peak is due to the freezing of the capillary condensed component of the pore filling whereas the 66 K peak signals the freezing of the mobile top part of the adsorbed film on the pore walls. The maximum area of the 66 K peak corresponds to the entropy of fusion of an amount of material equivalent of about one monolayer. From the f-value pertaining to this situation, it is clear that this is the third monolayer which resides on top of two dead layers. The 66 K-peak disappears at higher f where more and more solidified capillary condensate is present. The independent solidification of the third layer occurs only in those pore sections that are still void of solidified capillary condensate. The 66 K-anomaly does not reappear on heating. The material of the third monolayer delayers upon solidification, transforms into capillary-condensate-like plugs and then melts as such on subsequent heating. Such delayering processes have been predicted theoretically and observed experimentally for multilayer adsorption on planar substrates in a special situation termed triple point wetting [45]. There the delayering means transformation of a liquid monolayer into the bulk solid.

Isosteres in the μ–T plane show not only a kink but steplike discontinuity at 66 K [10], suggesting that by the time the liquid third layer is cooled down to 66 K, it is no longer an equilibrium but a metastable state with a chemical potential higher than that of the crystallized capillary condensate.

The cooling run of the first thermal cycle of an $f = 0.62$ sample shows (Fig. 8) an additional splitting of the high-T freezing peak, suggesting delayering not only of the third, but also of the fourth monolayer, with a freezing peak 74 K in addition to the solidification of a small amount of capillary condensate at 76 K. The results also indicate that for partial fillings, $f_c < f < 1$, the distribution of the liquid right after preparation of the pore filling by condensation is different from what is obtained when returning to the liquid state after the first thermal cycle.

This is also illustrated by optical transmission results, see Fig. 9. The transmission of a fresh $f = 0.60$ sample prepared by adsorption is high. This is because of fine partitioning of liquid. Upon solidification the transmission drops by several orders of magnitudes down to τ values comparable to those of

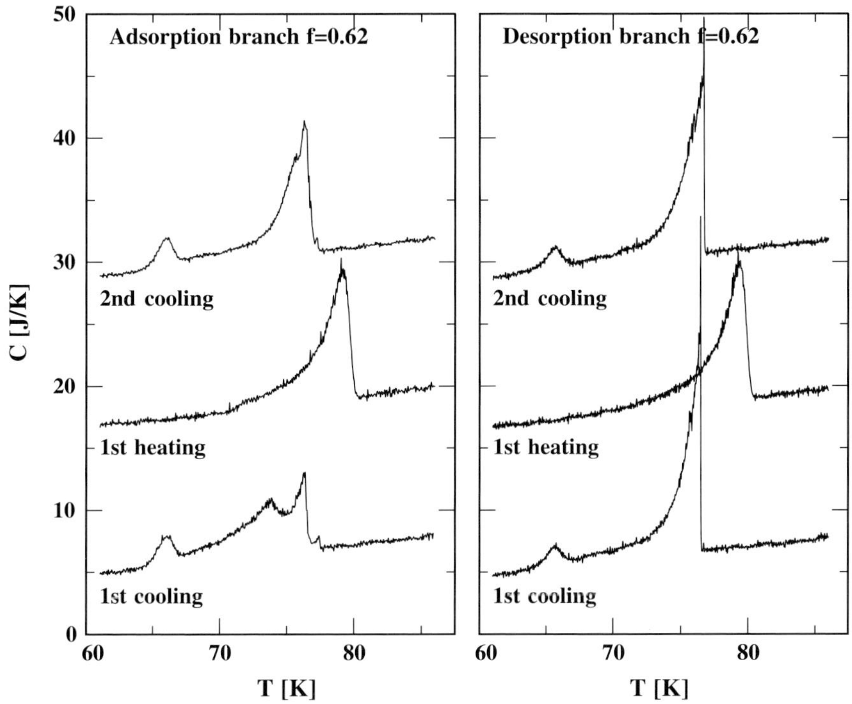

Fig. 8. Analogous to Fig. 6, showing that for a partial filling (here $f = 0.62$) the results of the first cooling run directly after preparing the sample by the condensation of the vapour into the pores in the liquid regime at 86 K are different from the results of other sample histories. From Ref. [10].

the desorption branch (see Fig. 1). Obviously the partition coarsens upon solidification, the crystallites are larger in size and fewer in number than the original liquid parcels. The vehicle of mass transport is presumably not only distillation but also diffusion within the mobile third and still liquid top adsorbed layer. In fact the decrease of τ comes to a halt at about 66 K when solidification is completed after the delayering of the third layer (Fig. 9). The driving force for the coarsening is presumably the crystalline anisotropy of interfacial energies between the solid and the vapour or more likely – since more abundant – between the solid and the amorphous dead layers.

In any case favourably oriented crystallites (see below) grow at the expense of others.

On heating τ increases, but does not reach the high values of the initial state (Fig. 9). A coarse arrangement remains. The third layer in the "empty" part of pore space is restored, the layer number four is not, as can be seen from the comparison of the first and second cooling run in Fig. 8. This different thermal behaviour is not surprising since the sorption isotherms of the

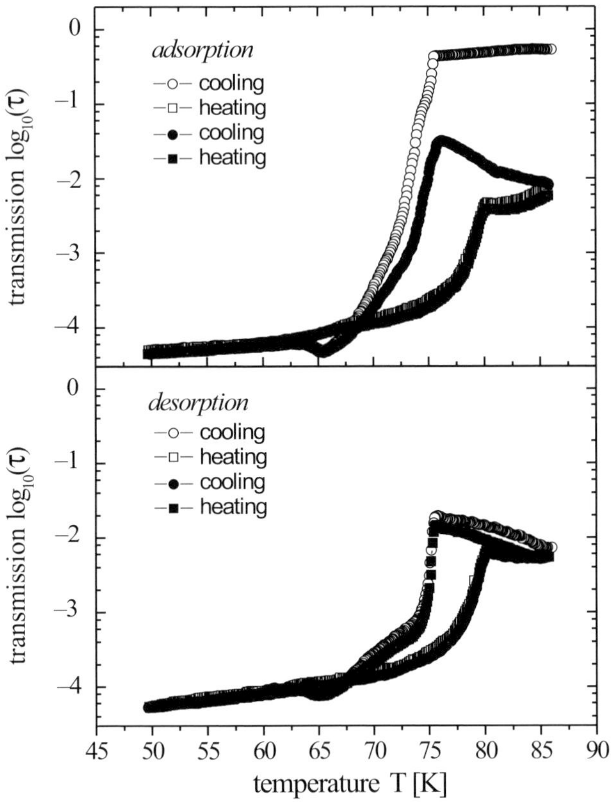

Fig. 9. The optical transmission of a fractional filling of Ar in Vycor glass. The samples have been prepared by adsorption up to $f = 0.6$ (upper frame) and by desorbing material from a completely filled sample down to $f = 0.6$ (lower frame) in the liquid regime at 86 K and have been subsequently cooled down to 50 K and heated back up to 86 K twice. From Ref. [11].

liquid regime suggest that higher layers such as the fourth one are metastable and can be prepared by adsorption only, and are therefore absent along the desorption branch of a sorption isotherm. Hence it is plausible that they are not restored in a coarse arrangement that is quite similar to the distribution of material obtained by desorption.

Above we referred to favourable orientations of the crystallites. Such preferred orientations of the pore solid with respect to the pore axis have indeed been observed, but only when complete pore fillings are solidified by cooling. For Ar in Vycor the evidence is somewhat indirect since it is based on the anisotropy of the coherence length [8], however in the oriented pores of Si it directly comes from pole figures [16, 17]. Ar has a relatively weak preference for one of the $\langle 111 \rangle$ directions to be parallel to the pore axis, but for hcp N_2 the preference for the unique hexagonal axis is strong. The texture is presumably due to the fact that the velocity of the advancing solidification front depends on the crystallographic direction. In a pore this leads to the selection process that is exploited for the Bridgeman technique of single crystal growth. Preferred orientations are absent, even for $f = 1$, when the pore solid is prepared directly by the condensation of the vapour.

The n-alkane $C_{17}H_{36}$ solidified in porous Si shows a preferred orientation not only with respect to the pore axis, but also with respect to the azimuthal angle about the pore axis [7]. This results from the highly anisotropic, namely layered crystal structure of the alkane in combination with the single-crystalline state of the Si substrate.

In summary, the experiments on the freezing transition of Ar in Vycor show that different parts of the pore filling have to be distinguished. The dead layers do not show a melting transition, the higher layers of the adsorbed film freeze in at temperatures different from the freezing of the capillary condensed component, but this behaviour is only observed as long as these layers are in contact with the vapour in the pore centre. In case the centre is filled they have lost their identity and are part of the capillary condensate, at least in Vycor, the following results on Ar in SBA-15 tell a slightly different story.

The calorimetric measurements on Ar in SBA-15 have not been completed yet [46], but first results are shown in Fig. 10. As for Vycor the first amount of material that has condensed in the pores at low values of p does not support the melting transition, again consistent with the idea of dead layers next to the pore walls. Melting gives rise to a single anomaly of asymmetric shape suggesting again that upon solidification mobile layers in the empty pore sections have been removed and have been converted into solid plugs by the delayering process described above. Hence after completion of solidification the crystallized component is structurally homogeneous to such an extent that there are no separate melting transitions of individual strata.

As to freezing, the low-T peak (at 65 K, see Fig. 10) is again due to the freezing and delayering of the top mobile part of the wall coating, as indicated by the peculiar f-dependence of the peak area. The freezing of the capillary condensate is however different from what is observed in Vycor. The saw-tooth shaped freezing anomaly observed in Vycor is replaced by a broad anomaly that comprises a relatively sharp peak at 67 K and a plateau extending down to this peak, with an onset at 71 K.

We interpret this complex freezing behaviour in terms of metastable states. In case a parcel of pore liquid is disconnected from other pieces of material, that have already solidified at higher temperatures, it remains in the liquid state down to the lower spinodal temperature of the interfacial freezing/melting scenario introduced above. At this temperature it eventually solidifies *via* homogeneous nucleation of the crystalline phase in the pore centre. Since there is little variation of the pore size in SBA-15, this happens at about the same temperature, namely 67 K, in all pores. If on the other hand, the parcel of li-

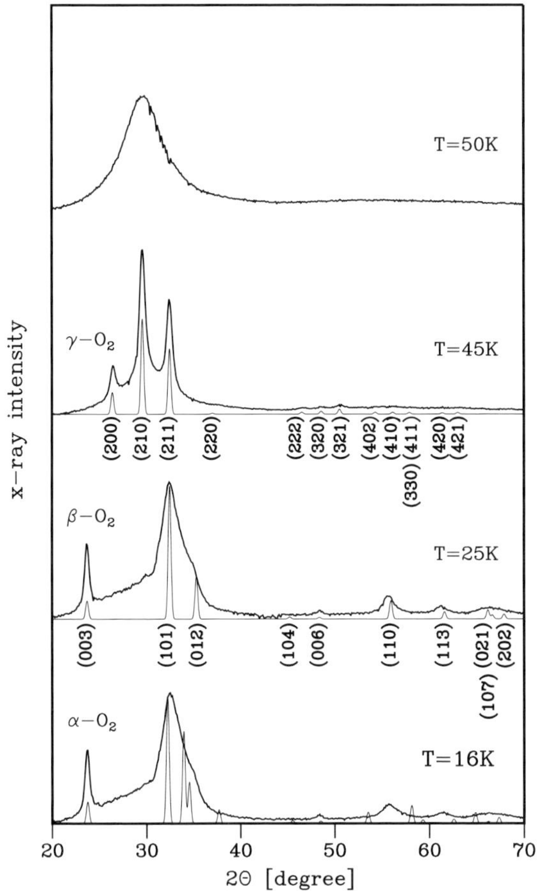

Fig. 11. X-ray diffraction patterns on O_2 in a mesoporous xerogel, for complete filling at selected temperatures representing the four phases, liquid, γ, β, α. Also shown are powder patterns calculated for the three crystalline phases of bulk O_2. From Ref. [14].

The diffraction patterns of the four phases encountered in O_2 are shown in Fig. 11. The capillary condensed component still shows the γ–β transition known from the bulk state. The phase transformation is complete in pores with average diameters of 7.5 nm and 13 nm, but incomplete in 5 nm-pores. Here only 40% of capillary condensate manage to transform into the β phase. See Fig. 12 that also gives an impression on the hysteresis and the temperature width of the liquid–γ and the γ–β coexistence on cooling and heating.

In the diffraction patterns of the bulk state, splittings of the β peaks signal the transition into the low-T α phase. In the pores such splittings are absent but some Bragg peaks, in particular the (110) peak, broaden significantly, suggesting that the spontaneous lattice distortion of the α with respect to the β phase

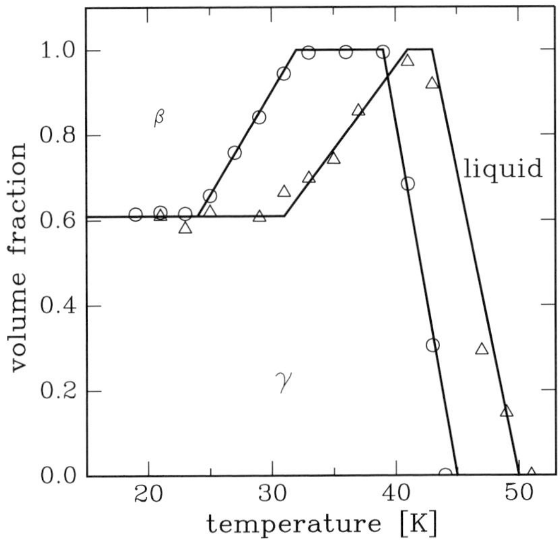

Fig. 12. The temperature dependence of the volume fractions of the phases liquid, γ, β of O_2 (complete filling) in a xerogel with an average pore diameter of 5 nm. Circles are for cooling, triangles for heating. From Ref. [14].

does indeed take place. This distortion is however not homogeneous, but rather characterized by a broad distribution of local distortions, induced by abundant lattice defects of the pore solid.

In pore confined N_2 the transition from the β into the low-T α phase is suppressed [13]. The same observation has been made in bulk Ar : N_2 solid solutions with N_2-concentrations less than 0.8 [48]. The idea is that the substitutional disorder introduces random strain fields that block the lattice transformation, thereby preserving the orientational disorder of the β phase. At lower T, the thermally excited molecular reorientations freeze in randomly, a situation that has been called orientational glass [52]. In the present case the random strain fields arise from lattice defects (not only stacking faults but also defects such as grain boundaries and dislocations that are not directly visible in a powder diffraction pattern) that are necessary to match the solid to the pore geometry, but the final low-T state is presumably similar. (He in Vycor is another example where a reconstructive phase transformation, here bcc to hcp, is suppressed in pores [21]).

Pore confined CO reproduces the liquid–β–α phase sequence known from the bulk state, of course with hysteresis, wide coexistence ranges, and reduced transition temperatures. See Fig. 13 for diffraction patterns of the three phases. A closer look shows however that this sequence is only obeyed in cooling runs. On heating an additional fcc phase intermediate to the α and the β phase shows up. This is illustrated in Fig. 14. Here the β-phase is represented by its (100) reflection, the Pa3 α phase by the (210) reflection,

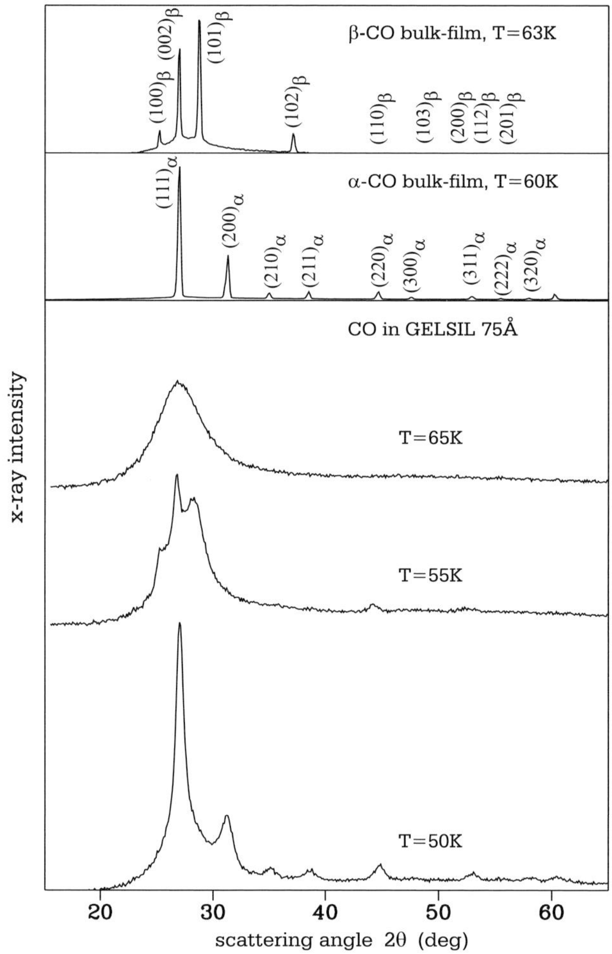

Fig. 13. X-ray diffraction patterns representing the three phases (liquid, β, α) of CO confined in a mesoporous xerogel, for complete filling. The diffraction patterns of the two solid phases of a quasi-bulk evaporated CO film are shown for comparison. From Ref. [13].

a superlattice reflection relative to the fcc phase. The third reflection shown, (200), is common to both the α and the additional fcc phase. On heating the (210) and the (200) disappear at different temperatures. Between 59 K and 62 K the centre-of-mass lattice is still fcc (as in the α phase), but the (210) reflection representing the four-sublattice type orientational order is already gone. An intermediate orientationally disordered fcc is expected to occur in a situation where the fcc-to-hcp reconstruction of the centre-of-mass lattice and the orientational order–disorder transition no longer coincide. On heat-

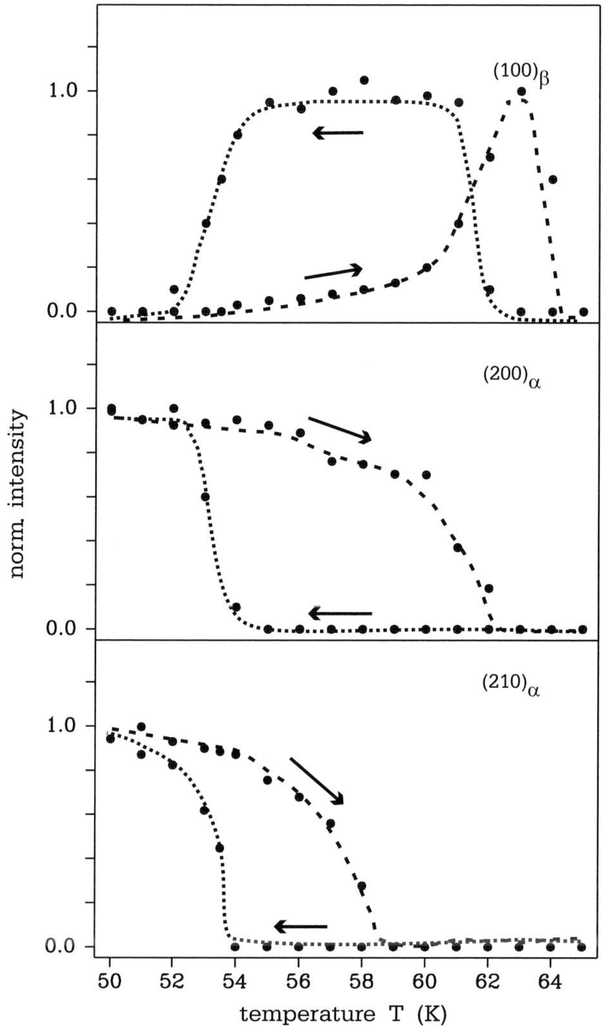

Fig. 14. The temperature dependence of the intensity of three selected Bragg peaks of CO confined in a mesoporous xerogel, for both cooling and heating. The intensities are measures of the volume fractions of the corresponding phases. On heating there is a T-range from about 58 K to 62 K in which the $(210)_\alpha$ reflection is already absent but the $(200)_\alpha$ is still present, suggesting an intermediate orientationally disordered phase with an fcc centre-of-mass lattice. From Ref. [13].

ing the reconstruction is delayed so much by lattice strains that the orientational ordering takes place in the fcc centre-of-mass lattice of the low-T state. Analogous observation have been made in bulk Ar : N_2 for N_2 concentrations around 0.9 [49].

Thus the three examples pertain to quite different situations. For O_2 pore confinement does not change the phase sequence known from the bulk state, in N_2 the low-T phase of the bulk state is suppressed, whereas in CO a new (but not totally unexpected) phase appears.

7. Summary

As exemplified above many of our conclusions are drawn from, and rely on, a proper characterization of both the thermodynamic as well as the structural state of the pore condensates under investigation. To a large extent this was possible due to measurements which combined sorption isotherms with a variety of other experimental techniques, ranging from specific heat, dielectric spectroscopy, and optical transmission to wide and small angle X-ray diffraction. In particular the extension of the sorption isotherm technique, a method well-established for the characterization of the liquid/vapour coexistence of pore condensates, towards the exploration of the vapour/solid coexistence, accomplished in the early stages of this project, turned out to be particularly fruitful for all but a few studies to follow. It allowed the first observation of capillary sublimation in mesopores [8, 10, 13] and subsequently the establishment of a phase diagram of mesopore confined Ar, one of the most simple mesoconfined system imaginable.

Over the years we had to learn, however, that the phase behaviour of even such a simple van-der-Waals system is more complex than captured in a phase diagram spanned by the variables T, p, and f. Depending on the preparation and the history of the sample a reorganisation of the confined liquid as well as solid can occur, a phenomenon which we could explore in a rigorous manner, and to the best of our knowledge for the first time, by systematic measurements on partial pore fillings. To some extent we could trace the complex behaviour encountered to the interplay of a radial arrangement of the condensate, a clear partitioning in two components, *i.e.* the material in the proximity of the wall and the material in the pore center, and the quenched disorder introduced by the porous matrices employed, most prominently the pore size distribution.

Therefore, one may be inclined to believe that the nowadays available mesoporous hosts with narrow pore size distributions and simpler geometry such as carbon nanotubes and aligned bundles thereof [53], tailored pores in silicon [12, 54], silica [27, 55] and aluminum [56] may lead to a thermodynamics which is more easily to grasp. Note, however, also in these systems the structural gradient along the radial direction will be of importance, due to the reduced disorder compared to tortuous pores with wide pore diameter distributions presumably even in a more distinct manner. In fact, the measurements on the freezing and melting of Ar in SBA-15, presented above, give first hints for a complex behaviour despite of a seemingly most simple confining geometry.

While studying mesopore confined condensates, we frequently encountered phenomena which were not only intermediate with respect to the scale of the system considered, but also at the border between surface and bulk physics. Therefore, we profited a lot from the well established microscopic models and theories for van-der-Waals systems on planar substrates [45, 50, 51] along with the well known and the comparably elementary bulk physics of these species [39].

After having achieved a better understanding of simple molecules and atoms, we felt recently encouraged to explore more complicated molecules as well, such as chain- or rod-like alkanes in mesopores [7, 22–24]. For the future the investigation of even more complex molecules and macromolecules such as proteins in mesopores may be envisioned and of interest [57–59]. For such studies the findings presented here may set the stage for a better understanding.

Finally we would like to note that the intimate relationship between thermodynamics and structure explored and documented in this project should not only be important for the equilibrium state of these systems. We believe it is also crucial for an understanding of non-equilibrium phenomena, such as lubrication and adhesion mediated by thin films of liquids [61, 62], or transport diffusion processes within [25, 31, 32, 60, 63] and flow of, liquids in extreme spatial confinement, a scientific and technological field just emerging [64] and which for flow in pores of 10 nm diameter can be truly termed nanofluidics [65, 66].

Acknowledgement

We enjoyed our collaboration with T. Hofmann, C. Schäfer, A. Henschel, A. Kityk, V. Soprunyuk, R. Ackermann, J. Albers, M. Enderle, M. Rheinstaedter and P. Kumar who contributed to the project as students, colleagues or guest scientists. We appreciated the financial support of the Deutsche Forschungsgemeinschaft.

References

1. J. M. Drake and J. Klafter, Phys. Today **43** (1990) 46.
2. H. K. Christenson, J. Phys. Condens. Matt. **13** (2001) R95.
3. L. D. Gelb, K. E. Gubbins, R. Radhakrishnan, and M. Sliwinska-Bartkowiak, Rep. Prog. Phys. **62** (1999) 1573.
4. C. Alba-Simionesco, B. Coasne, G. Dosseh, G. Dudziak, K. E. Gubbins, R. Radhakrishnan, and M. Sliwinska-Bartkowiak, J. Phys. Condens. Matt. **18** (2006) R15.
5. D. Wallacher, V. P. Soprunyuk, A. V. Kityk, P. Huber, and K. Knorr, Phys. Rev. B **71** (2005) 052 101; D. Wallacher, T. Hofmann, K. Knorr, and A. V. Kityk, Phys. Rev. B **71** (2005) 224 202.
6. D. Wallacher, P. Huber, and K. Knorr, J. Low Temp. Phys. **113** (1998) 13.
7. A. Henschel, T. Hofmann, P. Huber, and K. Knorr, Phys. Rev. E **75** (2007) 021 607.

8. P. Huber and K. Knorr, Phys. Rev. B **60** (1999) 12 657.
9. T. Hofmann, D. Wallacher, P. Huber, R. Birringer, K. Knorr, A. Schreiber, and G. H. Findenegg, Phys. Rev. B **72** (2005) 064 122.
10. D. Wallacher and K. Knorr, Phys. Rev. B **63** (2001) 104 202.
11. V. P. Soprunyuk, D. Wallacher, P. Huber, and K. Knorr, Phys. Rev. B **67** (2003) 144 105.
12. D. Wallacher, N. Künzner, D. Kovalev, N. Knorr, and K. Knorr, Phys. Rev. Lett. **92** (2004) 195 704.
13. P. Huber, D. Wallacher, and K. Knorr, Phys. Rev. B **60** (1999) 12 666.
14. D. Wallacher, R. Ackermann, P. Huber, M. Enderle, and K. Knorr, Phys. Rev. B **64** (2001) 184 203.
15. P. Huber and K. Knorr, Mater. Res. Soc. Symp. Proc. E **876** (2005) R3.1.
16. T. Hofmann, D. Wallacher, P. Huber, and K. Knorr, J. Low Temp. Phys. **140** (2005) 91.
17. K. Knorr, D. Wallacher, P. Huber, and V. P. Soprunyuk, Eur. Phys. J. E **12** (2003) 51.
18. V. P. Soprunyuk, D. Wallacher, P. Huber, R. Ackermann, and K. Knorr, J. Low Temp. Phys. **134** (2004) 1043.
19. D. Wallacher, V. P. Soprunyuk, A. V. Kityk, and K. Knorr, Phys. Rev. B **66** (2002) 014 203.
20. D. Wallacher, V. P. Soprunyuk, K. Knorr, and A. V. Kityk, Phys. Rev. B **69** (2004) 134 207.
21. D. Wallacher, M. Rheinstaedter, T. Hansen, and K. Knorr, J. Low. Temp. Phys. **138** (2005) 1013.
22. P. Huber, D. Wallacher, T. Hofmann, and K. Knorr, J. Phys. Condens. Matter **14** (2003) 309.
23. P. Huber, D. Wallacher, J. Albers, and K. Knorr, Europhys. Lett. **65** (2004) 351.
24. P. Huber, V. P. Soprunyuk, and K. Knorr, Phys. Rev. E **74** (2006) 031 610.
25. S. Gruener, Diploma thesis, Saarland University (2006).
26. P. Levitz, G. Ehret, S. K. Sinha, and J. M. Drake, J. Chem. Phys. **95** (1991) 6151.
27. D. Zhao, J. Feng, Q. Huo, N. Melosh, G. H. Fredrickson, B. F. Chmelka, and G. D. Stucky, Science **279** (1998) 548.
28. V. Lehmann and U. Gösele, Appl. Phys. Lett. **58** (1991) 856; V. Lehmann, R. Stengl, and A. Luigart, Mater. Sci. Eng. B **69** (2000) 11; L. T. Canham, Appl. Phys. Lett. **57** (1990) 1046.
29. W. F. Saam and M. W. Cole, Phys. Rev. B **11** (1975) 1086.
30. S. J. Gregg and K. S. W. Sing, *Adsorption, Surface Area and Porosity.* Academic Press, New York (1982); Y. C. Yortsos, *Methods in the Physics of Porous Media.* Academic Press, San Diego (1999).
31. R. Paul and H. Rieger, J. Chem. Phys. **123** (2005) 024 708.
32. R. Valiullin, S. Naumov, P. Galvosas, J. Kärger, H.-J. Woo, F. Porcheron, and P. A. Monson, Nature **443** (2006) 965.
33. G. Mason, Proc. R. Soc. London A **390** (1983) 47.
34. M. Alava, M. Dubé, and M. Rost, Adv. Phys. **113** (2004) 83.
35. B. Abeles, L. F. Chen, J. W. Johnson, and J. M. Drake, Isr. J. Chem. **31** (1990) 99.
36. J. H. Page, J. Liu, B. Abeles, H. W. Deckman, and D. A. Weitz, Phys. Rev. Lett. **71** (1993) 1216.
37. E. Kierlik, P. A. Monson, M. L. Rosinberg, L. Sarkisov, and G. Tarjus, Phys. Rev. Lett. **87** (2001) 055 701; T. Lambert, C. Gabay, L. Puech, and P. E. Wolf, J. Low Temp. Phys. **134** (2004) 293.
38. R. Guegan, D. Morineau, C. Loverdo, W. Beziel, and M. Guendouz, Phys. Rev. E **73** (2006) 011 707.
39. M. L. Klein, J. A. Venables (Eds.), *Rare Gas Solids.* Academic Press, London (1977).

40. J. Z. Larese, Q. M. Zhang, L. Passell, J. M. Hastings, J. K. Dennison, and H. Taub, Phys. Rev. B **40** (1989) 4271.
41. J. Hoffmann and P. Nielaba, Phys. Rev. E **67** (2003) 036 115; K. Nishio, J. Koga, T. Yamaguchi, and F. Yonezawa, Phys. Rev. B **69** (2004) 214 201.
42. M. Brun, A. L'Allemand, J. F. Quinson, and C. Eyfraud, Thermochim. Acta **21** (1977) 59.
43. E. Molz, A. P. Y. Wong, M. H. W. Chan, and J. R. Beamish, Phys. Rev. B **48** (1993) 5741.
44. J. W. M Frenken and J. F. Van der Veen, Phys. Rev. Lett. **54** (1985) 134.
45. R. Pandit and M. E. Fisher, Phys. Rev. Lett. **51** (1983) 1772; S. Dietrich, In: Phase Transitions and Critical Phenomena, C. Domb and J. L. Lebowitz (Eds.), Vol. 12, Academic Press, London and Orlando (1988), p. 1.
46. C. Schäfer, Diploma thesis, Saarland University (2007), unpublished.
47. R. Ackermann and M. Enderle, Europhys. Lett. **64** (2003) 260.
48. H. Klee and K. Knorr, Phys. Rev. B **42** (1990) 3152.
49. H. Klee and K. Knorr, Phys. Rev. B **43** (1991) 8658.
50. G. B. Hess, *Phase Transitions in Surface Films 2*. H. Taub, G. Torzo, H. J. Lauter, and S. C. Fain, Jr., (Eds.). Plenum, NewYork (1991).
51. P. G. de Gennes, Rev. Mod. Phys. **57** (1985) 827.
52. U. Höchli, K. Knorr, and A. Loidl, Adv. Phys. **39** (1990) 405.
53. S. Iijima, Nature **354** (1991) 56; B. J. Hinds, N. Chopra, T. Rantell, R. Andrews, V. Gavalas, and L. G. Bachas, Science **303** (2004) 62; J. K. Holt, H. G. Park, Y. Wang, M. Stadermann, A. B. Artyukhin, C. P. Grigoropoulos, A. Noy, and O. Bakajin, Science **312** (2006) 1034.
54. C. T. Black, K. W. Guarini, G. Breyta, M. C. Colburn, R. Ruiz, R. L. Sandstrom, E. M. Sikorski, and Y. Zhang, J. Vac. Sci. Technol. B **24** (2006) 3188; C. C. Striemer, T. R. Gaborski, J. L. McGrath, and P. M. Fauchet, Nature **445** (2007) 749.
55. C. T. Kresge, M. E. Leonowicz, W. J. Roth, J. C. Vartuli, and J. S. Beck, Nature **359** (1992) 710; P. D. Yang, T. Deng, D. Y. Zhao, P. Y. Feng, D. Pine, B. F. Chmelka, G. M. Whitesides, and G. D. Stucky, Science **282** (1998) 2244; K. Morishige, N. Tateishi, and S. Fukuma, J. Phys. Chem. B **107** (2003) 5177; M. Thommes, B. Smarsly, M. Groenewolt, P. I. Ravikovitch, and A. V. Neimark, Langmuir **22** (2006) 756.
56. K. J. Alvine, O. G. Shpyrko, P. S. Pershan, K. Shin, and T. P. Russell, Phys. Rev. Lett. **97** (2006) 175 703.
57. M. Beiner and H. Huth, Nature Mater. **2** (2003) 595.
58. A. Schoenhals, H. Goering, C. Schick, B. Frick, and R. Zorn, Colloid Polym. Sci. **282** (2004) 882.
59. N. E. Chayen, E. Saridakis, and R. P. Sear, Proc. Nat. Acad. Sci. **103** (2006) 597.
60. J. Baumert, B. Asmussen, C. Gutt, and R. Kahn, J. Chem. Phys. **116** (2002) 10 869.
61. M. Urbakh, J. Klafter, D. Gourdon, and J. Israelachvilli, Nature **430** (2004) 525; H. Bock, D. J. Diestler, and M. Schoen, Phys Rev. E **64** (2001) 46 124.
62. S. Herminghaus, Adv. in Phys. **54** (2005) 221; G. Huber, H. Mantz, R. Spolenak, K. Mecke, K. Jacobs, S. N. Gorb, and E. Arzt, Proc. Nat. Acad. Sci. **102** (2005) 16 293.
63. A. Meller, L. Nivon, E. Brandin, J. Golovchenko, and D. Branton, Proc. Nat. Ac. Sci. **97** (2000) 1079; C. Dekker, Nature Nanotechn. **2** (2007) 209.
64. P. Barthelemy, M. Ghulinyan, Z. Gaburro, C. Toninelli, L. Pavesi, and D. S. Wiersma, Nature Phot. **1** (2007) 172.
65. J. C. T. Eijkel and A. van den Berg, Microfl. Nanofl. **1** (2005) 249; M. Whitby and N. Quirke, Nature Nanotechn. **2** (2007) 87.
66. P. Debye and R. L. Cleland, J. Appl. Phys. **30** (1959) 843; P. Huber, S. Grüner, C. Schäfer, K. Knorr, and A. V. Kityk, Eur. J. Phys. Spec. Topics **141** (2007) 101.

Precursorchemistry with Metalalkoxides and their Use for Nano-Scaled Materials

By Michael Veith[1,2,*]

[1] Institute for Inorganic Chemistry, Saarland University, D-66041 Saarbrücken, Germany
[2] Leibniz-Institute for New Materials, Campus D2 2, D-66123 Saarbrücken, Germany

Single Source Precursors / Biphasic Nano-Composites / Cermets / Core-Shell Particles / Metastable Solid Phases

The synthesis of nano-scaled materials using several variations of the single source precursor concept in sol–gel and chemical vapour deposition techniques is demonstrated. Mono- and biphasic ceramics and cermets (metallic phases incorporated in a metal oxide phase) are addressed. The materials prepared are all nanocrystalline in a range from 5 to 50 nm. The ceramics belong *inter alia* to the spinel and perowskite-type. In the biphasic ceramic systems, especially $NdAlO_3$ in Al_2O_3 is highlighted. In the cermet family Ge, Sn and Pb in either their oxide matrices or in perowskite matrices ($M'MO_3$, $M' = Ca$, Sr, Ba; $M = Ge$, Sn, Pb) and Al and Ga in Al_2O_3 or Ga_2O_3 matrices are discussed. Two general concepts are presented in more detail: 1) the arrangement of mutually dependent ligands in the precursor molecule, to facilitate intra-molecular cascade or domino reactions, and 2) the role of metastable phases for physical patterning. The temperature and pressure dependent formation of aluminum/aluminum oxide composites and their structural consequences leading to zero-dimensional ball-shape spheres or to one-dimensional nano-wires is shown as well as the core shell structures of such composites.

1. General introduction to molecular precursors in the gas-phase and in sol–gel processes

As generally accepted, the synthesis of materials is tightly connected to the specific process and the starting materials. The more the materials to be produced have to be chemically pure, the more the effectiveness of the process and the purity as well as the quantitative reactivity of the precursors have to be considered. A well known example is GaAs, which can be made by reac-

* Corresponding author. E-mail: veith@mx.uni-saarland.de

Z. Phys. Chem. **222** (2008) 287–305

tion of trimethylgallium with arsenhydride [1, 2]. Since the band gap of this semiconductor is very much dependent on the purity of the material (Li or Cl should be less abundant than $10^{-4}\%$), trimethylgallium follows a specific treatment before being used in the reaction. Also from precipitation reactions it is well known that not only the components or the precipitate have to be as pure as possible, but also the solvent should be chosen in such a way, that no contamination or inclusion of solvent in the solid product is observed [3, 4].

Sol–gel as well as gas-phase techniques like the chemical vapour deposition (CVD) have been proven to be adequate processes for material synthesis [5–7]. While in the first method a liquid medium is used to bring the precursor entities together, which can either consist of complex ions or of molecules, and which react together following a condensation reaction (mostly by elimination of water), in the second method the precursors are in a gas-phase (mostly under reduced pressure) reacting together by loss of ligands. The particular structures of materials obtained by both processes may be one-, two- or three-dimensional (particles, wires, surfaces) [8–10]. With the choice of the process, the results of coating are quite different. So it is widely accepted that a complete coverage of a three dimensional surface, characterized by a rough and undulated landscape with small dimensions and tiny bumps and holes, is best be performed by gas-phase techniques as the molecular scales of the gas flow adopt nicely to the small dimensions. On the other side, smooth and large areas can be easily and cost efficiently covered by wet techniques using dip or spray coatings [11, 12].

There has been a debate on the role of single source precursors for material syntheses. In a broader sense a single source precursor is a system which has been assembled by chemical interaction and which has a definite composition prior to its use in a sol–gel or gas-phase process [13–15]. It therefore could consist of several molecules which interact by chemical bonds or van-der-Waals forces, the structure of which is often not known. In a more restricted definition a single source precursor is a molecular entity which is characterized by a distinct composition, a molecular mass, a distinct volatility and a precise molecular structure [16–19].

In the framework of this second more restricted definition, the precise composition and bonding within the precursor may apparently influence the chemical behavior. We have defined several possibilities for the formation of materials starting with such a single source precursor (Eqs. (1)–(3)) [20].

$$AB_nXY \rightarrow AB_n + X^{\bullet} + Y^{\bullet} \tag{1}$$

$$AB_nXY \rightarrow AB_n + X{-}Y \tag{2}$$

$$AB_nXY \rightarrow AB_{n-m} + B_m + X{-}Y \tag{3}$$

In these examples AB_nXY is a molecular precursor which in the first case (Eq. (1)) reacts to the compound AB_n releasing the radical species $X^{\bullet}$ and $Y^{\bullet}$. In Eq. (2) the same precursor again may form the phase AB_n, this time releasing a molecular species X–Y without formation of radicals and additionally in Eq. (3) AB_n is unstable under the reaction conditions forming two phases B_m and AB_{n-m} [20]. These principles, which strictly hold for gas-phase reactions, may also be adopted to sol–gel processes.

In the following review, we would like to describe our results using this single source precursor approach for the synthesis of new materials with nano-dimensional structures and with singular phase mixing. The key point in these procedures is the precursor which is composed in such a way that by heating or by hydrolysis certain bonds preferentially break or recombine, so that definite products are obtained. Such products could even be thermodynamically metastable.

2. The design of molecular precursors and the products

In this chapter, we describe several principles which can be used for the efficient synthesis of mono- or biphasic materials and which illustrate the single source precursor approach. The examples are exclusively taken from the work of our group, and as the starting point in all cases is a molecule, the targeted material is growing up to a nano- or micro-meter scale depending on the time and conditions of the reaction. This procedure is often addressed as bottom-up-approach.

2.1 Metal/metaloxide composites with Ge, Sn and Pb

Alkoxides of divalent germanium, tin and lead $M(OR)_2$ ($M = Ge, Sn, Pb$; $R =$ organic ligand) are normally oligomeric and even with bulky substituents like the tert-butyl group they are dimeric in the case of germanium and tin and trimeric in the case of the lead derivative [21–25]. As the metallic elements in these compounds are in an intermediate oxidation state (for Ge and Sn the $+4$ state is more stable), they may be activated by simple heat treatment to disproportionation into the oxidation states 0 and $+4$. Moreover, by using the tert-butyl group as a ligand on oxygen, iso-butene may form by thermolysis leaving behind the hydroxy group. In Eq. (4), the overall process of the thermolysis of bis(tert-butoxy)germanium, -tin and -lead is summarized.

$$2M[OC(CH_3)_3]_2 \rightarrow M + MO_2 + 2HOC(CH_3)_3 + 2H_2C{=}C(CH_3)_2 \tag{4}$$

By this process, two solid phases (the metallic M and the oxide MO_2) are formed at the same time liberating simultaneously tert-butanol and iso-butene. Both organic products are volatile and have been identified by on-line mass-spectrometry. The stoichiometric ratio of metal (0) and metal (IV) is strictly 1 : 1. The formation of tert-butanol can be explained by the abstraction of the proton of M–OH by the tert-butoxy group of $M{-}OC(CH_3)_3$.

The formation of two solid phases is not only restricted to the simple alkoxides of the divalent group-14-metals, but may also be applied, when

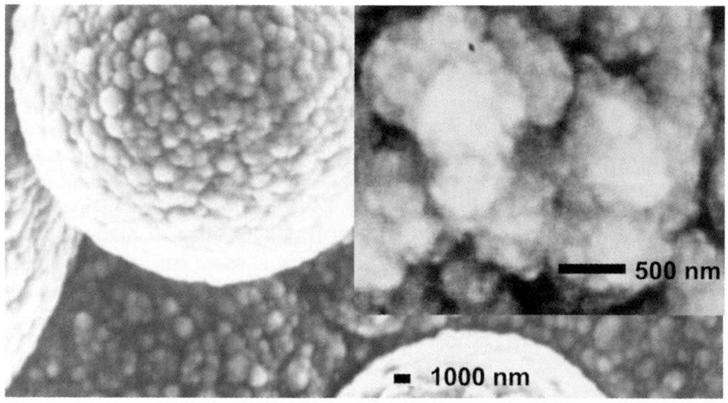

Fig. 1. Ball-like structures of $Sn \cdot BaSnO_3$ at two different scales. The inserted picture (SEM) shows the same surface enlarged by a factor of 10 [72].

these metal-alkoxides are parts of more complex entities like in the trimetallic complexes $M'M_2[OC(CH_3)_3]_6$ ($M' = Ca, Sr, Ba$; $M = Ge, Sn, Pb$) as may be deduced from Eq. (5).

$$M'M_2[OC(CH_3)_3]_6 \rightarrow M + M'MO_3 + 3HOC(CH_3)_3 + 3H_2C{=}C(CH_3)_2 \tag{5}$$

In this reaction, besides the metallic phase M, the perowskite $M'MO_3$ ($M' = Ca, Sr, Ba$; $M = Ge, Sn, Pb$) is formed, the process otherwise being very similar to Eq. (4). Again the two solid phases form stoichiometrically.

We have also used $BaSn_2[OC(CH_3)_3]_6$ in a sol–gel process using ethanol/water to activate the condensation reaction. Presumably the first step in this reaction is the displacement of tert-butanolate by ethanolate and by hydroxide. Anyway, by heating the gel up to 500 °C, the same phases Sn (0) and $BaSnO_3$ are obtained as in the CVD process (Eqs. (6) and (7)).

$$BaSn_2[OC(CH_3)_3]_6 \xrightarrow[-(CH_3)_3COH]{+C_2H_5OH,\ H_2O} BaSn_2(OC_2H_5)_x(OH)_{6-x}\text{-gel} \tag{6}$$

$$BaSn_2(OC_2H_5)_x(OH)_{6-x}\text{-gel} \xrightarrow{500\,^\circ C} Sn + BaSnO_3 + xC_2H_5OH + yH_2O \tag{7}$$

The consequences for the single source processes discussed so far are manifold. Because of the simple liberation of the organic ligands (in the gas-phase and sol–gel process) the contamination with carbon in the final solid products is negligible (0.5% C). The biphasic product is reproducibly obtained without much effort and the stoichiometric ratio of the phases (metal/metaloxide) is fixed. Presumably due to the common molecular source and the continuous mass-flow, the two phases cannot be separated quickly enough, and they form interpenetrating structures. In Fig. 1 SEM pictures of material obtained from

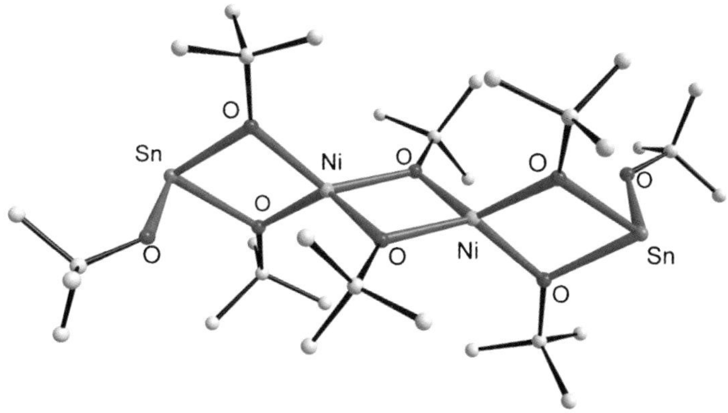

Fig. 2. Molecular structure (X-ray) of the mixed metal alkoxide $Ni_2Sn_2[OC(CH_3)_3]_8$, a single source precursor to Ni_3Sn_4 in SnO_2; the hydrogen atoms are omitted for clarity.

$BaSn_2[OC(CH_3)_3)]_6$ at different scales are shown, illustrating ball-like structures with fractal features, as every ball consists of smaller balls which are again made up of smaller ones and so on . . .

With surface sensitive techniques like XPS it can be shown that the balls have a core-shell structure with the metal occupying the core and the metal oxide occupying the shell [26].

Changing the reaction conditions a little bit, we have also succeeded to produce nano-scaled alloy particles in a metal oxide environment [27]: Ni_3Sn_4 clusters in a SnO_2 matrix are obtained by CVD of $Ni_2Sn_2[OC(CH_3)_3]_8$ [28]. Again XPS techniques have been used to discriminate between the different possible alloys which might be formed in this reaction as Ni_3Sn, Ni_3Sn_2 or Ni_3Sn_4 [27]. The molecular structure of the precursor molecule is displayed in Fig. 2.

Using a classical bimolecular redox-approach in organic solvents, the same metal clusters like Ge_n, Sn_n and Pb_n or even Ni_n at a nano-scale may also be prepared, with the choice of different matrices [29–32]. These can be organic solvents (colloids) or structured mesoporic materials like Al_2O_3. Metal clusters of aluminum or gallium, again synthesized from single source precursors, are addressed to in Sect. 2.3.

2.2 Metal oxide phases and metal oxide composites

2.2.1 Single oxide phases by single source precursors

Before discussing the simultaneous synthesis of two metal oxide phases using a single precursor molecule, we would like to present our approach to metal-oxide phases using different and complementary ligands at the metal atom.

We had very good experiences combining the tert-butoxy group with the hydride ligand. Whereas the hydrogen at the metal atom is negatively charged because of the polarity of the M–H bond, the hydrogen atoms on the methyl groups of the tert-butyl group are positively charged. The two opposite charges in the molecule can of course "balance each other" forming dihydrogen and thus behave as starting point of a cascade reaction, which is shown for $HAl[OC(CH_3)_3]_2$ in Eqs. (8)–(10) [33].

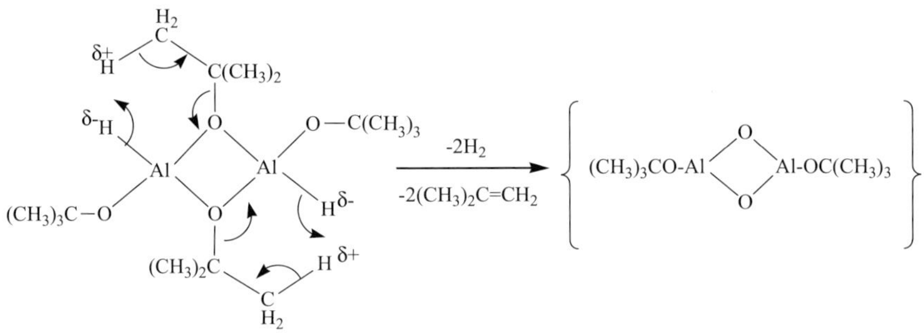

(8)

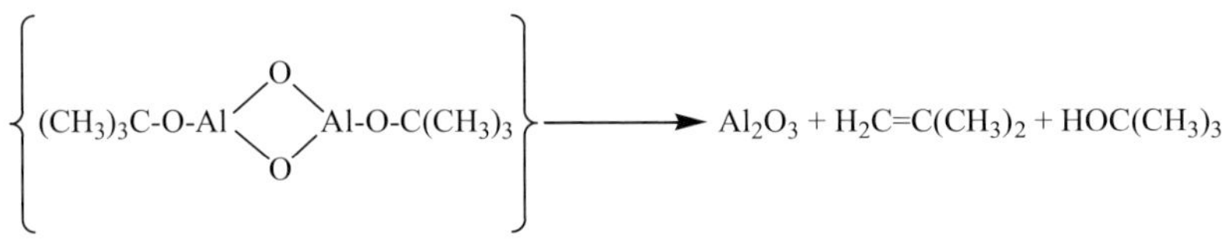

(9)

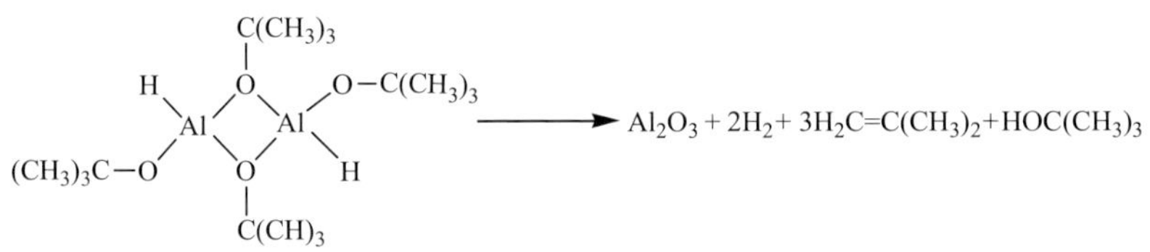

(10)

The overall reaction is assembled in Eq. (10) while some probable mechanism details are found in Eqs. (8) and (9) with the transient intermediate in brackets. We have been able to get some more insight in the reaction path using $[(CH_3)_3C{-}O]_2Al{-}D$ instead of $[(CH_3)_3C{-}O]_2Al{-}H$: in the mass spectrum, which has been obtained on-line during this reaction, only HD and no D_2 has been detected besides the mass peaks and fragments of iso-butene and tert-butanol. This means that, as indicated in Eq. (8), the source of the dihydrogen is in equal parts from the hydride ligand and the tert-butyl group. Anyway, by this method dense and pure Al_2O_3 coatings on metallic (Fe, Ni, Cu, Pt) or other substrates (Si, SiO_2, glass) may be obtained, which at a temperature below 400 °C are X-ray amorphous and above 450 °C mostly belong to the γ-Al_2O_3 phase.

Of course, this process, with two different ligands at the aluminum atom, described for Al_2O_3, can also be adopted to other oxide systems containing apart from aluminum other metallic elements [34–36]. In Eq. (11) the reaction of $[(CH_3)_3CO]_4H_4Al_2Mg$ to spinel $MgAl_2O_4$ is summarized.

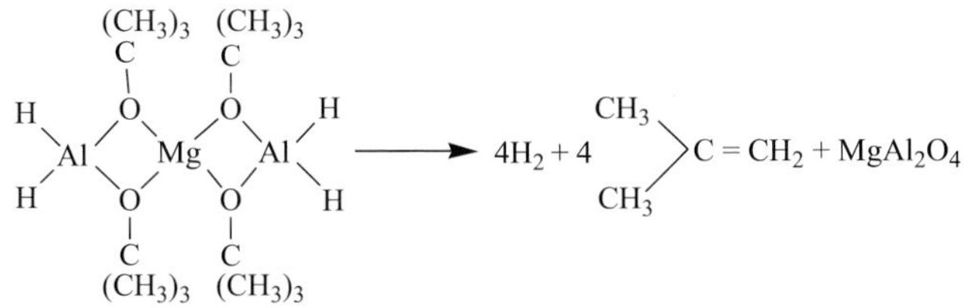

(11)

The spinel obtained in this reaction is chemically pure (carbon content less than 0.5%), especially when a gas-phase method is used. As the spinel phase is built up by a bottom up process, the size of the crystalline particles can be easily adjusted by the temperature and the running time of the process. Typically at temperatures around 500 °C, crystals of 5–20 nm diameters are obtained, but also thin films on metallic substrates (like steel) with a high amorphous ratio may be produced [34]. The ratio of normal to inverse spinel (some of the magnesium atoms also occupying octahedral sites) has been checked by solid state NMR spectroscopy and has been found to be approximately 70 : 30%. The superiority of the hydride/alcoholate ligand system compared to ordinary alkoxides can be shown by using $[(CH_3)_3CO]_4H_4Al_2Mg$, $[(CH_3)_2HCO]_8Al_2Mg$ or $[(CH_3)_3CO]_8Al_2Mg$ in a CVD process under comparable conditions: the smallest particles of spinel (less than 10 nm in diameter) and at quite moderate conditions (400–450 °C compared to 550 °C with the pure alkoxides) are found for the hydrido/alcoholate precursor [36]. Moreover, after annealing, the deposits obtained from the hydrido/alcoholate precursor show the highest sheet resistance [36].

Spinels of the general formula MAl_2O_4 (M = Co, Ni, Cu) can be prepared at the nano-scale (5–45 nm) using single source precursors of the type $MAl_2(OR)_8$ (R = iso-propyl, tert-butyl) in a microemulsion assisted sol–gel process [37, 38]. The hydrolysis/condensation process is run in water droplets, their diameters being tunable by the chain length of the surfactant molecules. There is a correlation between the initial droplet size and the crystallite size of the resulting spinel. In the solid state ^{27}Al-NMR, the almost complete inverse nature of the obtained spinel $NiAl_2O_4$ can be deduced from the chemical shift, whereas in the other cases mixtures seem to be present. A test with a stoichiometric mixture of cobalt and aluminum alcoholate in the microemulsion droplets reveals less homogeneity in the ion distributions compared to the corresponding single source precursor.

Besides the nano-scaled spinel phases, a large number of oxide ceramics have been synthesized at the nano-scale and have been extensively characterized. Most of these processes have been performed in solution (sol–gel) or in the gas-phase (CVD) using again the single source precursor concept. In the case of perowskites, a precursor to $BaTiO_3$ and $BaZrO_3$ has been developed [39] and even PZT-like phases ($BaTi_{0.5}Zr_{0.5}O_3$) by molecular mixing using sol–gel techniques have been addressed to [40]. An optimization of the synthetic procedure to nano-crystalline $Y_3Al_5O_{12}$, using again different precursors, has been performed [41]. Some of these synthetical studies were connected with the physical properties and the aim to adjust them by control of the particle size and the high purity of the material. In this respect, $ZnAl_2O_4$ for optical uses [42], iron oxides with singular magnetic properties [43–47] and $ZnFe_2O_4$ due to its magnetic behavior in the nano-regime were synthesized [48].

2.2.2 Composites of metal oxide phases from a single source

As pointed out in Sect. 1 and in Eqs. (1)–(3), the molecular precursor may react to a single phase or may be assembled in such a way that not only one solid product is obtained but, in contrast, two phases are produced simultaneously [12]. We have shown this for the composites of a metallic phase and its oxide for Ge, Sn and Pb in Sect. 2.1 and we will demonstrate this for aluminum and gallium phases in their oxide matrices in Sect. 2.3. In this chapter, we show that the stoichiometric ratio of metals in the precursor assembly may also be used to create two different oxide phases at the same time. We illustrate this in detail for the $NdAlO_3/AlO_3$ system [49–51], but also other mixtures are possible [52] or have been studied [53].

In Eqs. (12) and (13) the overall reactions of the gas-phase as well as the sol–gel processes of two hetero metal alkoxides, $\{NdAl[O{-}CH(CH_3)_2]_6[HO{-}CH(CH_3)_2]\}_2$ and $\{NdAl_3[O{-}CH(CH_3)_2]_{12}[HO{-}CH(CH_3)_2]\}$, are assembled.

$$\{NdAl[O{-}CH(CH_3)_2]_6[HO{-}CH(CH_3)_2]\}_2 \rightarrow 2NdAlO_3 + \text{products} \quad (12)$$

$$\{NdAl_3[O{-}CH(CH_3)_2]_{12}[HO{-}CH(CH_3)_2]\} \rightarrow NdAlO_3/Al_2O_3 + \text{products} \quad (13)$$

Whereas in the gas-phase process, the solid ceramics are formed together with volatiles like iso-propanol, hydrogen, propene and acetone [49], in the sol–gel procedure the hydrolysis releasing the alcohol is followed by a condensation process at higher temperatures [50, 51]. Clearly, there is a correlation between the stoichiometric ratio of neodymium and aluminum in the products and in the precursor, as the excess of aluminum in the tetrametallic precursor leads to the simultaneous formation of neodym aluminum oxide and aluminum oxide. This formation is independent of the process used (gas-phase or solution) and differs only in the process parameters, the CVD product being formed at lower temperatures and therefore with a smaller particle size. In Fig. 3 the precursor molecules $\{NdAl[O{-}CH(CH_3)_2]_6[HO{-}CH(CH_3)_2]\}_2$ and $\{NdAl_3[O{-}CH(CH_3)_2]_{12}[HO{-}CH(CH_3)_2]\}$ are depicted, as found out by X-ray diffraction analysis, whereas in Fig. 4 a transmission microscopy image

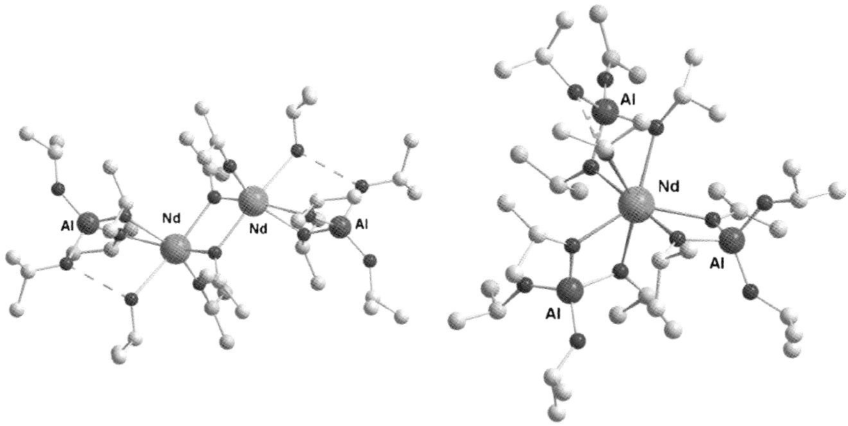

Fig. 3. The molecular structures of $\{NdAl[O{-}CH(CH_3)_2]_6[HO{-}CH(CH_3)_2]\}_2$ and $\{NdAl_3[O{-}CH(CH_3)_2]_{12}[HO{-}CH(CH_3)_2]\}$ from X-ray diffraction analysis (see also caption of Fig. 2). The dashed lines stand for hydrogen bridges.

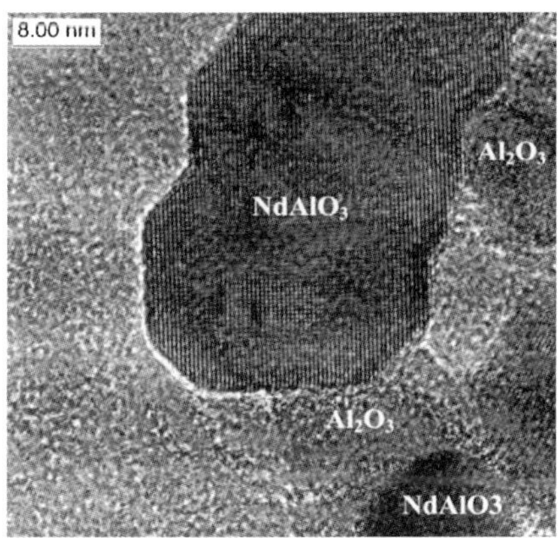

Fig. 4. High resolution TEM picture of $NdAlO_3/Al_2O_3$ composite obtained at 1400 °C; the crystallinity of $NdAlO_3$ may be seen from the lattice spacings.

(TEM) from the sol–gel $NdAlO_3/Al_2O_3$ is shown, sintered at 1400 °C. From those images as well as from powder-X-ray diffraction and electron diffraction studies, it can be concluded that at temperatures below 1200 °C only the $NdAlO_3$ is crystalline while Al_2O_3 is amorphous (X-ray), only crystallizing above this temperature (γ-, δ-, κ-phases). In the TEM images (Fig. 4) the crystallite sizes of $NdAlO_3$ are of the order of 50 nm (sol–gel) or 300 nm (CVD) while the Al_2O_3-grains are less than 10 nm. The single precursor process to $NdAlO_3/Al_2O_3$ can be compared to a classical solid-state reaction between Al_2O_3 and Nd_2O_3, first grinded and then heated at 1000–1400 °C.

Contrarily to the results described above, only $NdAlO_3$ and Nd_2O_3 can be characterized by their X-ray diffraction patterns together with an amorphous background.

The homogeneous distribution of the two oxide phases in the ceramic composite $NdAlO_3/Al_2O_3$ follows from electron microscopy, solid state ^{27}Al-NMR, XPS-spectroscopy and especially from the optical behavior of the composite [51, 53]. When a photoluminescence spectrum of $NdAlO_3$ ($Nd\,4f^4\,F_{3/2} \rightarrow {}^4I_{9/2}$) is compared to that of the composite $NdAlO_3/Al_2O_3$ (excitation at 4 K by 351 nm radiation of an Ar laser) the intensity is enhanced by a factor of 35! The most important difference between the two samples could be the quenching effect in $NdAlO_3$ because of the higher probability of $Nd^{3+} - Nd^{3+}$ contacts which is of course less in $NdAlO_3$ "diluted in an Al_2O_3 matrix". Another effect could be the energy transfer to Nd^{3+} by the Al_2O_3 matrix [51]. These findings seem to be of general importance and applicability as they may be adopted for neodymium doped yttrium aluminum garnets [53].

2.3 Metal/metal oxide composites with Al and Ga

2.3.1 Metastable phases in the hydrido-oxo-aluminum system

As described in Sect. 2.2.1, the cascade reaction between the ligands H and $O{-}C(CH_3)_3$ in $H{-}Al[O{-}C(CH_3)_3]_2$ can be frequently used to synthesize high-purity nano-scaled γ-Al_2O_3. By changing the ligand ratios to $H_2Al[O{-}C(CH_3)_3]$ another precursor is available which is also dimeric like $H{-}Al[O{-}C(CH_3)_3]_2$ [54]. This dihydrido compound in the gas phase at temperatures as low as 270 °C loses hydrogen and iso-butene (compare also Eq. (11)) to form the ternary solid phase HAlO (Eq. (14)) [55–58].

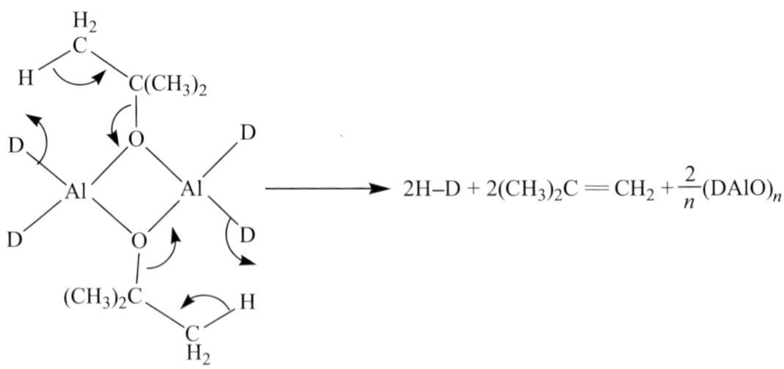

(14)

The deuterated compound in Eq. (14) has been used to demonstrate that the formed dihydrogen (HD) has its origin in the hydride (deuteride) on aluminum and the protic hydrogen of the tert-butyl group. In the on-line mass-spectrum not only the exclusive formation of HD can be recognized (no D_2 detected) but also the simultaneous formation of iso-butene (compare also Sect. 2.3.1).

The new phase hydrido aluminum oxide (found independently also by another group [59]) is amorphous, colorless and glass-like and may be produced as a coating on metallic and semi-conductor surfaces (Fe, Cu, Ni, Pt, Si *etc.*). Freshly prepared, it has a 1 : 1 : 1 stoichiometry with respect to aluminum, hydrogen and oxygen. The hydrogen is bonded to aluminum and not to oxygen as found *inter alia* in IR-spectra taken from the hydride and its deuterated derivative [58]. From theoretical calculations using different models, it can be concluded that the structural skeleton is made up of aluminum–oxygen bonds with hydrogen bonded to the aluminum atoms at a terminal position or generally speaking at the periphery [60, 61].

The solid glass-like phase HAlO is metastable as may be deduced from heating experiments. Indeed, very slowly at a temperature above 300 °C or quickly at 550 °C, HAlO (or DAlO) lose hydrogen (deuterium) and transform to biphasic Al/Al_2O_3 (Eq. (15)) [55–58].

$$3\mathrm{HAlO} \xrightarrow{\Delta T} \frac{3}{2}\mathrm{H_2} + \mathrm{Al} + \mathrm{Al_2O_3} \tag{15}$$

The two solid phases are mixed and interpenetrating and cannot be separated from one another by heating to induce phase separation. The physical and structural consequences of this behavior are addressed to in Sect. 2.3.2 in more detail. The composite Al/Al_2O_3 is, as far as we know, almost stoichiometric and can also be obtained from $H_2Al[O{-}C(CH_3)_3]$ without the intermediate formation of HAlO, when the dihydride precursor is heated up to 550 °C [55–57].

The formation of the composite Al/Al_2O_3 and the stoichiometric ratio of 1 Al^0 to 2 Al^{3+} can be explained by the instability of {AlO} which should remain after elimination of hydrogen from HAlO as expressed in Eq. (16).

$$\begin{aligned} 3\mathrm{HAlO} &\rightarrow \frac{3}{2}\overset{0}{\mathrm{H}}_2 + 3\{\overset{+2}{\mathrm{Al}}\mathrm{O}\} \\ 3\{\overset{+2}{\mathrm{Al}}\mathrm{O}\} &\rightarrow \overset{0}{\mathrm{Al}} + \overset{-3}{\mathrm{Al}}_2\mathrm{O}_3 \end{aligned} \tag{16}$$

The entire formation process of the two interpenetrating phases Al and Al_2O_3 from kinetic and thermodynamical points of view can be expressed in the qualitative diagram displayed in Fig. 5.

Within the diagram in Fig. 5, we were able to isolate the metastable HAlO; the {AlO} species could be transient or really intermediate. Until now, we were not able to define the nature of this {AlO} state clearly [33].

The metastability of HAlO makes this ceramic coating an excellent candidate for structuring using single laser beams or interferencing beams. Instead of transforming the whole surface of a substrate by general heating, laser optics allow to modify only such parts of the surface where the beam is interacting. In Fig. 6, the pattern of such a local transformation is shown by using a Nd:YAG laser in the interference mode [62].

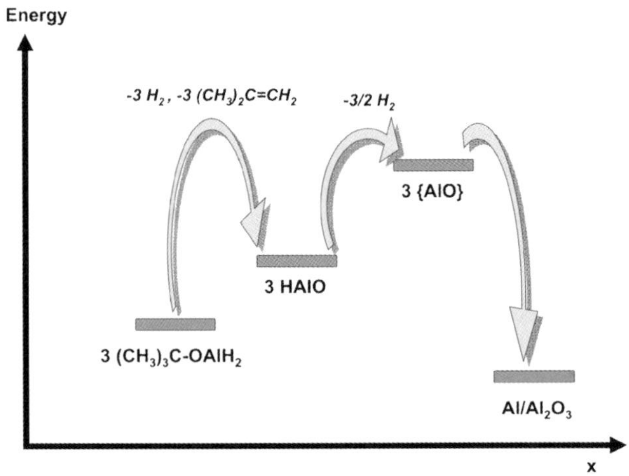

Fig. 5. Possible reaction coordinate x from $(CH_3)_3C{-}OAlH_2$ to Al/Al_2O_3 and metastable phases.

Fig. 6. Micro patterning of a HAlO coated surface with a Nd-YAG laser (SEM). The equally spaced lines, which consist of Al/Al_2O_3 phases, have been obtained by splitting the beam and by interference. In between, the HAlO phase remains unchanged.

A detailed characterization of the Al/Al_2O_3 lines obtained by a laser treatment of the HAlO coated surface reveals that, apart from the chemical transformation, also some melting processes have to be considered [63].

We have also tried to use the $(CH_3)_3C{-}OGaH_2$ precursor instead of $(CH_3)C{-}OAlH_2$ [64]. Interestingly, in this case we were not able until now to isolate a hydrido-oxo phase of the composition HGaO similar to HAlO. Instead, we observed the formation of gallium droplets together with Ga_2O_3.

The droplets seemed to have an "oxide skin" because they were stable up to 170 °C without changing their shape. In contrast to the aluminum system, the two phases gallium and gallium oxide seem to separate from each other easily, which may be attributed to the low melting point of gallium which is around 30 °C.

2.3.2 *The Al/Al_2O_3 composite and its structural forms*

As pointed out in the former chapter, the composite Al/Al_2O_3 can be either prepared from $(CH_3)_3C{-}OAlH_2$ directly at temperatures > 450 °C or from HAlO. When using the hydrido alkoxy precursor, the pressure and the gas flow play a role in the chemical purity of the composite, as with high pressure (> 0.3 atm) and a moderate gas flow the product is contaminated by Al_4C_3 (aluminum carbide) as found out from powder X-ray diffraction [33]. This might give a hint to the reactivity of the aluminum clusters present at an early stage in the phase mixture, which in contact with hydrocarbon entities (like iso-butene) could partly transform to aluminum carbide under the experimental conditions [33].

When the flow of the gases is high or the organic by-products are efficiently pumped off, the material of phase mixture is no longer contaminated by carbides. At temperatures between 430–470 °C at a pressure of 0.01 atm of $(CH_3)_3C{-}OAlH_2$ on metallic targets as well as on non-conducting targets like glass which are indirectly heated in a field of microwaves by direct contact to a graphite body, ball-like particles grow which resemble those obtained from germanium alkoxides or tin alkoxides (see Sect. 2.1) As may be seen from Fig. 7, again the particles are fractal and composed of hierarchies of balls. The fractal dimension has been established from neutron diffraction to be approximately between 2 and 3 [65]. The particles seem to have a core-shell structure, with the aluminum phase occupying the center of the ball and the oxide phase occupying the shell. This explains nicely that the XPS spectra of the phase mixtures display only signals for Al^{3+} while the Al^0 signals are only visible after scratching or continuous argon sputtering. It also explains that these conglomerates are stable on air although they contain Al^0-clusters of small dimensions (0.5–5 nm) which should instantly inflame if they were not protected.

We were very much surprised that these ball-like structures changed to nano-wires when the deposition temperature for the Al/Al_2O_3 composite was raised to over 500 °C. The zero-dimensional structures changed to one-dimensional with an aspect ratio of 20 nm diameter to several μm length. Clearly, the biphasic material seems to grow presumably driven by a phase separation due to phase ordering through crystallisation.

From high-resolution TEM studies, it can be shown that the wires are also of the core-shell type and the inner part of the wire, displaying under certain angles a higher contrast, consists of crystalline aluminum (electron and X-ray diffraction). The outer shell of the wire is made of several tiny crystals

Fig. 11. Branched Al/Al_2O_3 wires obtained by gase-phase deposition of $(CH_3)_3C{-}OAlH_2$ at 530 °C/0.1 atm.

Fig. 12. A droplet of water on an Al/Al_2O_3 coating showing ultrahydrophobic properties.

corundum phase. It seems that the aluminum regions in the composite, which, because of the nano-scale, have a high surface area, react under the laser destruction of the composite very efficiently with oxygen. The liberated energy is trapped in the system and is high enough to drive the transformation to the high temperature α-Al_2O_3 form (see also Fig. 9).

The Al/Al_2O_3 surface, either made of ball-like structures or of wires, under further exposion to the precursor $(CH_3)_3C{-}OAlH_2$ may grow onward in different ways. The structures may form bigger micrometer clusters or islands or the wires are branching (Fig. 11) [70]. This tertiary structure has a dramatic effect on the general properties of the surfaces, as may be shown by wettability experiments [71]. We observed in these experiments that a range of hydrophilic (contact angles: 2–3°) to ultrahydrophobic (contact angle: 178°) regions is found (see Fig. 12), depending on the nano- and micro-structure ratio and arrangement on the surface.

3. Conclusion

The adequate chemical composition of a precursor molecule used in a sol–gel or chemical vapour deposition process, has a high impact on the purity of materials obtained and allows to control the phases, the primary and secondary structures and therefore also the physical properties of the materials. Our approach using simple intra-molecular "cascade reactions" (also known in organic chemistry as "domino reactions") and the choice of the molecular ratio of the metallic components in the precursor allows to obtain metastable and mixed solid phases. Whereas the metastability allows subsequent transformations using physical methods (heat, laser, electron-beam, light), the molecular approach of biphasic materials leads to nano-structures of high reproducibility and structural control. This last aspect may be described as self-assembly of solid phases and will be an exciting platform of research and application in the future.

Acknowledgement

This report would not have been possible without the high engagement of my co-workers, especially S. Mathur (see citations), and the cooperation within the SFB 277, especially with my colleagues S. Hüfner, R. Hempelmann, H. P. Beck, M. Springborg, K. Jakobs, and U. Hartmann. The Deutsche Forschungsgemeinschaft is not only acknowledged for its financial aid, but also for its spiritual support through the experts and their contributions. Some work cited here has also been made in the framework of the international Graduiertenkolleg 532. Finally, I would like to thank the Fonds der Chemischen Industrie for its continuous support.

References

1. H. M. Manesevit and W. I. Simpson, J. Electrochem. Soc. **116** (1969) 1725.
2. S. Ito and T. Shinohara, J. Electrochem. Soc. **120** (1973) 1419.
3. J. D. Wright and N. A. J. M. Sommerdijk, *Sol-Gel Materials Chemistry and Applications*. CRC Press (2001).
4. U. Schubert and N. Hüsing, *Synthesis of Inorganic Materials*. 2. edn., Wiley-VCH, Weinheim (2004).
5. K. T. Ranjit, I. Martyanov, D. Demydov, S. Uma, S. Rodrigues, and K. J. Klabunde, J. Sol-Gel Sci. Technol. **40** (2006) 335.
6. G. Frenzer and W. F. Maier, Ann. Rev. Mater. Res. **36** (2006) 281.
7. K. Ariga, A. Vinu, and M. Miyahara, Curr. Nanosci. **2** (2006) 197.

8. N. Kimizuka, J. Synth. Org. Chem. Japan **62** (2004) 440.
9. K. Sada, M. Takeuchi, N. Fujita, M. Numata, and S. Shinkai, Chem. Soc. Rev. **36** (2007) 415.
10. M. Willander, O. Nur, Y. E. Lozovik, S. M. Al-Hilli, Z. Chiragwandi, Q. H. Hu, Q. X. Zhao, and P. Klason, Microelectron. J. **36** (2005) 940.
11. C. J. Brinker and G. W. Scherer, *Sol-Gel Science: The Physics and Chemistry of Sol-Gel Processing*. Academic Press, Boston (1990).
12. C. J. Brinker, A. J. Hurd, P. R. Schunk, G. C. Frye, and C. S. Ashley, J. Non-Cryst. Sol. **147/148** (1992) 424.
13. C. D. Chandler, M. J. Hampden-Smith, and R. W. Schwartz, Mater. Res. Soc. Symp. Proc. **310** (1993) 357.
14. J. P. Carpenter, C. M. Lukehart, S. B. Milne, S. R. Stock, and J. E. Wittig, Inorg. Chim. Acta **251** (1996) 151.
15. B. D. Fahlman, Curr. Organic Chem. **10** (2006) 102.
16. N. L. Pickett and P. O'Brien, Chem. Record **1** (2001) 467.
17. S. Mathur, M. Veith, H. Shen, and S. Hüfner, Mater. Sci. Forum **386–388** (2002) 341.
18. P. Werndrup, M. Verdenelli, F. Chassagneux, S. Parola, and V. G. Kessler, J. Mater. Chem. **14** (2004) 344.
19. V. G. Kessler, J. Sol-Gel Sci. Technol. **32** (2004) 11.
20. M. Veith, J. Chem. Soc., Dalton Trans. (2002) 2405.
21. M. Veith, S. Weidner, K. Kunze, D. Käfer, J. Hans, and V. Huch, Coord. Chem. Rev. **137** (1994) 297.
22. C. Sanchez, G. J. D. A. Soler-Illia, F. Ribot, T. Lalot, C. R. Mayer, and V. Cabuil, Chem. Mater. **13** (2001) 3061.
23. M. Veith, S. Hill, and V. Huch, Z. Anorg. Allg. Chem. **627** (2001) 1495.
24. M. Veith, C. Mathur, and V. Huch, J. Chem. Soc. Dalton Trans. (1997) 995.
25. H. Reuter and M. Kremser, Z. Anorg. Allg. Chem. **598** (1991) 259.
26. M. Veith, S. J. Kneip, A. Jungmann, and S. Hüfner, Z. Anorg. Allg. Chem. **623** (1997) 1507.
27. M. Veith, N. Lecerf, S. Mathur, H. Shen, and S. Hüfner, Chem. Mater. **11** (1999) 3103.
28. M. Veith, D. Käfer, J. Koch, P. May, L. Stahl, and V. Huch, Chem. Ber. **125** (1992) 1033.
29. M. Veith, O. Schütt, J. Blin, S. Becker, J. Frères, and V. Huch, Z. Anorg. Allg. Chem. **628** (2002) 138.
30. M. Veith, S. Mathur, P. König, C. Cavelius, J. Biegler, A. Rammo, V. Huch, H. Shen, and G. Schmid, Compt. Ren. Chimie **7** (2004) 509.
31. M. Veith, J. Frères, P. König, O. Schütt, V. Huch, and J. Blin, J. Eur. Inorg. Chem. (2005) 3699.
32. J. Hesse, H. Bremers, O. Hupe, M. Veith, E. W. Fritscher, and K. Valtchev, J. Magn. Magn. Mater. **212** (2000) 153.
33. M. Veith, Top. Organomet. Chem. **9** (2005) 81.
34. M. Veith, A. Altherr, and H. Wolfanger, Chem. Vapor Depos. **5** (1999) 87.
35. R. Winter, M. Quinten, A. Dierstein, R. Hempelmann, A. Altherr, and M. Veith, J. Appl. Cryst. **33** (2000) 507.
36. S. Mathur, M. Veith,T. Rügamer, E. Hemmer, and H. Shen, Chem. Mater. **16** (2004) 1304.
37. F. Meyer, R. Hempelmann, S. Mathur, and M. Veith, J. Mater. Chem. **9** (1999) 1755.
38. F. Meyer, M. Veith, A. Dierstein, C. Beck, W. Härtl, R. Hempelmann, and S. Mathur, Nanostruct. Mater. **12** (1999) 71.
39. M. Veith, S. Mathur, V. Huch, and T. Decker, Eur. J. Inorg. Chem. (1998) 1327.
40. M. Veith, S. Mathur, N. Lecerf, V. Huch, T. Decker, H. P. Beck, W. Eiser, and R. Haberkorn, J. Sol-Gel Sci. Technol. **15** (2000) 145.

41. M. Veith, S. Mathur, A. Kareiva, M. Jilavi, M. Zimmer, and V. Huch, J. Chem. Mater. **9** (1999) 3069.
42. S. Mathur, M. Veith, M. Haas, N. Lecerf, R. Haberkorn, H. P. Beck, M. Jilavi, and S. Hüfner, J. Am. Ceram. Soc. **84** (2001) 1921.
43. S. Mathur, M. Veith, V. Sivakov, H. Shen, and H.-B. Gao, J. Phys. IV France **11** (2001) 487.
44. S. Mathur, M. Veith, V. Sivakov, H. Shen, V. Huch, U. Hartmann, and H.-B. Gao, Chem. Vapor Depos. **8** (2002) 277.
45. G. F. Goya, M. Veith, R. Rapalavicinte, H. Shen, and S. Mathur, Appl. Phys. Lett. **79** (2004) 118.
46. S. Mathur, M. Veith, R. Rapalavicinte, H. Shen, G. F. Goya, W. L. M. Filho, and T. S. Berquo, Chem. Mater. **16** (2004) 1906.
47. G. F. Goya, M. Veith, R. Rapalavicinte, H. Shen, and S. Mathur, Appl. Phys. A **80** (2005) 1523.
48. M. Veith, M. Haas, and V. Huch, Chem. Mater. **17** (2005) 95.
49. M. Veith, S. Mathur, N. Lecerf, K. Bartz, M. Heintz, and V. Huch, Chem. Mater. **12** (2000) 271.
50. S. Mathur, M. Veith, H. Shen, N. Lecerf, and S. Hüfner, Scripta Mater. **44** (2001) 2105.
51. M. Veith, S. Mathur, H. Shen, N. Lecerf, S. Hüfner, and M. Jilavi, Chem. Mater. **13** (2001) 4041.
52. M. Veith, C. Mathur, S. Mathur, and V. Huch, Polyhedron **17** (1998) 1005.
53. S. Mathur, M. Veith, H. Shen, R. Rapalavicinte, and T. Agne, J. Am. Ceram. Soc. **89** (2006) 2027.
54. M. Veith, S. Faber, H. Wolfanger, and V. Huch, Chem. Ber. **129** (1996) 381.
55. M. Veith, S. Kneip, S. Faber, and E. Fritscher, Mater. Sci. Forum **269–272** (1998) 303.
56. M. Veith, Mater. Sci. Forum **343–346** (2000) 531.
57. M. Veith and K. Andres, J. Metastable Nanocryst. Mater. **15/16** (2003) 279.
58. M. Veith, K. Andres, S. Faber, J. Blin, M. Zimmer, Y. Wolf, H. Schnöckel, R. Köppe, R. de Masi, and S. Hüfner, Eur. J. Inorg. Chem. (2003) 4387.
59. S. Liu, U. Fooken, C. M. Burba, M. A. Eastman, and R. J. Wehmschulte, Chem. Mater. **15** (2003) 2803.
60. Y. Dong, M. Springborg, M. Burkhart, and M. Veith, J. Phys. Chem. B. **109** (2005) 22 820.
61. Y. Dong, M. Springborg, M. Burkhart, and M. Veith, Lect. Ser. Comput. Comput. Sci. **4** (2005) 1010.
62. M. Veith, K. Andres, C. Petersen, C. Daniel, C. Holzapfel, and F. Mücklich, Adv. Eng. Mater. **7** (2005) 27.
63. M. Veith and C. Petersen, unpublished results; C. Petersen, Dissertation, University of Saarbrücken 2007.
64. M. Veith and E. Sow, unpublished results.
65. M. Veith, S. Faber, R. Hempelmann, S. Janssen, J. Prewo, and H. Eckerlebe, J. Mater. Sci. **31** (1996) 2009.
66. M. Veith, E. Sow, C. Petersen, U. Werner, and C. Aktas, paper in preparation.
67. A. Morales and C. Lieber, Science **279** (1998) 208.
68. F. Zhang, R. Barrowcliff, G. Stecker, W. Pan, D. Wang, and S.-T. Hsu, Japan. J. Appl. Phys. **44** (2005) L398.
69. M. Veith, Y. Wolf, and C. Aktas, unpublished results.
70. M. Veith and C. Petersen, unpublished results.
71. M. Veith, C. Petersen, and K. Jacobs, unpublished results.
72. M. Veith and S. Kneip, J. Mater. Sci. Lett. **13** (1994) 335.

One-Dimensional Semiconductor Nanostructures: Growth, Characterization and Device Applications

By Sanjay Mathur[1,2,*] and Sven Barth[1,2]

[1] INM Leibniz-Institut für Neue Materialien gGmbH, D-66123 Saarbrücken, Germany
[2] Department of Chemistry, Würzburg University, Am Hubland, D-97074 Würzburg, Germany

Molecular Precursor / Chemical Vapor Deposition / Semiconductor / Nanowires / Sensing

One dimensional (1D) inorganic materials are gaining increasing attention because of their unique structural features and interesting functional properties. Given the structural stability, they show promising application potential in vacuum as well as in oxidizing atmospheres, which provides them a competitive edge over their carbon-based counterparts. A number of synthetic procedures have been developed and demonstrated for 1D nanostructures that have led to intriguing morphological variations (wires, tubes, belts, rods, *etc.*), however the control over radial and axial dimensions remains a continuing challenge. In addition, the choice of material is rather limited. We have developed a generic approach for the size-selective and site-specific growth of nanowires by combining vapor–liquid–solid (VLS) approach with molecule-based chemical vapor deposition. The synthesis of nanowires (NWs) is based on the decomposition of discrete molecular species, which allows growing nanowires at low temperatures with a precise control over their diameter and length. The precursor chemistry can be tuned to facilitate the stripping of organic ligands and to achieve complete decomposition that is critical for maintaining the gas phase super-saturation necessary for 1D growth. High-yield synthesis of elemental (Ge) and compound semiconductors (SnO_2, Fe_3O_4, V_2O_5, In_2O_3) was performed by the chemical vapor deposition of appropriate metal-organic precursors. Axial and radial dimensions of the NWs were varied by adjusting the precursor feedstock, deposition temperature, and catalyst size. Finally, the device potential of these building blocks as photo- and gas sensors was investigated by integrating individual nanowires in electrical circuits using focussed ion beam (FIB) assisted nano-lithography.

---

* Corresponding author. E-mail: s.mathur@uni-wuerzburg.de

Z. Phys. Chem. **222** (2008) 307–317

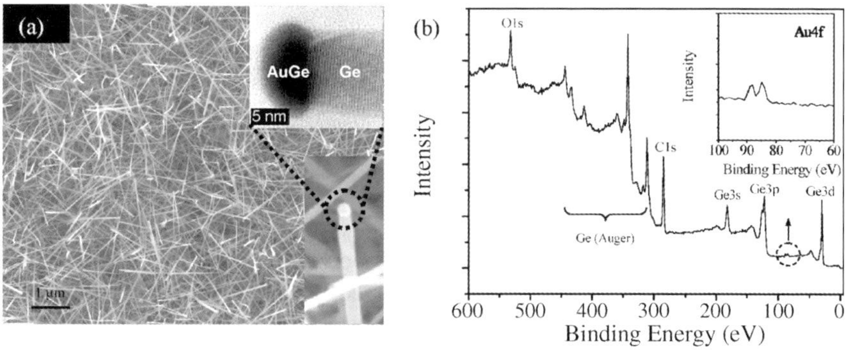

Fig. 1. (a) SEM and TEM images and (b) XPS spectra of as-prepared Ge NWs.

Fig. 2. TEM images of single crystalline Ge nanowires (a and b) and Ge core-shell nanowires (c and d).

The high-resolution TEM image of an individual NW (Fig. 2a) revealed the defect-free single crystalline nature of the Ge nanostructure with a highly ordered surface, however few NWs displayed crystal defects (twinning) and local disordered regions (Fig. 2b). The diameter of the nanowire can be tuned

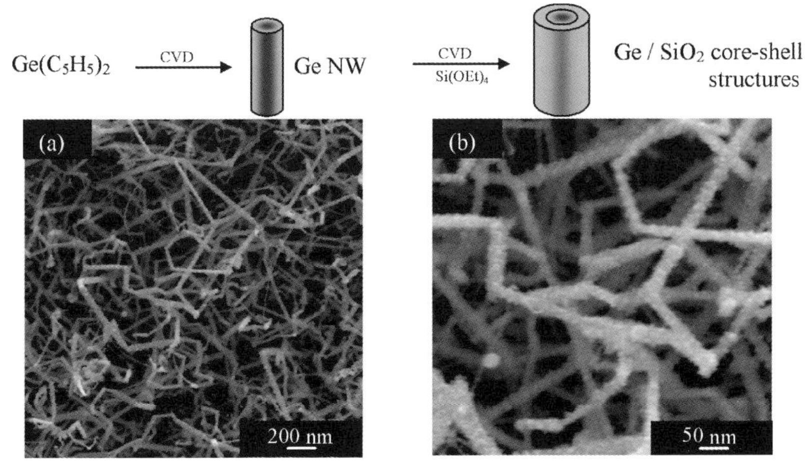

Fig. 3. Schematic representation of the experimental procedures used to fabricate core-shell nanostructures and corresponding SEM images of Ge/SiO_2 core-shell nanowires.

by changing the size of the gold catalyst, which could be varied using gold colloids of different sizes or by varying the thickness of sputtered gold films [15]. The Au film was kept thin in order to avoid the formation of undesired large Au aggregates produced upon thermal annealing.

Single crystalline Ge nanowires with diameters of less than 5 nm were successfully synthesized by adjusting appropriate precursor flux and deposition temperature. Germanium nanostructures of these dimensions far below the Bohr radius of Ge (24.3 nm) are expected to exhibit prominent quantum effects [16]. Figure 2 shows TEM images of single, twinned and core-shell 1D germanium nanostructures, which could be obtained under specific parameter combinations. The porosity and thickness of amorphous Ge overlayers could be controlled by tuning the gas phase saturation of $[Ge(Cp)_2]$ precursor. Increased precursor flux at low process temperatures resulted in the formation of amorphous Ge shells on single crystalline Ge cores. The surface of as-deposited Ge nanowires can be modified and/or protected by passivating the surface with inert shells, such as a SiO_2 coating. Figure 3 shows the morphology of Ge/SiO_2 core-shell composites obtained by the CVD of $[Si(OEt)_4]$ on Ge NWs obtained by the CVD of $[Ge(C_5H_5)_2]$.

Since the size and arrangement of catalytic metal droplets determine the density and dispersity of 1D structures, selective growth of Ge nanowires on patterned substrates was performed in order to generate arrays of nanostructures for device applications. Figure 4 shows the SEM images of patterned Ni dots on a silicon substrate and selective deposition of Ge nanostructures in which the Ge 'flowers' composed of bundles of Ge nanowires are located orderly on the original periodicity of Ni catalysts.

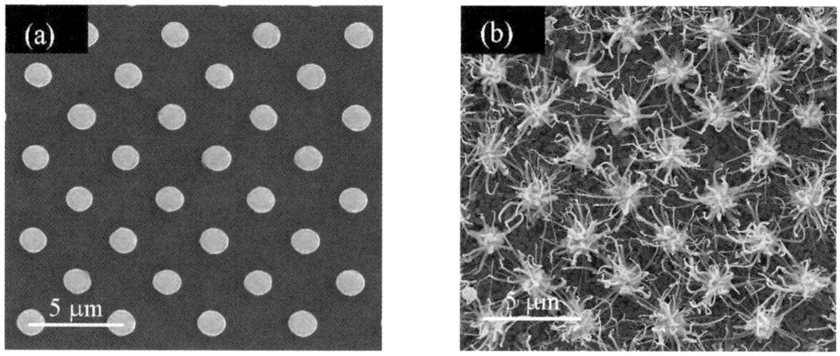

Fig. 4. SEM images of (a) patterned Ni dots on silicon and selective deposition of Ge 'flowers'.

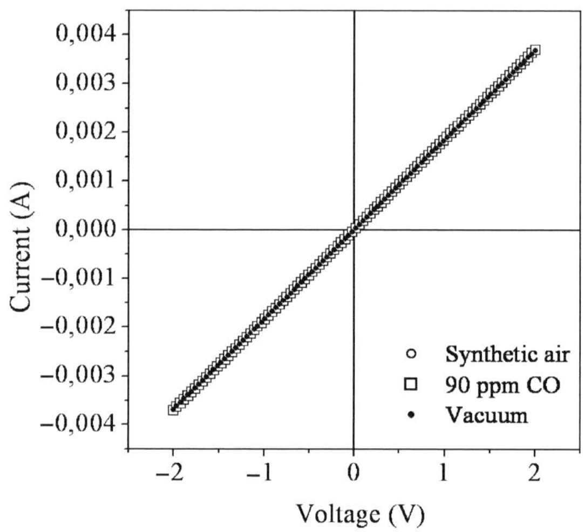

Fig. 5. Room temperature *I–V* characteristics of Ge nanowire films in different atmospheres.

Nanowires can be used not only as interconnecting units but also as functional sensing elements due to their response to external chemical, optical or electrical stimulus. In this context, the charge carrier transport in nanowires depends on their surface state. *I–V* characteristics of as-prepared Ge nanowires were studied under various atmospheres (synthetic air, 90 ppm CO in air and vacuum) in order to check the influence of external gas species on the charge carrier transport behavior.

Figure 5 shows the current–voltage plots of Ge nanowires in different environments, indicating low sensitivity of Ge NWs towards different gases at room temperature.

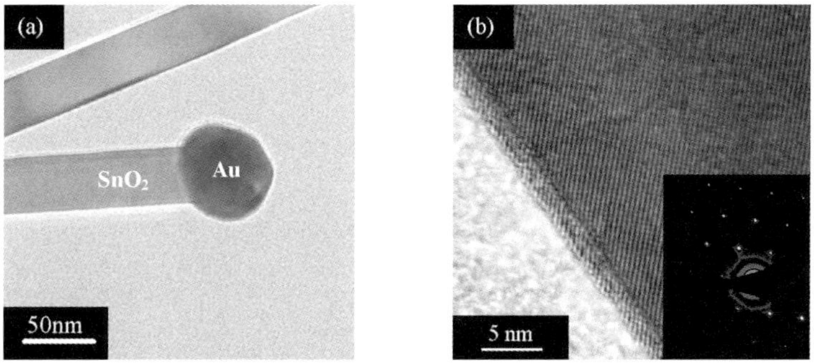

Fig. 6. (a) TEM and (b) HR-TEM images of SnO_2 nanowire grown at 700 °C. Corresponding SAED pattern of a nanowire is shown in inset.

The extraordinary chemical stability of Ge NWs is possibly due to surface passivation, for instance, through chemisorption of atomic hydrogen, produced during the thermolysis of $[Ge(Cp)_2]$, on the nanowire surface. The stable intrinsic and surface properties such as oxidation resistance are highly interesting for device applications.

2.2 Oxide nanowires

Semiconducting metal oxide nanowires are gaining increasing attention due to their numerous potential applications [17]. In this context high aspect ratio SnO_2 nanostructures have been synthesized by various gas phase and solution techniques [18–21]. We have grown tin oxide NWs by CVD of $[Sn(OBu^t)_4]$ on gold coated substrates (alumina, quartz, silicon, *etc.*) in a cold-wall reactor [22]. The chemical source contained preformed Sn–O bonds, which allowed NW synthesis at moderate temperatures (550–900 °C) without additional oxygen source. The decomposition of the tin(IV)-*tert*-butoxide lead to the formation of *iso*-butene and *tert*-butyl alcohol, which were identified by online mass spectroscopic analysis (Eq. (2)). XRD measurements (not shown here) of tin oxide NW coatings revealed pure tetragonal SnO_2 with a preferred growth direction $\langle 100 \rangle$ [9]. Globular particles at the end of each wire indicated a gold particle catalysed NW growth (Fig. 6a). Nano-probe EDX revealed gold as the sole constituent of the catalytic particle, whereas merely tin and oxygen were detected in the NW body. This indicated a catalyst-assisted growth mechanism, where no intermixing of components in the molten metal tip is observed. However a recent study showed compositional variation of the catalyst tip during growth [23]. We assume an initiated nanowire growth by surface diffusion of precursor species/fragments to a thermodynamically favoured growth site at the interface between the substrate and the liquefied catalyst. Growth may continue by adsorption of phase forming building blocks on the growing

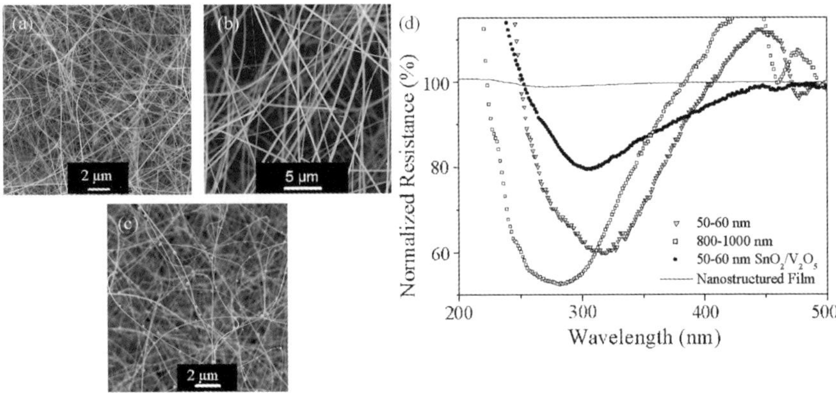

Fig. 7. SEM images of SnO_2 nanowires with average radial dimensions of (a) 50 nm and (b) 500 nm. (c) shows vanadium oxide decorated SnO_2 NWs as in (a), and (d) displays photoconductance values of pure SnO_2 NWs of different diameters, V_2O_5 decorated tin oxide wires and an nanograined film.

nanowire and/or the catalyst surface followed by diffusion to a low-energy surface (100).

$$[Sn(OBu^t)_4] \xrightarrow{CVD} SnO_2 + C_4H_9OH + C_4H_8\,. \tag{2}$$

The HR-TEM images of nanowires and sharp SAED pattern showed single crystalline SnO_2 wire bodies with low dislocation and defect density (Fig. 6b). SEM images of as grown products revealed randomly oriented nanowires in high areal density with tunable diameter in the range of 20–1000 nm and length of several micrometers. Two SnO_2 nanowire coated alumina substrates of different diameters are shown in Fig. 7a (50 nm) and b (500 nm), respectively.

Nanowire diameter was found to correspond to the size of gold particles and the precursor gas phase saturation during growth. Diameter control in SnO_2 NWs was used to investigate the influence of the modulation of wire diameter on the intrinsic properties of SnO_2. Illumination of semiconducting metal oxides with UV photons causes (i) desorption of ionosorped oxygen species (O^{2-}, O^- and O^{2-}) combined with release of surface-bound electrons in the native depletion layer and (ii) the exitation of charge carriers from the valence band into the conduction band [24].

Both effects reduce the electrical resistance by providing a higher density of charge carriers. Photo-conductance estimation of individual SnO_2 nanowires was performed on quartz substrates containing prefabricated Au comb-electrodes. A nanowire suspension was dropped on the device elements whereby SnO_2 wires bridged the electrodes.

A significant blue-shift of the main conduction value (photo-response peak, Fig. 7d) with shrinking wire diameter from average radial dimensions of

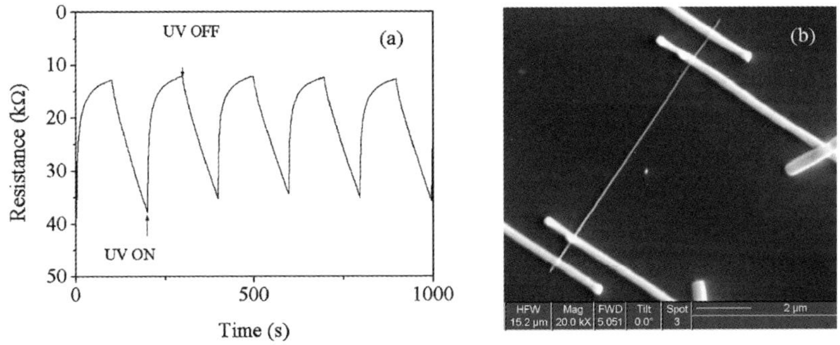

Fig. 8. (a) Periodical photoresponse of SnO_2 nanowires (diameter ~ 900 nm) recorded with and without illumination and (b) single SnO_2 NW contacted by FIB nanolithography.

900 nm to 50 nm was observed. This effect might be attributed to a quasi-one-dimensional confinement of the charge carriers, which is described in literature for a series of nanowire diameters [9]. However, the influence of surface area and possible stoichiometric defects, especially those located on the surface, cannot be ruled out. Prefabricated tin oxide NWs were decorated with vanadium oxide nanocrystals by CVD of $[VO(OPr^i)_3]$. SEM images showed the presence of the second phase, where the decoration of wires with nanocrystals (150–400 nm) is evident (Fig. 7c). The photo-response spectra revealed a shift in the photoresponse for SnO_2/V_2O_5 NWs to larger wavelength compared to pure SnO_2 (Fig. 7d), which suggests lowering of the band gap value. Vanadium oxide on SnO_2 nanostructures can influence the band-structure on the surface and/or the depth of the depletion layer when compared to pure SnO_2.

Photoswitching behaviour in SnO_2 NWs was demonstrated (Fig. 8a) by turning the UV source on and off, whereby the response alternated between high (on) and low (off) conductivity states. In these experiments, the UV-source was switched off at 90–95% of R_{min} (minimal resistance) value identified by flattening of the slope of resistance drop. Finally nanowire-based devices were fabricated by electrically contacting single nanowires using electron- and ion-beam assisted focused ion beam (FIB) nanolithography to evaluate electrical and sensing properties [25]. Sensors based on nanowires showed two orders of magnitude lower detection limit (5 ppm) towards CO when compared to literature reports [26]. In addition, oriented growth of SnO_2 NWs and 1D SnO_2/V_2O_5 hierarchical nanostructures was achieved and will be published elsewhere [27].

3. Conclusions

We described herein a low-temperature molecule-based CVD approach for controlled growth of nanowires. Combining catalytic growth mode, with single molecular sources allowed precise control over dimensions, site-specific growth, surface states and transport behaviors. Low decomposition temperature and clean ligand stripping of chemically designed precursors resulted in highly pure single crystalline Ge nanowires devoid of any native oxide. Increasing the precursor feedstock in the gas phase produced core-shell nanostructures in which single crystalline germanium nanowire cores were coated with an amorphous Ge overlayer. Further a two-step strategy was developed to produce Ge/SiO_2 heterostructures. Such core-shell nanocables will play an important role in the device applications due to surface modification and passivation treatment. Single crystalline tin oxide nanowires were synthesized in high yield at low temperatures by catalyst assisted CVD of $[Sn(OBu^t)_4]$ and showed a size- and surface-dependent photo-conductance. CO sensing behaviour of FIB contacted individual wire devices showed a detection limit of a few ppm indicating the sensitivity of nanowire based sensors.

In summary we have been able to demonstrate the strength of chemical methods in the controlled growth of one-dimensional nanostructures and modulation or preservation of their functional properties by surface functionalization. Finally, the device potential of semiconductor nanowires as building blocks for future devices has been demonstrated.

Acknowledgement

Thanks are due to the German Science Foundation (DFG) for supporting this work in the frame of the priority programme on nanomaterials – *Sonderforschungsbereich 277* – at the Saarland University, Saarbruecken, and for providing a Ph.D. fellowship to SB. Authors are thankful to the Saarland state and central government for providing the financial assistance. The infrastructure facilities and scientific guidance provided by Prof. M. Veith is gratefully acknowledged.

References

1. Y. Xia, P. Yang, Y. Sun, Y. Wu, B. Mayers, B. Gates, Y. Yin, F. Kim, and H. Yan, Adv. Mater. **15** (2003) 353.
2. Y. Wu, H. Yan, M. Huang, B. Messer, J. H. Song, and P. Yang, Chem. Eur. J. **8** (2002) 1261.
3. Y. Huang, X. Duan, and C. M. Lieber, Small **1** (2005) 142.
4. D. Yoffe, Adv. Phys. **42** (1993) 173.
5. M. Law, J. Goldberger, and P. Yang, Ann. Rev. Mater. Sci. **34** (2004) 83.
6. Y. Wu and P. D. Yang, Chem. Mater. **12** (2000) 605.
7. M. Batzill and U. Diebold, Prog. Surf. Sci. **79** (2005) 47.
8. D. Zhang, C. Li, S. Han, X. Liu, T. Tang, W. Jin, and C. Zhou, Appl. Phys. A **77** (2003) 163.
9. S. Mathur, S. Barth, H. Shen, J. C. Pyun, and U. Werner, Small **1** (2005) 713.

10. Y. Xia, P. Yang, Y. G. Sun, Y. Y. Wu, B. Mayers, B. Gates, Y. D. Yin, F. Kim, and Y. Q. Yan, Adv. Mater. **15** (2003) 353.
11. S. Mathur, H. Shen, V. Sivakov, and U. Werner, Chem. Mater. **16** (2004) 2449.
12. S. Mathur. S. Barth, J. C. Pyun, and H. Shen, Mater. Res. Soc. Symp. Proc. **910E**, Warrendale, 0901-Rb15-02 (2006).
13. R. S. Wagner and W. C. Ellis, Appl. Phys, Lett. **4** (1964) 89.
14. L. J. Lauhon, M. S. Gudiksen, D. Wang, and C. M. Lieber, Nature **420** (2002) 57.
15. M. S. Gudiksen, F. Wang, and C. M. Lieber, J. Phys. Chem. B **105** (2001) 4062.
16. Y. Maeda, Phys. Rev. B **51** (1995) 1658.
17. S. M. Sze, *Semiconductor Devices*. Wiley, New York (2001).
18. Z. R. Dai, J. L. Gole, J. D. Stout, and Z. L. Wang, J. Phys. Chem. B **106** (2002) 1274.
19. Z. Liu, D. Zhang, S. Han, C. Li, T. Tang, W. Jin, X. Liu, B. Lei, and C. Zhou, Adv. Mater. **15** (2003) 1754.
20. X. C. Jiang, Y. L. Wang, T. Herricks, and Y. N. Xia, J. Mater. Chem. **14** (2004) 695.
21. L. Vayssieres and M. Graetzel, Angew. Chem. Int. Ed. **43** (2004) 3666.
22. S. Mathur, M. Veith, V. Sivakov, H. Shen, V. Huch, U. Hartmann, and H. B. Gao, Chem. Vapor. Depos. **8** (2002) 277.
23. J. C. Harmand, G. Patriarche, N. Péré-Laperne, M.-N. Mérat-Combes, L. Travers, and F. Glas, Appl. Phys. Lett. **87** (2005) 203103.
24. P. Camagni, G. Faglia, P. Galinetto, C. Perego, G. Samoggia, and G. Sberveglieri, Sens. Actuators B **31** (1996) 99.
25. (a) F. Hernandez-Ramirez, A. Tarancon, O. Casals, J. Rodriguez, A. Romano-Rodriguez, J. R. Morante, S. Barth, S. Mathur, T. Y. Choi, D. Poulikakos, V. Callegari, and P. M. Nellen, Nanotechnology **17** (2006) 5577; (b) F. Hernandez-Ramirez, A. Tarancon, O. Casals, E. Pellicer, J. Rodríguez, A. Romano-Rodriguez, J. R. Morante, S. Barth, and S. Mathur, Phys. Rev. B, accepted.
26. F. Hernandez-Ramirez, A. Tarancon, O. Casals, E. Pellicer, J. Rodriguez, A. Romano-Rodriguez, J. R. Morante, S. Barth, and S. Mathur, Adv. Funct. Mater. (submitted).
27. S. Mathur, S. Barth, and H. Shen, Small (submitted).

Moderne Festkörperphysik

Siegfried Hunklinger
Festkörperphysik
2007 | 595 S. | Br.
€ 44,80
ISBN 978-3-486-57562-0

Eine moderne und didaktisch elegante Darstellung der Festkörperphysik.

In dem ausgefeilten und klar strukturierten Lehrbuch werden alle wichtigen Teilgebiete der Festkörperphysik behandelt und anschaulich die grundlegenden Gesetzmäßigkeiten und die für die Festkörperphysik typische Betrachtungsweise eingeführt. Konsequente Berücksichtigung finden zudem die amorphen Festkörper, die in unserer Umwelt eine wichtige Rolle spielen und in der Wissenschaft zunehmend an Bedeutung gewinnen. Zur Illustration von experimentellen Ergebnissen werden nicht nur schematische Darstellungen präsentiert, sondern in erster Linie Originaldaten herangezogen. Hierdurch sollen nicht zuletzt auch die Schwierigkeiten verdeutlicht werden, denen ein Experimentalphysiker in der Praxis gegenübersteht. An die Kapitel schließen sich Übungsaufgaben an, die die unmittelbare Überprüfung des Gelernten ermöglichen.

Prof. Dr. Siegfried Hunklinger ist Professor am Kirchhoff-Institut für Physik der Universität Heidelberg.

Oldenbourg

oldenbourg-wissenschaftsverlag.de

Nanocrystalline Metals Prepared by Electrodeposition

By H. Natter and R. Hempelmann*

Universität des Saarlandes, Physikalische Chemie, Geb. B2 2, D-66123 Saarbrücken, Germany

Pulsed Electrodeposition / Nanocrystalline / Nano Metals / Fuel Cell Catalysts / Ionic Liquids

An overview is presented about the preparation of nanocrystalline metals by pulsed electrodeposition out of aqueous electrolytes and ionic liquids. Appropriate selection of the bath chemistry and the plating parameters provide the flexibility to control the crystallite size. For selected examples we demonstrate how these electrochemical methods can be used for the preparation of nanostructured bulk metals and nanoparticle for fuel cell applications. Electrochemical analytical methods like cyclic voltammetry and chronoamperometry are used to fundamentally understand the deposition mechanisms, the effect of bath additives on the micro- and nanostructure of the deposits and to characterize the catalytic activity of the electrocatalysts. Some improved physical properties of electrodeposited nanocrystalline metals are discussed, namely thermal stability of n-Fe, magnetisation of n-Ni and hardness of n-Al.

1. Introduction

Nanostructured materials, *i.e.*, materials with ultra-fine crystals, less than 100 nm in size, were initially introduced as interfacial materials in the 1980's [1–3]. Traditional synthesis routes for nanostructured materials include physical and chemical vapour phase processing, severe plastic deformation as well as chemical methods which typically yield nano-powders. Electrodeposition has been used in the synthesis of bulk, compact nanocrystalline metals for over twelve years [4–8]. Electrochemical processing techniques offer the ability to nanoscale a large number of metals [9], alloys, and composite materials in various bulk forms and as improved innovative coatings [10]. Simultaneously,

* Corresponding author. E-mail: r.hempelmann@mx.uni-saarland.de

Z. Phys. Chem. **222** (2008) 319–354

sired the deposition should be performed at ambient temperatures or below in order to decrease the mobility of the adatoms and the metal ions in the electrolyte.
4. The ionic species (complex formation), the bath composition, the pH-value, the hydrodynamic conditions and also the use of special current shapes are further possibilities to control the crystallite size.

2.2 Variation of the crystallite size

A double-walled plating cell with a volume of 400 mL is suited for most of the electrochemical experiments. The temperature is kept constant using a thermostatic unit. The galvanostatic mode is preferred for pulse plating experiments because the potentiostatic mode with its three-electrode setup has problems with the voltage regulation for high pulse repetition rates: to instantaneously obtain a desired potential at the electrode the current should start theoretically from an infinite value, which is not feasible because of the limitations of the electronic equipment. In addition, in the potentiostatic mode a short reverse pulse would be necessary to achieve the initial rest potential, which could cause passivation problems.

All samples are deposited onto stainless steel, titanium, copper, or iron electrodes. The coating thickness (several microns to 2 cm) can be varied by the deposition time. Typically the deposits can be mechanically removed from the electrode. Otherwise, the temporary electrode substrate has to be dissolved in diluted acid. TEM (transmission electron microscopy) and XRD (X-ray diffraction) are used for crystallite size determination. We derive crystallite size, crystallite size distribution and micro strain from the X-ray patterns by means of our modified Warren–Averbach technique [27–30].

Organic compounds interact with a metal surface *via* free electron pairs (O, N, S) or π-electron systems of aromatic compounds. For this reason the effect of different additives on the nanostructures of a deposit has been studied. The bath composition and the experimental details are given in [9]. We consider a compound with free amino groups (butanediamine), a complex former (diammonium-EDTA) and a compound with a functional sulphur group (saccharin). For electrolytes without any grain refiners crystallite sizes in the range of 100 nm are obtained. After the addition of a very small amount of additives a reduction of the crystallite size by a factor of 2 is observed, see Fig. 1. Further addition of grain refiners decreases the crystallite size.

Butanediamine shows the strongest effect whereas diammonium-EDTA or saccharin lead to crystallite sizes not smaller than 40 and 54 nm, respectively. This behaviour is a proof for the assumed mechanism that free electron pairs interact with the metal surface. The nitrogen atoms in butanediamine interact strongly with the gold surface and therefore the inhibiting effect is very strong. Similar results are obtained for the system nano-copper/citric acid [16], nano-nickel/saccharin [7] and nano-Pd/Na_2EDTA [8]. A detailed description

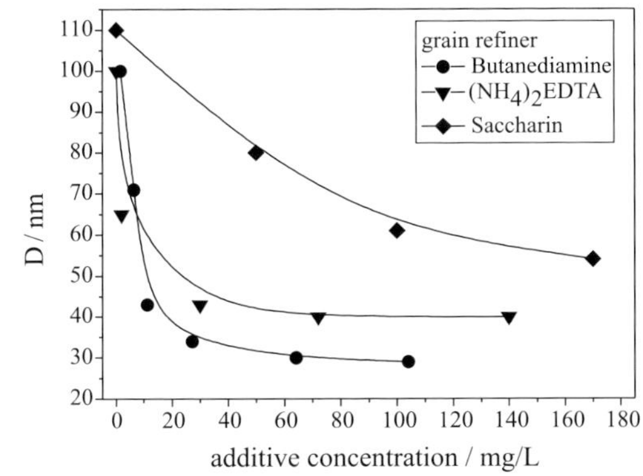

Fig. 1. The influence of different grain refiners on the nanostructure of gold deposits.

of the influence of organic additives on the microstructure of metal deposits is given by Fischer [31]. The reversibility of the adsorption process was confirmed with an ICP-OES analysis (electrolyte with 130 mg/L butanediamine, I_{pulse}: 500 mA/cm^2, t_{on}: 2 ms, t_{off}: 78 ms). The concentration of metallic impurities is below the detection limit in the case of nanostructured palladium [8]. The content of light elements is: 590 wt. ppm carbon, 130 wt. ppm hydrogen, 236 wt. ppm nitrogen and 47 wt. ppm oxygen. This analysis shows that the organic additives are not occluded in the deposit.

As demonstrated by Georgiev *et al.* [32] water soluble polymers are also suited as grain refining additives. The use of polyethylene glycol, poly-*N*-vinylpyrrolidone, starch, gum arabic, sodium alginate, quarternized guar gum and polydimethylaminoethylmethacryloleylpropanesulfonate leads to nickel deposits with crystallite sizes below 10 nm.

Alternatively, the crystallite size can be controlled by physical means, namely by the variation of t_{on} and t_{off} as will be demonstrated in the following example of nano-palladium. The off-time has been varied between 1 and 200 ms at constant on-time (3 ms) and at constant average current density (0.037 A/cm^2). The deposition at a constant average current density demands the variation of two pulse parameters (t_{off}, I_{pulse}): an increase in off-time is combined with an increase in pulse current density. The electrolyte in this example consists of 112 mmol/L $PdCl_2$, 378 mmol/L ammoniumsulfate, and 224 mmol/L Na_2EDTA. The pH is adjusted to 6–7 with $(NH)_4OH$. In this way the grain size can be varied between 20 nm and 128 nm, see Fig. 2.

As shown by Volmer [33] the energy of grain nucleus formation depends on the cathodic overpotential. A large cathodic overpotential reduces the energy of nucleus formation, and for this reason the number of nuclei increases. We clearly demonstrated in previous publications [8, 9, 16] that deposits prepared with a high pulse current density exhibit smaller grain sizes than those prepared with low pulse current densities. In our view the explanation for the

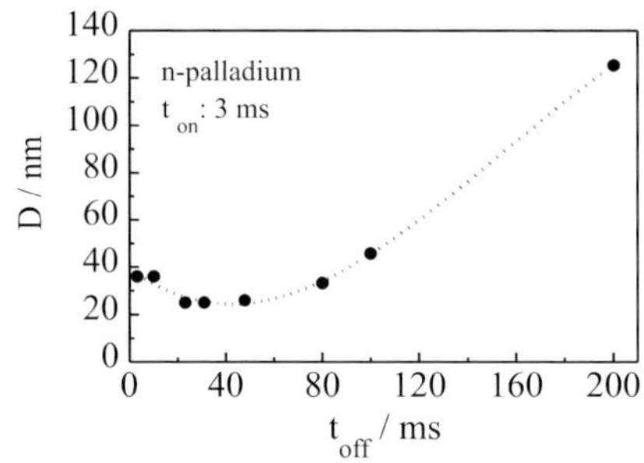

Fig. 2. Crystallite size variation of nanostructured palladium by the t_{off}-time.

Table 1. The influence of the bath temperature on the crystallite size of copper deposits.

$T/°C$	D/nm
20	< 10
30	10
40	14
50	44
60	121
80	microcryst.

grain growth at long off-times is that a high pulse current density causes an increase in the number of nuclei formed during a pulse. These small initial nuclei are thermodynamically instable and tend to form larger grains (recrystallization). In addition the recrystallization process is impeded by the adsorption of inhibiting molecules (EDTA) which block the growth sites, and the result is a fine grained deposit. Similar results have been obtained for Ni [7], Cu [16] and Fe [27].

It is well known from the literature [34] that electrochemical processes are influenced by different diffusion processes: the diffusion of metal ions in the electrolyte, the movement of adatoms on the electrode surface and also the mobility of the grain refiners. All these processes, therefore, depend on the temperature and should influence the nanostructuring of the deposits. For the experimental verification of the influence of the temperature on the nanostructure a copper electrolyte with citric acid additive has been used. Experimental details and parameters are given in [16]. The crystallite size increases from less than 10 nm to microcrystalline copper between 20 and 80 °C deposition temperature (Table 1) which indicates a strong temperature dependence of the nanocrystal size. Small metal grains have a high interface energy and a high tendency for its reduction by crystallite growth. This result is in accordance with the temperature dependence of the processes described above. The optimal temperature range for the preparation of nanocrystals is 20–40 °C.

2.3 Kinetics of nanophase formation

Electrochemical techniques are a powerful tool to investigate in situ the electrocrystallization processes [9]. The basics of phase formation and growth during the early stages of electrocrystallization on a foreign substrate are now fairly well understood [26]. The growth of each individual nucleus is determined by the rate of incorporation of new atoms, *i.e.* ion transfer and/or diffusion. It is well known that the current due to the initial growth of nuclei depends on their geometry. Different models based on different geometrical forms, namely hemispheroids [35], paraboloids [36], spherical caps [37], right circular cones [38, 39], are minutely described in the literature. After the initial growth, overlap of growing centres takes place. In the case of charge transfer control the growth initiates the overlap of the growing nuclei, whereas, in the case of diffusion control, the covering of the diffusion zones has to be considered [38]. This approach is well described for hemispheroids [35] and right circular cones [40]. Furthermore, side reactions such as hydrogen evolution [41] and/or concurrent adsorption processes [42] must be considered when describing the early stages of the electrocrystallization. The utilization of rotating disc electrodes (RDE) allows to minimize/avoid the influence of diffusion; therefore, the analysis of current-time transients is rather simple based on existing theoretical models. On the other hand, in real industrial applications it is generally impossible to rotate the cathode and both charge-transfer and diffusion controlled processes must be taken into account.

In a recent work [43] the kinetics of electrocrystallization of nanostructured gold on stationary and rotating disc glassy carbon electrodes has been measured using chronoamperometry, and the kinetic parameters describing early stages of crystallization have been evaluated. The electrochemical system consists of an aqueous solution of tetrachloroaureate(III) acid, 3-mercapto-1-propanesulfonic acid sodium salt and 50% ammonium hydroxide solution and is kept at room temperature. This electrolyte is non-toxic and can replace toxic cyanide electrolytes, details have been described elsewhere [44]. Glassy carbon was used as working electrode which, compared to metallic substrates, has the advantage of being amorphous; therefore, the nanostructure and microstructure formation of the deposit is not biased by any preferred texture. The counter electrode was a platinum plate and the reference electrode was a standard calomel electrode. Gold was deposited on the glassy carbon surface by means of cyclic voltammetry, linear sweep voltammetry (LSV) and single step chronoamperometry. Figure 3 represents current time transients recorded on a rotating disc electrode with rotation speed of 2000 rpm at the potentials −1700 mV and −1800 mV *vs.* SCE.

The data can be well described by three-dimensional nucleation of right circular cones and charge-transfer controlled growth as is obvious from the corresponding curve fits in Fig. 3. The analysis of gold kinetics by using of RDE

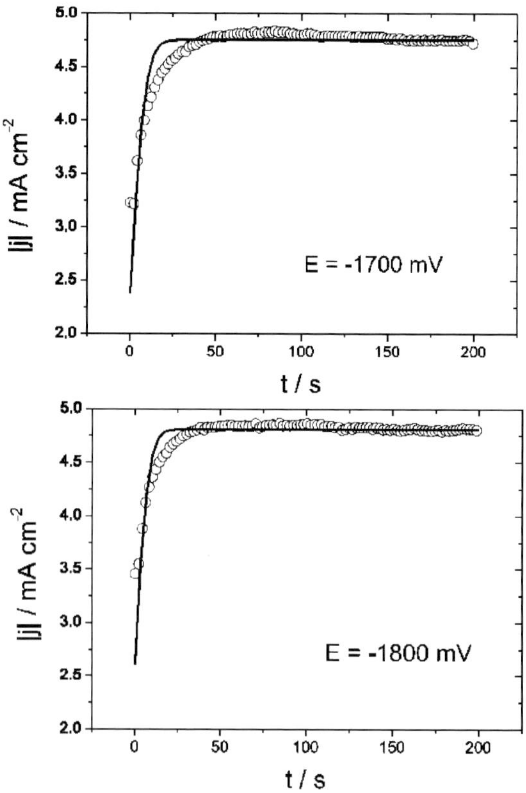

Fig. 3. Current–time transients registered on RDE at the potentials −1700 and −1800 mV *vs.* SCE with 2000 rpm rotation speed; the solid lines represent curve fits as explained in the text.

seems to be clear according to the existing theoretical models; at high rotation rates the influence of diffusion can be neglected.

Figure 4 illustrates a set of current–time transients in a wide range of potentials obtained using a stationary electrode. The potential is moved from −500 mV *vs.* SCE at $t = 0$ where gold deposition was not observed down to more negative potentials at $t > 0$. In the case of stationary working electrodes, diffusion cannot be neglected and the analysis of current–time transients becomes more complicated. The initial current decrease in Fig. 4 could be caused by different processes: a) a double layer charging current; b) a direct deposition of metal into the lattice without nucleation; and c) side reaction, such as hydrogen evolution. After an induction time, the nucleation process starts and gold crystals grow. This leads to an increase of the current with time as result of the increase of the electroactive surface area. The time, corresponding to the current minima extends to

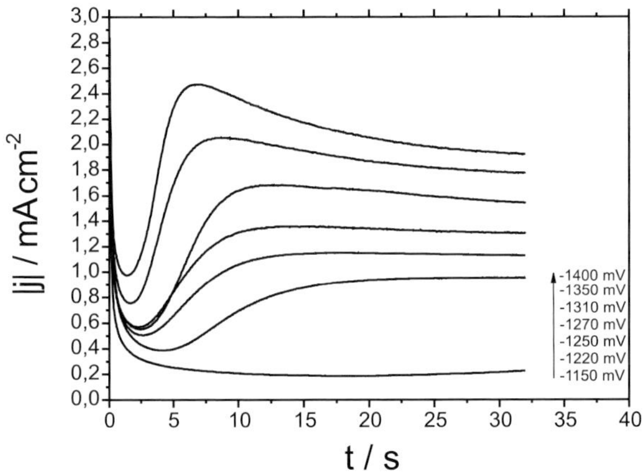

Fig. 4. Current–time transients registered on a stationary electrode in the potential range [−1150 mV; −1400 mV] *vs.* SCE.

a few seconds, depending on the potential applied. The estimated time of 6 ms required for the double layer charging is too short to contribute significantly to the current–time transients. Therefore it was assumed that the gold deposition process starts before the time corresponding to the current minimum is reached. The initial current minimum corresponds principally to a direct gold deposition without nucleation. This could be possible *via* an adsorption of gold complex followed by the electron transfer at more positive potentials, whereas at more negative potentials gold deposition can occur by direct charge transfer. It should be mentioned that no simple model of nucleation and growth could describe the experimental curves quantitatively. The process of nucleation and growth at the limiting conditions can be controlled by charge transfer at short times or by diffusion at longer times, when the concentration of active species in the vicinity of the cathode decreases. Probably, if the initial concentration of active species in the electrolyte is very low, the process of new phase formation can simultaneously be controlled by both charge transfer and diffusion processes [45]. Such mixed "charge transfer + diffusion" controlled growth was observed by [46] during the electrodeposition of copper on glassy carbon electrode and by [45] during the electrodeposition of cobalt-molybdenum on glassy carbon electrode [47]. Has shown that during copper electrocrystallization on oxidized ruthenium at long times of chronoamperometric experiments the current decays to a potential-dependent steady-state current that is consistent with mixed "diffusion + charge transfer" control. Analyzing all above described results and comparing them with the literature it is possible to suppose that gold electrodeposition on a stationary glassy carbon electrode, probably, occurs by mixed "charge transfer + diffusion" controlled nucleation and growth.

2.4 Crystallite growth of nanostructured iron

Of particular importance for any nanocrystalline material is its thermal stability. The onset of grain growth ultimately limits the maximum operating temperature [48]. A possibility to stabilize nanostructures is the addition of doping elements which act as diffusion barrier during the heating process. Boylan [49] showed for electrodeposited samples an increased thermal stability compared to metals prepared by inert gas condensation and attributed this behaviour to codeposited impurities. The pulse plating method can be used to incorporate so-called retarding elements like oxygen, nitrogen or sulphur into metallic samples. In the case of nanocrystalline Ni, with an intentionally increased oxygen content of 6000 wt. ppm [9], starting from an initial crystallite size of 19 nm only a slight grain growth up to 35 nm was observed at 400 °C at long times. High purity samples prepared, *e.g.*, by the inert gas condensation method, show a fast crystallite growth at this temperature within the first few minutes of annealing, resulting in a microcrystalline state. However, for nanocrystalline metals prepared by pulsed electrodeposition from an aqueous electrolyte with the concomitant incorporation of some oxygen into the grain boundaries (which thus are stabilized), at 400 °C and 500 °C the grain size initially increases rapidly by a factor of 2–3 but thereafter essentially remains constant and independent of the annealing time. A similar behaviour was observed for a Ni-S alloy at 300 °C [50, 51]. Several approaches including microalloying can be used to substantially increase the thermal stability [52].

The kinetics of grain growth has been studied in detail on the example of nanocrystalline iron by means of *in-situ* real-time X-ray synchrotron experiments [27]. Nanocrystalline iron samples with an initial crystallite size of 19 nm were prepared from a citric acid bath containing 50 g/L $(NH_4)_2Fe(SO_4)$, 220 g/L citric acid trisodium salt, 10 g/L citric acid and 40 g boric acid. The plating parameters are: $t_{on} = 2$ ms, $t_{off} = 8$ ms and a pulsed current density of 0.4 A/cm^2. The plating temperature was 30 °C. The porosity of the samples ($< 2\%$) was measured by mercury porosimetry. The crystallite size distribution is characterized by a σ-parameter of 1.5 (in terms of the log normal distribution). Chemical analysis of the resulting nano-Fe yields (in wt. ppm/wt. %): boron 538 ppm, Cr 8.3 ppm, Mn 9.9 ppm, Co 190 ppm, Ni 541 ppm, Cu 33 ppm, Mo 4.4 ppm, Ag 1.0 ppm, Cd 1.2 ppm, and Sn 1.8 ppm, hydrogen 0.09%, carbon 0.87%, oxygen 1.5%, and nitrogen 0.20% (wt. %). The metallic impurities probably originate from the Fe salt.

It should be noted that crystallite sizes larger than about 100 nm cannot reliably be determined using a laboratory X-ray diffractometer because the peak broadening resulting from the nanostructured sample is substantially smaller than the instrumental resolution (FWHM: 0.07–0.1°). A better resolution and thus access to larger grain sizes can be achieved by the use of monochromatic synchrotron radiation with a negligible beam divergence. The instrumental resolution for our experiment at the ESRF, Grenoble, beamline BM 16, is

0.006° (FWHM). Further advantages of synchrotron radiation are the high photon flux and the adjustable wavelength which for a real-time experiment results in a very good time resolution with, nevertheless, good data statistics. To apply the Warren–Averbach method at least two harmonic peaks of the same order have to be recorded. Since the area sensitive detector only covers a 2Θ-range of 19°, the incoming wavelength has been decreased to $\lambda = 0.4898$ Å such that the 2Θ-range of the instrument (in the present case $11^\circ \leq 2\Theta \leq 30^\circ$) comprises all the X-ray diffraction peaks necessary for the evaluation. The nanostructured iron sheet was crushed and filled into quartz capillaries (∅ 0.7 mm). The Θ–2Θ mode (transmission technique) was used for all experiments. To avoid texture effects the samples were rotated. Diffractograms were recorded in time steps of one minute isothermally at different temperatures (663, 683, 703, 739, 753, and 783 K). The sample temperature was controlled with a hot air blower. The instrumental resolution was calibrated with a NIST-LaB_6 standard sample.

The recorded patterns were fitted with one symmetric Pearson VII-function because the line profiles obtained from the synchrotron experiment do not show any asymmetries. From the χ^2-values (quality of a least squares fit), which do not change as function of time, it is obvious that the sample consists of a monomodal crystallite size distribution. A constant σ-behaviour indicates a normal crystallite growth. All nanostructure parameters like crystallite size, size distribution and micro strain were determined by our modified Warren–Averbach procedure from the 110 and 220 peak profiles.

Three different kinetic models of crystallite growth are used for the data evaluation:

1. Generalized parabolic grain growth model [53–56]

$$D(t)^n - D_0^n = k_1 t \tag{2}$$

(D_0 is the initial grain size, k_1 means the temperature dependent rate constant and n is an empirical grain growth exponent)

2. Grain growth model of Burke with impediment [57, 58]

$$k_2 t = \frac{D_0 - D(t)}{D_\infty} + \ell n\left(\frac{D_\infty - D_0}{D_\infty - D(t)}\right) \tag{3}$$

(D_∞ means the final crystallite size, a_2 and b_2 denote retarding terms connected to D_∞ by $k_2 = b_2^2/a_2 = a_2/D_\infty^2$)

3. Modified Burke model with a size-dependent retarding term [59, 60]

$$D(t) = \sqrt{\frac{a_3}{b_3} + \left(D_0^2 - \frac{a_3}{b_3}\right)\exp\left(-2b_3 t\right)} \tag{4}$$

($k_3 = 2b_3 = 2a_3/D_\infty^2$, the constants $a_{1,2,3}$ contain the product of specific interface energy γ and grain boundary mobility M).

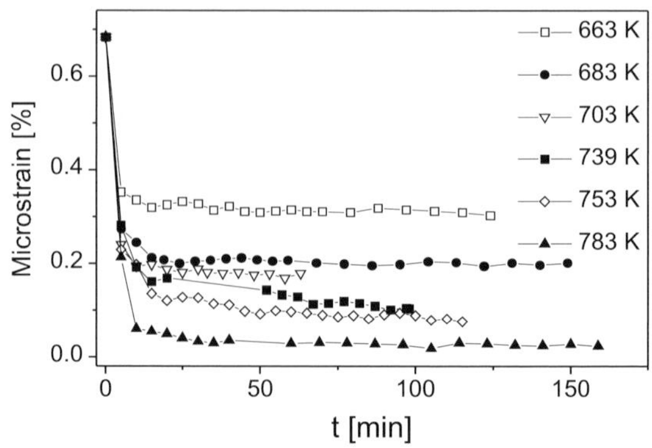

Fig. 5. Time evolution of the microstrain of nanostructured iron for different temperatures.

In the following only a short description of the models is given. Detailed information can be found in the cited literature. The generalised parabolic grain growth model introduces a parabolic relationship for crystallite growth kinetics. For grain boundary migration we assume an atom transport across the boundary under a pressure due to the surface curvature. A characteristic feature of this model is the continuous growth which does not lead to a crystallite size limit. Burke and later Grey and Higgins took account of the experimental observation of a limiting grain size: in their model, as result of grain growth, the driving pressure vanishes at a certain stage; consequently Eq. (2) has to be supplemented by a growth retarding term. An expression for $D(t)$ can only numerically be obtained from Eq. (3), and therefore in this grain growth model with impediment for the data evaluation it is convenient to fit the reverse function $t(D)$ to the experimental data instead of $D(t)$.

An improvement of model 2 was suggested by Michels *et al.* [60] who assume that the retardation constant b_2 should be a function of the grain size in grain growth processes of nanocrystalline metals. They argue that in these systems the impurities in the grain boundaries are more and more enriched when in the course of grain growth the grain boundary volume fraction decreases. The activation energies of the grain boundary diffusion processes can be obtained from the slopes of Arrhenius plots of Tk_1 or $Tk_{2,3}D_{\infty}^2$, respectively (γ is assumed to be temperature independent). Figure 5 shows the evolution of the microstrain with time and temperature. For nano-iron, as prepared, the microstrain amounts to 0.65%; in the isothermal grain growth experiments this value decreases extremely rapidly right at the beginning and then remains constant at limiting values between 0.3% at 663 K and 0.1% at 753 K; at the highest temperature, 783 K, the microstrain essentially disappears.

The volume-weighted average diameters resulting from the different isothermal measurements, D_{vol}, are displayed in Fig. 6. At high temperatures they show a fast increase of the grain size at the beginning of each experiment,

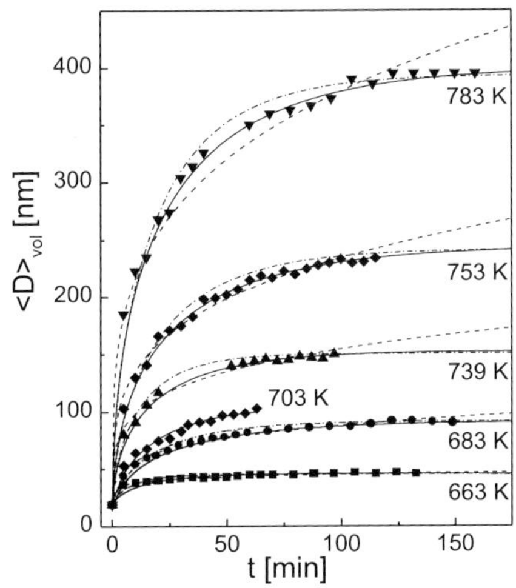

Fig. 6. The time evolution of the crystallite sizes of n-Fe for different temperatures. The symbols show the experimental data. Curve fits with the generalized parabolic law (dashed line), the Burke model (dashed-dotted line) and according to Michels *et al.* (straight line) are displayed.

and evidently a limiting grain size value at long times, D_∞, which exhibits a pronounced temperature dependence. At less elevated temperatures only a moderate and comparatively smooth grain growth is observed which stops after a short time period. Curve fits with the three kinetic models outlined above are compared in Fig. 6 with the experimental data.

The dashed lines represent the results of curve fits with model 1. The resulting grain growth parameters, n, decrease from about 12 at the lowest temperature to about 3 at the highest temperature of our experiments. The dash-dotted lines originate from curve fits with the growth model 2. The resulting growth retarding parameter b_2 is temperature-independent for $663\,\mathrm{K} \leq T \leq 793\,\mathrm{K}$, but then, for 753 and 783 K, clearly increases with temperature. For this reason two temperature regimes are apparently distinguishable. The solid lines are the results of curve fits with model 3. The resulting b_3 values amount to about $0.02\,\mathrm{s}^{-1}$ for all temperatures except for the lowest one where it is about twice that value. Conventional Arrhenius evaluations of the rates $a_1 = k_1/2$ and $a_{2,3} = k_{2,3} D_\infty^2$ yield activation energies of $Q_1 = 221\,\mathrm{kJ/mol}$, $Q_2 = 121\,\mathrm{kJ/mol}$, and $Q_3 = 118\,\mathrm{kJ/mol}$. There is, however, evidence for deviations from straight-line behaviour of a_2 and a_3. The extended Arrhenius plots of these rates are displayed in Fig. 7.

As can be seen in Fig. 7 the results from model 2 and 3 are in good agreement. One can clearly distinguish different activation energies in the low and high temperature regimes, respectively. For the kinetic model 2 we obtain in the high temperature regime $Q_2^{\mathrm{HT}} = 173 \pm 10\,\mathrm{kJ/mol}$ and in the low tempera-

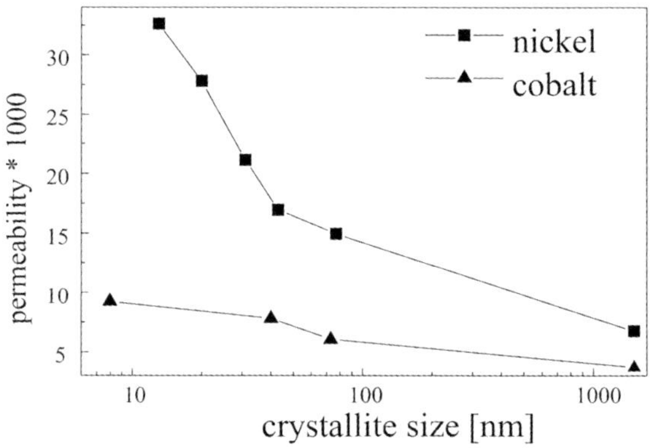

Fig. 9. Permeability of nanostructured Ni and Co as function of the crystallite size.

small crystallite sizes the local magnetocrystalline anisotropies are randomly averaged out. The exchange length for Ni is 26 nm and for Co 50 nm, these data fit well with the experimentally observed maxima in the coercivity, which are 3 times larger for Ni and 5 times larger for Co compared to the respective reference material. In the case of n-Ni we could not confirm the data of Aus *et al.* [63]. These investigators, also using electrodeposited n-Ni, found a minimum in coercivity at grain sizes near to the exchange length without being able to explain this behaviour.

The maximum permeability increases monotonically with decreasing grain size as shown in Fig. 9. For the permeability it is known that it decreases with decreasing grain size and reaches a minimum at a grain size near the exchange length. For smaller grain sizes it rises by several orders of magnitude [67]. In our work we investigated the maximum permeability, but to our knowledge there are no theoretical predictions for this behaviour. For large grains the magnetization can follow the easy magnetic axis in the single grains, and domains can be formed within the grains. For very small grains the ferromagnetic exchange interaction more and more forces the magnetic moments to align parallel, thus impeding the magnetization to follow the easy directions of each individual grain, and so it is easier to magnetize the nanocrystalline samples.

In the literature the same behaviour like in our experiments was observed for ball milled n-Ni and n-Fe [64] and also for inert gas condensed n-Fe [68].

3. Electrochemical deposition of less noble metals and alloys from ionic liquids

The use of aqueous electrolytes restricts the electrochemical deposition of nanostructured metals and alloys to those metals which have a higher standard reduction potential than hydrogen. Metals with slightly lower standard potential can still be electrodeposited if they exhibit a high overvoltage for hydrogen formation. Less noble materials like aluminium, magnesium, and the refractory metals (Ta, W, Nb) as well as semiconductors (Si, Ge) cannot be electrodeposited from aqueous solutions. To extent the electrochemical window organic solvents and compounds can be used [69–74]. The deposition from these organic electrolytes shows some negative effects:

1. The organic solvents can easily be decomposed during the electrochemical deposition and then the decomposition products are found as impurities in the deposits.
2. The extremely high vapour pressure, the flammability and the explosiveness make these solvents dangerous and make it difficult to integrate these electrolytes in an industrial process.

A further possibility for the electrodeposition of less noble metals is the use of an eutectic mixture of two inorganic salts with a relatively low melting point. For titanium and titanium-aluminium deposition a $NaCl/AlCl_3$ mixture can be used as solvent which melts at 195 °C. Details are described in [75]. However the preparation of nanostructures from these electrolytes is not possible because the high temperature of this process causes crystallite growth and therefore the product eventually is of microcrystalline structure.

Excellent electrolytes for the deposition of less noble metals are ionic liquids, IL. Ionic liquids have a single- or multi-component composition of organic- or organic/inorganic substances. Features of ILs are a broad electrochemical window (up to 6 V [76]), a sufficient solubility for metal ions, a good ionic conductivity and a very low vapour pressure [77]. Most of the ILs are non-toxic and not inflammable and thus environmentally friendly in the sense of "green chemistry". From their first description in 1914 [78] up to now three groups of ILs were developed. ILs of the first generation consist of imidazolium- or pyrrolidinium cations and halide anions (complex anions like $[AlCl_4]^-$ are also possible [76, 79–82]). The disadvantage of these ILs is the extreme moisture and oxygen sensitivity what makes them interesting only for academic research. ILs of the second generation are characterized by a good compatibility with moisture. Examples are choline chloride based ILs and metal salts mixed with urea [83]. ILs of the third generation are strongly hydrophobic resulting from fluorine rich anions [84]. The bis(trifluoromethylsulfonyl)imide (TFSI) and the trifluoromethanesulfonate anions combined with pyrrolidinium, imidazolium or trimethyl-n-hexylammonium cations are examples for such ILs [85, 86]. From these ILs metals like In, Se, Si, Ge and also Ta [87] have been deposited.

For the electrodeposition of aluminium and aluminium alloys from chloroaluminate based ionic liquids in nanoscale microstructure chemical additives like thiourea or nicotinic acid are suited to prevent Ostwald ripening in ILs. Grain sizes down to approximately 100 nm can in many cases be achieved either by pulsed electrodeposition from an electrolyte without additives or by dc electrodeposition with active additives, *i.e.*, those which also act as complex

Fig. 10. Bulk aluminium sample with a crystallite size of 15 nm deposited from $AlCl_3$/[EMIm]Cl mixture (65 : 35 mol-%) with 3 wt. % nicotinic acid additive [88].

formers and in this way require/enable an enhanced overvoltage. Grain sizes below approximately 10 nm, however, can in most cases only be achieved by a combination of grain refiners and high overvoltage.

3.1 Experimental details of electrodeposition

Nanostructured aluminium has been prepared from a Lewis acid $AlCl_3$/butyl-methyl-imidazolium chloride ([BMIm]Cl) mixture (65 mol % $AlCl_3$, 35 mol % [BMIm]Cl) whereas palladium alloys could only be deposited from a mixture of 45 mol % $AlCl_3$ and 55 mol % [BMIm]Cl. For alloy deposition we use an $AlCl_3$/[BMIm]Cl (53 : 47 wt. %) system with addition of the corresponding metal salts (*e.g.* 5.5 wt. % $InCl_3$ or $MnCl_2$). The aluminium content in the alloys can be controlled by the composition of the electrolyte [89, 90]. For example, aluminium rich alloys were prepared from ionic liquids with $AlCl_3$ contents between 20 and 39 mol %. Mixtures of $AlCl_3$/immidazolium salts are very sensitive to water and for this reason all experiments have to be performed in a glove box containing a nitrogen atmosphere with water content below 1 ppm. The electrochemical cell and all parts which are in contact with the electrolyte were built from glass, PTFE or other inert materials. The cathode (60 mm × 15 mm) consists of glassy carbon or copper. To avoid a decrease of the aluminium concentration, a sacrificial aluminium anode was used. The bath temperature (40 °C) was controlled with a thermostat. Temperatures above 60 °C lead to crystallite growth during the deposition and have to be avoided. A magnetic stirrer is used to avoid ion concentration gradients in the electrolyte. Grain refining additives like nicotinic or benzoic acid and also saccharin have been used for crystallite size reduction because these organic substances exhibit a very good solubility in ionic liquids. The deposits can be removed from the copper cathode mechanically (Fig. 10) using a sharp knife or

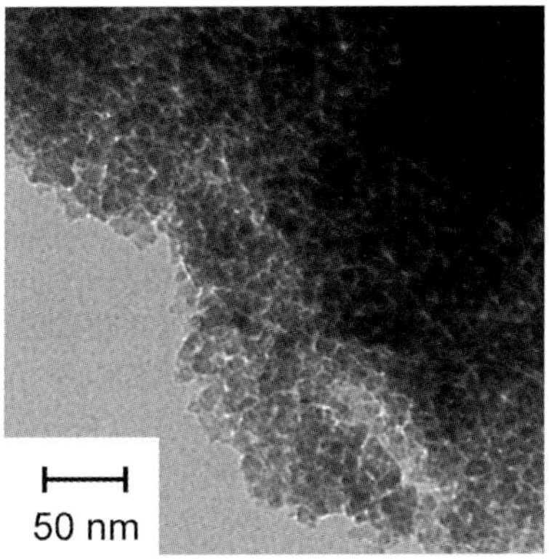

Fig. 11. Transmission electron micrograph of a nanostructured aluminium sample prepared from 5.0 g 1-ethyl-3-methyl-1H-imidazolium chloride ([EMIm]Cl) and 8.8 g absolute dry aluminium chloride, 3 wt. % benzoic acid, dc current density: 3 mA/cm^2.

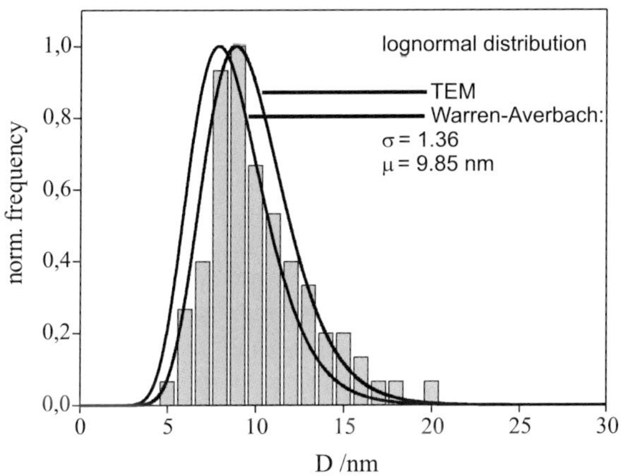

Fig. 12. Crystallite size distribution of the sample shown in Fig. 11.

by dissolving the copper substrate. Adsorbed electrolyte can be washed from the metal deposit with dried toluene or benzene. The nanostructural parameters like crystallite size, crystallite size distribution and microstrain content were determined by XRD. TEM (Fig. 11) and SEMicrographs were additional recorded to observe the bulk- and the surface nanostructure.

A systematic study of physical properties of these materials as function of the crystallite size is possible because this method allows a variation of the crystallite size with a narrow size distribution (Fig. 12).

The influence of the pulse current parameters on the nanostructure of the deposits has been studied first. All samples were deposited from a mixture of

Table 3. The crystallite size of the aluminium deposits as function of the pulse parameter (electrolyte 5.0 g 1-ethyl-3-methyl-1H-imidazoliumchlorid ([EMIm]Cl) and 8.8 g absolute dry aluminiumchloride). In the case of grain refiner addition a concentration of 2.5 g/L benzoic acid was used.

$I_{average}$ mA/cm²	I_{peak} mA/cm²	t_{on} ms	t_{off} ms	additive g/L	D nm
11.9	310	2	50	0	140
21.2	550	2	50	0	129
60.1	1570	2	50	0	122
60.1	1570	2	50	2.5	7

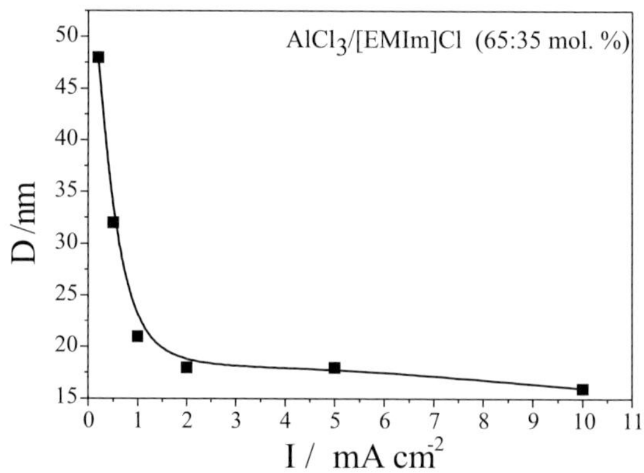

Fig. 13. Crystallite size of aluminium deposits for different dc-current densities. The electrolyte contains 3 wt. % benzoic acid.

5.0 g 1-ethyl-3-methyl-1H-imidazoliumchlorid ([EMIm]Cl) and 8.8 g absolute dry aluminium chloride without any additives. As summarized in Table 3 the pulsed current density was increased from 310 to 1570 mA/cm² for constant t_{on} (2 ms) and t_{off} (50 ms) times. The X-ray analysis reveals that the resulting crystallite size decreases with increasing current density.

As shown in previous publications [80, 91] the ratio of 2 ms t_{on}-time and 50 ms t_{off}-time is appropriate to get homogeneous and smooth deposits from this electrolyte. To achieve a further reduction of the crystallite size the pulsed current would have to be further increased – but above 1570 mA/cm² a change in the colour of the electrolyte from colourless to light-brown indicates decomposition of the electrolyte. The next step to reduce the crystallite size is the use of additives. For that purpose 3 wt. % benzoic acid is added to a $AlCl_3$/[EMIm]Cl (65 : 35 mol %) electrolyte. To simplify the process and also for a better understanding a dc-current with current densities from 0.2 to 10 mA/cm² is applied. The result is shown in Fig. 13.

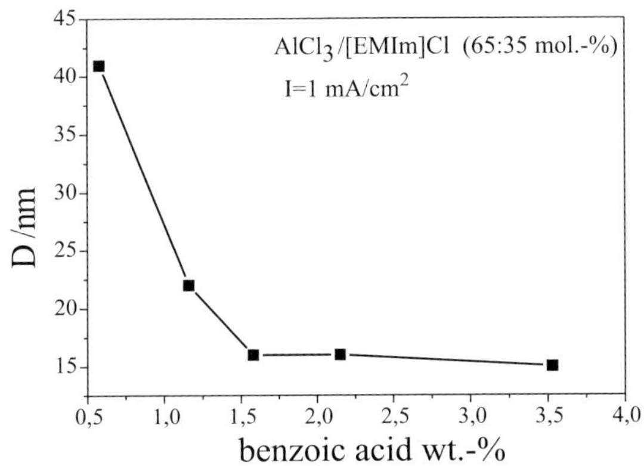

Fig. 14. Crystallite size of aluminium deposits for different additive amounts. A dc current of 1 mA/cm^2 was used.

Other possible organic additives are carboxylic acids, amines, and different aromatic and sulphur containing compounds which impede the growth of the nuclei by reversible adsorption on the electrode surface.

To verify the assumption that the crystallite size depends on the additive concentration and on the chemical structure of the additives, deposits have been prepared under constant physical process parameters with increasing amounts of benzoic acid (Fig. 14).

The crystallite size without any additive is 144 nm. Already an addition of a very small amount of 0.6 wt. % decreases the crystallite to 41 nm. This trend continues up to a concentration of 1.6 wt. % benzoic acid. Above this concentration a more or less constant behaviour can be observed with only a little change in crystallite size. Exactly the same behaviour could be observed for nanostructured Cu [16], Pd [8] and Au deposits [44].

The influence of the molecular structure of the grain refiner on the microstructure of the deposits was studied keeping the physical parameters (temperature, current parameters) constant. The composition of the basic electrolyte was 63 mol % absolute dry $AlCl_3$ and 37 mol % [EMIm]Cl. The dc-current density was 5 mA/cm^2 and the additive concentration was kept constant at 4 wt. % for all additives. The grain refiners were selected from four groups of organic molecules with different functional groups (Table 4):

1. Aromatic and aliphatic carboxylic acids with one or two carboxylic groups.
2. Aromatic carboxylic acids with chlorine substituents in different positions (2-, 3- and 4-chlorobenzoic acid).
3. Carboxylic acids with one or two hydroxy substituents
4. Substances with a sulphur containing functional group (benzoic acid sulfimide, sodium butanesulfonate, sodium dodecylsulfate).

50 °C only a slight increase in the crystallite size from 23 to 29 nm can be observed followed by a strong increase up to 72 nm at 60 °C. Obviously, for high temperatures fewer molecules are adsorbed at the surface compared with ambient temperatures. This leads to an enhanced Ostwald ripening resulting in fast growth of nuclei and large crystallites.

In this series of experiments it could be demonstrated, that the best conditions for the deposition of nanostructured aluminium from this kind of electrolyte are:

1. High current densities
2. Low temperatures (40–50 °C)
3. Use of an aromatic grain refiner molecule containing a functional group with sulphur or nitrogen atoms.

Ionic liquids are also good electrolytes for the deposition of the hydride forming metal palladium which upon electrodeposition from an aqueous electrolyte would absorb the concomitantly formed hydrogen what could lead to embrittlement and crack formation. Actually, it is very difficult to prepare thin palladium foils without any cracks from aqueous solutions. From ionic liquids as electrolyte hydrogen codeposition does not occur and therefore crack free deposits result.

Sun and Hussey demonstrated that Pd can be deposited from a Lewis basic [EMIm]Cl/$AlCl_3$-electrolyte but unfortunately the solubility of $PdCl_2$ is not sufficient to prepare large samples. They could show that the reducible species in this melt is $PdCl_4^{2-}$ [95]. For this reason we use a mixture of [EMIm]Cl/$PdCl_2$ (71.2 : 28.8 mol %) and 4.8 wt. % nicotinic acid as grain refiner. At a temperature of 100 °C (in order to get mixing) and a dc-density of 8.3 mA/cm^2 nanostructured palladium foils are obtained with a crystallite size of 34 nm and a thickness of 50 microns.

Nanostructured iron forms immediately hydroxides or oxides, respectively, during the deposition process from aqueous solutions. The use of ionic liquids prevents the oxide formation. First studies were done by Sun *et al.* for an IL basid on $AlBr_3$/1-methyl-3-ethylimidazolium bromide [96]. Yamagata *et al.* studied the behaviour of iron complexes in hydrophobic ionic liquids [97, 98]. Aravinda and Freyland measured the deposition mechanism of Fe from a $AlCl_3$/1-methyl-3-butylimidazolium chloride IL [99]. We prepare nanostructured iron from an electrolyte consisting of 5.0 g [BMIm]Cl, 8.8 g absolute dry $AlCl_3$ and 0.5 g anhydrous $FeCl_3$. 40 g/L benzoic acid was used as grain refiner. The crystallite size of the deposits can be adjusted by the variation of the grain refiner content from 60 (for 80 g/L) to 82 nm (without additive). The current density in these experiments is 5 mA/cm^2 [92]. The deposition of Pb [100], Te [101], Sn [102], Si, Ge [80], Ga [103], Co [104], Ni [105], AlAg, AlCo, AlNi [106], AlTi [107] is described in the literature.

Alloys like Al_xMn_{1-x} and Al_xIn_{1-x} can be prepared by PED or dc procedures from Lewis neutral systems by the addition of a corresponding metal salt.

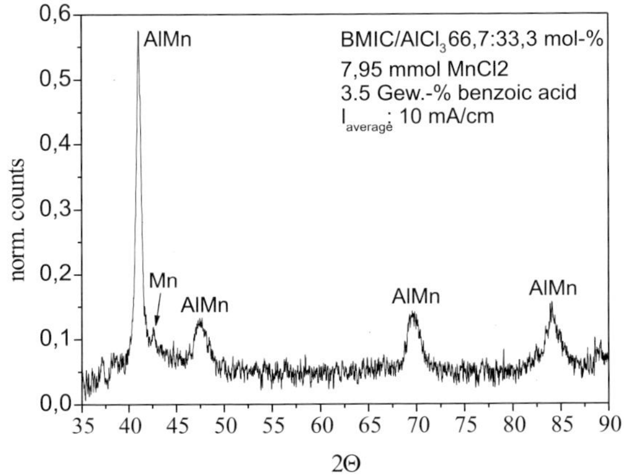

Fig. 16. X-ray diffraction pattern of an AlMn alloy prepared from butyl-methylimmidazolium chloride rich IL.

A IL consisting of $AlCl_3$/[BMIm]Cl (53 : 47 wt. %) with addition of 5.5 wt. % $InCl_3$ or $MnCl_2$ was used. The samples were deposited with a dc-density of 10 mA/cm^2 at 50 °C. In both cases nanostructured alloys with a crystallite size of 25 nm have been obtained (Fig. 16) [80]. The alloy composition can be controlled by the composition of the bath [89, 90, 108, 109]. A higher aluminium chloride surplus in the ionic liquid increases the aluminium content up to a 50 : 50 alloy composition.

3.2 Hardness of aluminium

Some properties of nanocrystalline metallic materials depend strongly on the crystallite size. Examples are: strength, ductility and hardness [110], wear resistance and coefficient of friction [51], electrical resistivity [111, 112], magnetism [63, 113–115], hydrogen solubility and diffusivity [18], resistance to localized corrosion and intergranular stress corrosion cracking [116, 117], as well as thermal stability [48, 118, 119].

Hall [121] was the first who found a relation between crystallite size and hardness (Hall–Patch relation). This model reveals that the hardness of a material is a function of the square root of the crystallite size. As a result of Hall–Petch strengthening, nanocrystalline materials display significant increases in hardness and strength relative to their coarse-grained counterparts. The reason for this behaviour is a blocking of the migrating dislocations during the deformation process by the grain boundaries. Experiments with Ni-P samples [110] confirm this behaviour for samples with crystallite sizes above 10 nm. Furthermore, it was reported in the literature that nanomaterials with crystallite sizes below 8 nm should show the so-called inverse Hall–Petch relationship [110]. Erb *et al.* observed a softening for Ni samples with crystallite

Table 6. The nanohardness of aluminium as function of the crystallite size [92]. The mechanical tests were done by nanoindentation using a Berkovich-Indenter (with a three-sided pyramid diamond tip). The values of the nanohardness were determined according to the Oliver–Pharr method [120].

D/nm	nanohardness/GPa
14	3.4
18	2.8
24	2.2
32	1.9
67	2.7
141	1.9
polycrystalline	1.4

sizes below 10 nm. It was assumed for nanostructures with a grain boundary content of more than 60% that a so-called "deformation-twinning" [122–124] is responsible for this behaviour. The energy of a migrating dislocation can be reduced by the formation of a new partial dislocation inside the grain which leads to a reorientation of the grain. Carsley *et al.* [125] developed a model with two terms, one representing the grains and the other one representing the atoms in the grain boundaries.

In order to study this effect we have prepared a set of nanostructured aluminium samples as described above. The crystallite size was varied between 14 and 141 nm. The results of the nanoindentation experiments are displayed in Table 6. As predicted from the Hall–Petch relation the hardness increases with decreasing crystallite size in a square root law. Our data, however, do not confirm the inverse Hall–Patch relationship because we did not succeed in preparing samples with crystallite sizes smaller than 14 nm [92].

4. Fuel cell catalysts prepared by pulsed electrodeposition

During the last two decades intensive research efforts have been undertaken in the field of electrocatalysis for polymer electrolyte membrane fuel cells (PEMFCs). Due to their high operational efficiencies and environmental acceptability this type of fuel cells seems to be a good alternative to existing mobile power sources (batteries, gasoline engines). The search for efficient and less expensive catalyst materials is a challenging task for PEMFCs because the commonly used catalyst material, platinum, has a high and permanently increasing price and a limited availability in the future. The mentioned problems also exist for direct methanol fuel cells (DMFCs) which work with a relatively high PtRu-catalyst loading of 4 mg/cm^2 for the anode. Ru is of even less frequent abundance than Pt, and meanwhile more expensive. In the course of commercialization, a distinct reduction of the costs, that means also a reduction of the catalyst load, of the PEMFCs is necessary. The conventional preparation techniques for catalyst layers of PEMFCs electrodes start from pressing catalyst particles such as platinum or platinum ruthenium supported on carbon black onto the polymer membrane. In this procedure a substantial fraction of catalyst particles are electrochemically inactive because they are not located [126] in direct contact to the three-phase boundary consisting of proton-conducting electrolyte, electron-conducting electrode (carbon net) and fuel [127]. Catalyst particles deposited within the porous carbon nanostructure may be inaccessible to the polymer electrolyte. Consequently a distinct reduction of the catalyst loading can be achieved by localizing the catalyst particles exclusively in the three-phase boundary regions. This is possible by electrochemical preparation routes. It is reported in the literature that catalyst particles were electrochemically deposited in a carbon-Nafion®-layer using an outer aqueous plating bath with platinum ions [128], but the disadvantage of this method is the contamination of the bath during the deposition process with Nafion® and carbon black. In the following we describe an alternative electrochemical preparation method which allows, starting from a platinum precursor salt, to deposit the catalyst particles directly in the gas diffusion layer [129]. The catalyst precursor salt is dissolved in solubilised polymer electrolyte (Nafion®), mixed with carbon black particles (size 50 nm) and sprayed onto a Nafion® membrane. Subsequently the catalyst is site-selectively reduced with a pulsed galvanostatic current at the three phase sites. Catalyst alloys can be obtained employing a mixture of catalyst precursor salts.

4.1 Experimental

This electrochemical deposition process for catalyst preparation consists of three steps: First, the catalyst precursor salt is dissolved in 5 wt. % solubilized Nafion® ionomer (Fluka). In order to prepare carbon supported catalyst layers the solution was mixed with high-surface-area carbon (Vulcan XC 72) and dispersed using an ultrasonic bath and an Ultraturrax™ homogenizer. H_2PtCl_6 was used for the preparation of carbon supported layers [130, 131]. For the preparation of platinum alloys, mixtures of hexachloroplatinum acid and the corresponding metal chlorides as precursor solution can be used. Second, the precursor solution/dispersion is sprayed onto the Nafion® membrane. Third, the catalyst precursor salt is electrochemically deposited in an electrochemical cell shown in Fig. 17.

The Nafion®-membrane is in contact with the cathode at the bottom of the cell. Platinum nets serve as electronic contact. In order to avoid a diffusion of the precursor salt out of the precursor layer, the electrolysis is performed in a humid atmosphere which is produced by bubbling air through warm water. After 30 min. in this atmosphere the current is switched on. The deposition time is varied between 30 s and several minutes depending on the used cur-

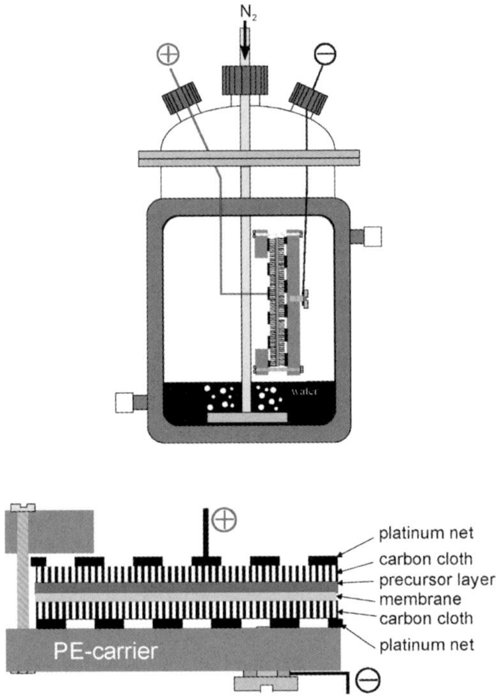

Fig. 17. Experimental setup for electrochemical catalyst deposition.

rent density and current shape. Direct current densities which show a good result are between 0.5–1 A/cm^2. For pulsed currents, as the result of statistically planed tests, $t_{on} = 1$ ms, $t_{off} = 1$ ms, $I_{peak} = 0.6\ A/cm^2$ turned out as the best combination of current parameters.

4.2 Characterisation of the catalysts

A number of samples were synthesized as described above and characterized by X-ray diffraction (XRD) to verify the alloy formation. In the case of alloy formation a shift of the Bragg peak positions can be observed resulting from the change in the lattice constant compared with the pure metals, according to Vegard's law. A standard material (KCl or LaB_6 powder) was measured simultaneously with the samples to determine absolute peak positions. The crystallite size of the deposited particles was determined by the Scherrer [132] and Warren–Averbach [119, 133] technique. In addition, the particle size, particle size distribution and degree of agglomeration of the deposited catalyst particles were investigated by high resolution transmission electron microscopy (HR-TEM). TEM samples were prepared by drying and pulverizing the carbon-Nafion®-layer. The powder samples were suspended

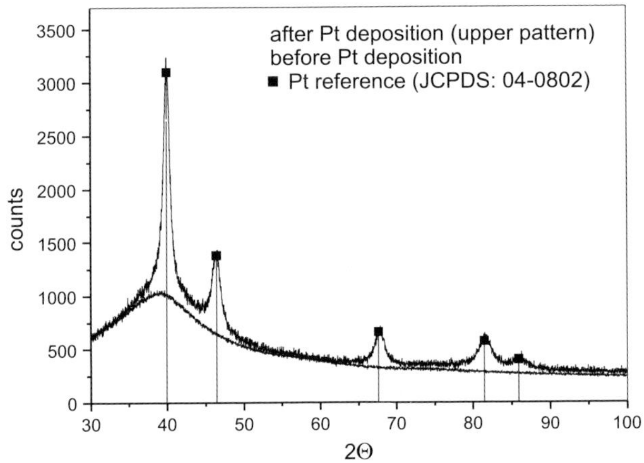

Fig. 18. X-ray diffraction pattern of the MEA before and after the electrochemical catalyst deposition.

in ethanol, deposited on a copper grid and dried. The deposits were electrochemically characterized by cyclic voltammetry. A standard three-electrode electrochemical cell was used for the measurements in pure 1M H_2SO_4 or 1M H_2SO_4/1M CH_3OH solution at room temperature, and the scan rate was 20 mV/s. A reversible hydrogen electrode (RHE) served as reference electrode; a Pt mesh was used as counter electrode.

4.2.1 Results of structural characterization

The deposition of carbon supported platinum metal in the precursor layers could be verified by comparing the XRD patterns of the membrane before and after electrolysis as shown in Fig. 18.

The measured X-ray diffraction pattern after the electrochemical catalyst deposition displays the metallic catalyst pattern, and the background represents the carbon support and the Nafion®. The Pt-pattern can be obtained by subtracting the background pattern from the pattern after deposition. A fit of the Pt [111]-peak with a Lorentzian function and a subsequent Scherrer analysis yields a volume-weighted average grain size of about 9 nm. In addition, the particle size distribution of the catalysts has been analysed employing the Warren–Averbach method. This Fourier method allows the separation of size and strain effects and enables the determination of the crystallite size distribution. Analysing the Pt [111] and Pt [200] peaks a log-normal distribution function was used yielding a median crystallite size of $\mu = 8.7$ nm and a geometrical standard deviation $\sigma = 1.2$ as shown in Fig. 19.

Platinum/ruthenium alloy electrodeposits were prepared employing a precursor mixture of hexachloroplatinic acid and ruthenium chloride (Pt : Ru

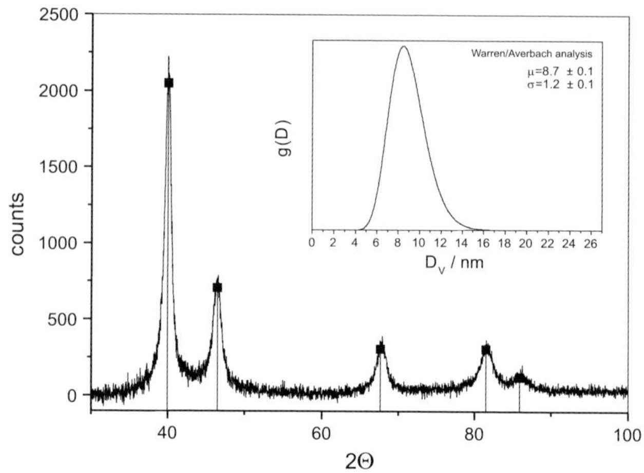

Fig. 19. Pure Pt-pattern deduced from Fig. 18, the Pt reference (JCPDS: 04-0802) and the particle size distribution (see inset).

50 : 50 at. % in the precursor solution). The diffraction pattern after electrochemical catalyst reduction did not show any lines corresponding to the ruthenium hexagonal structure indicating that a platinum/ruthenium alloy is formed as a substitutional solid solution, wherein the smaller Ru atoms replace the Pt atoms in the platinum cubic structure. This alloy formation results in the reduction of the platinum lattice constant and therefore in a peak shift to larger 2θ values. A fit of the shifted Pt [111]-peak with a Lorentzian function and a subsequent Scherrer analysis yielded a volume-weighted average grain size of about 8 nm. Using the Warren–Averbach analysis a median size of $\mu = 8.1$ nm and a geometrical standard deviation of $\sigma = 1.2$ were determined.

Furthermore, the average particle size and the particle size distribution of the electrodeposits were analysed using HR-TEM. Figure 20 shows an electron micrograph of a Nafion® membrane after the electrodeposition of platinum representing the platinum particles as dark spots on the carbon support (grey background). An EDX-spectrum of the catalyst layer confirmed the presence of platinum in addition to carbon, chloride and sulphur.

The data of Fig. 20 fit well to a log-normal distribution function (solid line), yielding a median size $\mu = 15.1$ nm, a geometrical standard deviation $\sigma = 1.51$ and thus a volume averaged size of 27.4 nm. The deviation of the μ value and the broadening of the log-normal distribution as compared to the Warren–Averbach results might be due to the agglomeration of smaller nanoparticles.

The characterization of membranes after the Pt/Ru electrodeposition by HR-TEM exhibited similar results. An EDX-spectrum verifies the presence of ruthenium in addition to platinum and carbon. The integration of the Pt-M and

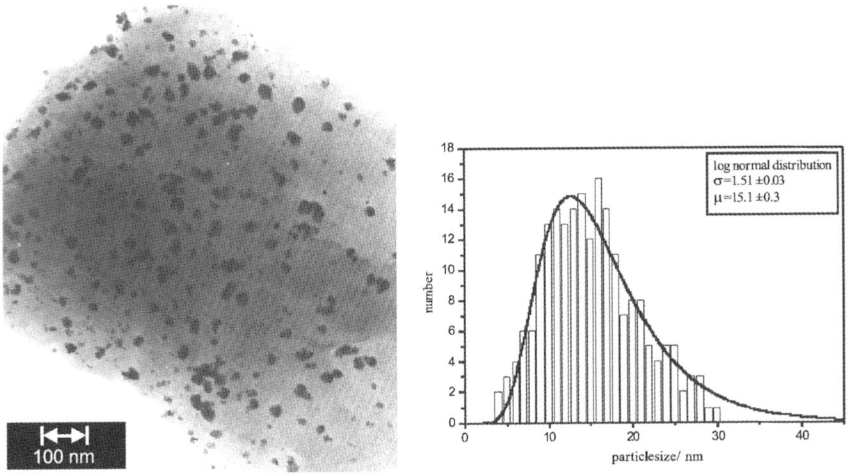

Fig. 20. Electron micrograph of a test electrode after electrochemical platinum deposition and particle size distribution of the platinum particles. Evaluating two hundred particles the size histogram shows a maximum at about 12 nm.

the Ru-L signal results in a Pt : Ru ratio of 54 : 46 at. % (precursor solution: 50 : 50 at. %).

4.2.2 Results of electrochemical characterization

The electrodeposits were electrochemically characterized by cyclic voltammetry measurements in sulphuric acid/1M CH_3OH solutions in order to verify the Pt/Ru alloy formation (see Fig. 21).

Alloying Pt with Ru results in a significant reduction of the overpotential of the methanol oxidation. Figure 21 indicates that the onset of the methanol oxidation peak was shifted from 0.95 V *vs.* RHE (Pt catalyst, curve 1) down to about 0.74 V *vs.* RHE (Pt/Ru catalyst, curve 2). The investigation of a commercial PtRu catalyst (commercial catalyst, Pt : Ru 50 : 50, curve 3) showed similar results.

It was demonstrated that platinum and platinum/ruthenium catalysts can be synthesized *in-situ* by the application of suitable catalyst precursor layers on Nafion® membranes and subsequent electrodeposition of the catalyst in a water vapour atmosphere. HR-TEM, SEM and XRD were used to characterize the catalyst layers. Both HR-TEM and XRD measurements verified the catalyst salt reduction and catalyst deposition as well as the nanocrystalline nature of the platinum and platinum/ruthenium catalyst particles with a volume weighted average grain size of typically 8–10 nm. The formation of platinum/ruthenium alloys could be verified by XRD and by CV measurements in methanol solution. An effort to reduce the catalyst particle size in

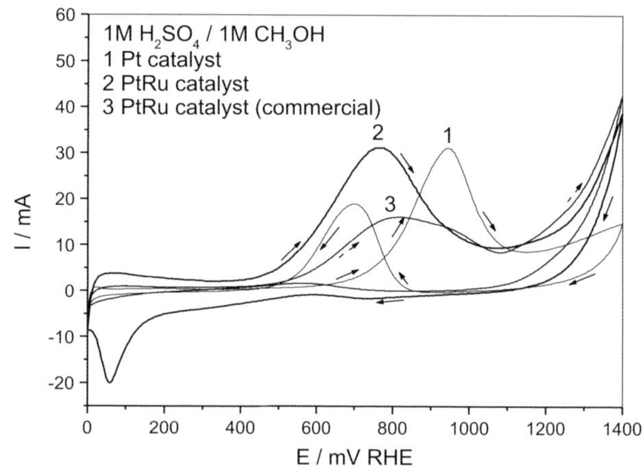

Fig. 21. Cyclic voltammograms of Pt and Pt/Ru electrodeposits in 1M CH_3OH.

the electrodeposition step by using pulsed electrodeposition (PED) to further enhance the catalytic activity is in progress.

5. Concluding remarks

Electrodeposition, as presented in this review, namely as a combination of sophisticated features of electrical engineering, of electroanalytical techniques, and – with respect to the electrolytes – of approaches towards green chemistry and sustainable use of resources, can be considered as a particularly innovative branch of electrochemical engineering. It has opened the possibility to synthesize a variety of nanomaterials and thus has provided access to surface coatings with innovative functionalities, to bulk materials with improved physical properties, and to nanoparticles with excellent electrocatalytic activities. A prerequisite for systematic progress in this field is, as has been demonstrated for a variety of examples, a fundamental understanding of the underlying atomic mechanism of nanocrystal formation. The transfer of this accumulated knowledge from academia into industrial applications, which in a few cases is already taking place, remains the great challenge for the near future.

Acknowledgement

The present studies were initiated and performed at the Universität des Saarlandes (Germany) in the framework of Sonderforschungsbereich 277 (Grenzflächenbestimmte Materialien). We gratefully acknowledge the financial support by the Deutsche Forschungsgemeinschaft. For stimulating discussions we thank Dr. M. Schmelzer, Dr. M.-S. Löffler, Dr. A. Shkourankov, Dr. J. Divisek, Dr. K. Wippermann, J. Mergel, Prof. J.-W. Schultze, Dr. U. König, Prof. C. E. Krill, Prof. R. Birringer, Dr. Krajewsky, Dr. H. Wolf, Dr. Z. Guan, and Prof. T. Wichert.

References

1. H. Gleiter, Prog. Mater. Sci. **33** (1989) 223.
2. H. Gleiter, Makromol. Chem. – Macromol. Symp. **50** (1991) 171.
3. H. Gleiter, Adv. Mater. **4** (1992) 474.
4. U. Erb, Can. Metall. Quarterly **34** (1995) 275.
5. A. M. Alfantazi, J. Page, and U. Erb, J. Appl. Electrochem. **26** (1996) 1225.
6. U. Erb, G. Palumbo, B. Szpunar, and K. T. Aust, Nanostruct. Mater. **9** (1997) 261.
7. H. Natter, M. Schmelzer, and R. Hempelmann, J. Mater. Res. **13** (1998) 1186.
8. H. Natter, T. Krajewski, and R. Hempelmann, Ber. Bunsenges. Phys. Chem. **100** (1996) 55.
9. H. Natter and R. Hempelmann, Electrochim. Acta **49** (2003) 51.
10. H. Natter, M. Schmelzer, S. Janssen, and R. Hempelmann, Ber. Bunsenges. Phys. Chem. **101** (1997) 1706.
11. E. Thiele, R. Klemm, L. Hollang, C. Holste, N. Schell, H. Natter, and R. Hempelmann, Mater. Sci. Eng. A **390** (2005) 42.
12. B. Yang, H. Vehoff, and R. Hempelmann, Int. J. Mater. Res. **97** (2006) 1220.
13. H. Gleiter, Acta Mater. **48** (2000) 1.
14. H. Konrad, T. Haubold, R. Birringer, and H. Gleiter, Nanostruct. Mater. **7** (1996) 605.
15. C. Politis and W. L. Johnson, J. Appl. Phys. **60** (1986) 1147.
16. H. Natter and R. Hempelmann, J. Phys. Chem. **100** (1996) 19525.
17. U. Erb, Nanostruct. Mater. **6** (1995) 533.
18. H. Natter, B. Wettmann, B. Heisel, and R. Hempelmann, J. Alloys Compd. **253** (1997) 84.
19. S. Z. El Abedin, E. M. Moustafa, R. Hempelmann, H. Natter, and F. Endres, Electrochem. Commun. **7** (2005) 1111.
20. S. Z. El Abedin, E. M. Moustafa, R. Hempelmann, H. Natter, and F. Endres, Chem. Phys. Chem. **7** (2006) 1535.
21. M. M. Viana, T. D. S. Mohallem, G. L. T. Nascimento, and N. D. S. Mohallem, Brazilian J. Phys. **36** (2006) 1081.
22. M. Veith, A. Altherr, and H. Wolfanger, Chem. Vapor Deposition **5** (1999) 87.
23. M. Veith, S. Kneip, S. Faber, and E. Fritscher, Mech. Alloyed Metastable Nanocrystal. Mater. **269** (1998) 303.
24. C. C. Koch, J. Mater. Sci. **42** (2007) 1403.
25. B. Gunther, G. Veltl, and H. D. Kunze, Philos. Mag. B **68** (1993) 825.
26. E. Budevski, G. Staikov, and J. W. Lorenz, in *Electrochemical Phase Transformation and Growth.* VCH, Weinheim (1996) p. 161.
27. H. Natter, M. Schmelzer, M. S. Loeffler, C. E. Krill, A. Fitch, and R. Hempelmann, J. Phys. Chem. B **104** (2000) 2467.
28. B. E. Warren and B. L. Averbach, J. Appl. Phys. **23** (1952) 497.
29. B. E. Warren and B. L. Averbach, Phys. Rev. **86** (1952) 656.
30. B. E. Warren and B. L. Averbach, E. Warekois, Phys. Rev. **91** (1953) 233.
31. H. Fischer, *Elektrolytische Abscheidung und Elektrokristallisation von Metallen.* Springer, Berlin, Göttingen, Heidelberg (1954).

102. X. H. Xu and C. L. Hussey, J. Electrochem. Soc. **140** (1993) 618.
103. P. Y. Chen, Y. F. Lin, and I. W. Sun, J. Electrochem. Soc. **146** (1999) 3290.
104. C. A. Zell and W. Freyland, Langmuir **19** (2003) 7445.
105. W. Freyland, C. A. Zell, S. Z. El Abedin, and F. Endres, Electrochim. Acta **48** (2003) 3053.
106. R. T. Carlin, H. C. De Long, J. Fuller, and P. C. Trulove, J. Electrochem. Soc. **145** (1998) 1598.
107. C. L. Aravinda, I. Mukhopadhyay, and W. Freyland, Phys. Chem. Chem. Phys. **6** (2004) 5225.
108. B. Grushko and G. R. Stafford, Scr. Metall. **23** (1989) 557.
109. G. R. Stafford, B. Grushko, and R. D. Mcmichael, J. Alloy. Compd. **200** (1993) 107.
110. U. Erb, G. Palumbo, R. Zugic, and K. T. Aust, in *Processing and Properties of Nanocrystalline Materials*. TMS, Warrendale (1996).
111. M. J. Aus, B. Szpunar, U. Erb, A. M. Elsherik, G. Palumbo, and K. T. Aust, J. Appl. Phys. **75** (1994) 3632.
112. I. Bakonyi, E. Toth-Kadar, J. Toth, T. Tarnoczi, and A. Cziraki, in *Processing and Properties of Nanocrystalline Materials*. C. Suryanarayana (Ed.), TMS, Warrendale (1996) p. 465.
113. M. J. Aus, B. Szpunar, A. M. Elsherik, U. Erb, G. Palumbo, and K. T. Aust, Scripta Metall. Mater. **28** (1993) 544.
114. M. J. Aus, C. Zheung, B. Szpunar, U. Erb, and J. Szpunar, J. Mater. Sci. Lett. **17** (1998) 1949.
115. B. Szpunar, M. Aus, C. Cheung, U. Erb, G. Palumbo, and J. A. Szpunar, J. Magn. Magn. Mater. **187** (1998) 325.
116. S. H. Kim, K. T. Aust, U. Erb, F. Gonzalez, and G. Palumbo, Scripta Mater. **48** (2003) 1379.
117. S. H. Kim, U. Erb, K. T. Aust, and G. Palumbo, Scripta Mater. **44** (2001) 835.
118. G. Hibbard, K. T. Aust, G. Palumbo, and U. Erb, Scripta Mater. **44** (2001) 513.
119. H. Natter, M. S. Loffler, C. E. Krill, and R. Hempelmann, Scripta Mater. **44** (2001) 2321.
120. W. C. Oliver and G. M. Pharr, J. Mater. Res. **7** (1992) 1564.
121. E. O. Hall, Nature **173** (1954) 948.
122. V. Yamakov, D. Wolf, S. R. Phillpot, and H. Gleiter, Acta Mater. **50** (2002) 61.
123. V. Yamakov, D. Wolf, S. R. Phillpot, and H. Gleiter, Acta Mater. **50** (2002) 5005.
124. V. Yamakov, D. Wolf, S. R. Phillpot, A. K. Mukherjee, and H. Gleiter, Nature Mater. **1** (2002) 45.
125. J. E. Carsley, J. Ning, W. W. Milligan, S. A. Hackney, and E. C. Aifantis, Nanostruct. Mater. **5** (1995) 441.
126. S. D. Thompson, L. R. Jordan, and M. Forsyth, Electrochim. Acta **46** (2001) 1657.
127. J. Mcbreen, J. Electrochem. Soc. **132** (1985) 1112.
128. E. J. Taylor, E. B. Anderson, and N. R. K. Vilambi, J. Electrochem. Soc. **139** (1992) L45.
129. M. S. Löffler, B. Gross, H. Natter, R. Hempelmann, T. Krajewski, and J. Divisek, Scripta Mater. **44** (2001) 2253.
130. W. Faul and B. Kastening, Galvanotechnik **68** (1977) 699.
131. R. Hempelmann, H. Schmitz, M.-S. Loeffler, H. Natter, and J. Divisek, *Method for Coating a Membrane Electrode Unit with a Catalyst and Device for Carrying out the Method.* EP 1 307 939 B1 (2001).
132. P. Scherrer, Göttinger Nachrichten **2** (1918) 96.
133. B. L. Averbach and B. E. Warren, Phys. Rev. **78** (1950) 644.

Investigation of Nanocrystalline Materials Using Radioactive Isotopes

By Th. Wichert*, H. Wolf, Z. Guan, and X. Li

Technische Physik, Universität des Saarlandes, D-66041 Saarbrücken, Germany

Hyperfine Interaction / Nanocrystal / Doped Semiconductor / Grain Boundary / Néel Temperature

The potential of detecting the hyperfine interaction at the nuclear site of radioactive isotopes for the investigation of nanocrystalline (nc) materials is discussed. The subject is illustrated with help of different examples for characterizing nc samples. Using the perturbed $\gamma\gamma$ angular correlation technique (PAC) for detecting the hyperfine interaction at the site of the radioactive probe $^{111}In/^{111}Cd$ experiments regarding the doping of nc semiconductors ZnO and InP, the shift of the Néel temperature in nc antiferromagnetic NiO, the structure of grain boundaries in Ni, and the homogeneity of nc Ni-Cu alloys are presented. The resulting information is compared with that obtained by Mössbauer spectroscopy and by standard techniques used in materials science.

1. Introduction

The properties of nanocrystalline (nc) materials are essentially determined by the size of the crystals and the structure of the grain boundaries of the respective material [1,2]. Having crystal sizes between 1 and 100 nm, a substantial fraction of atoms is located at surfaces or in grain boundaries (GB). Often forming non-equilibrium phases, nc materials are prone to atomic disorder, point defects and dislocations. Because of their special properties, they are of great current interest for many technological applications as well as for basic research [3–6].

The characterization of nc materials is performed by a multitude of different techniques well known in materials sciences. But, there is an obvious need for techniques that possess short range sensitivities in order to deliver separate information about crystallites and GB of the nc materials. Among other techniques, nuclear hyperfine interaction (hfi) methods have a very high local

* Corresponding author. E-mail: thw@tech-phys.uni-sb.de

Z. Phys. Chem. **222** (2008) 355–386

sensitivity as is illustrated in the compilation edited by G. S. Collins [7]. It turns out that a particularly sensitive way of characterizing materials is the employment of radioactive isotopes that are suited for detecting the respective hfi by Mössbauer spectroscopy (MS) or perturbed angular correlation of gamma rays (PAC). The application of the PAC technique for the investigation of nc semiconductors, metals, and alloys will be the subject of this contribution. The results will be discussed in context with those obtained by other techniques detecting the hfi, in particular MS, as well as by standard techniques used in materials sciences. To obtain information being complementary to that delivered by the hfi, these standard techniques are applied to the same samples investigated by PAC.

Taking into account that many techniques are sensitive to the global response of crystallites and GB forming the nc material under study, it will be shown that radioactive isotopes along with the hfi are able to deliver separate information about these two constituents. It turned out that the signals arising from crystallites and GB are particularly well distinguishable if materials that are studied possess internal electric field gradients or magnetic fields, as it is the case *e.g.* for the wurtzite lattice of ZnO or ferromagnetic Ni. It will be shown that Ni precipitates in nc NiCu alloys that are not resolvable by standard structural techniques like X-ray diffraction are still detectable *via* their characteristic magnetic hfi.

2. Common experimental aspects

Central for the characterization of the nc materials discussed in this contribution is the employment of radioactive probe atoms for obtaining local information on an atomic scale *via* the hfi that is measured at the nuclear site of the probe atom. In the present case, the hfi is detected by PAC. The technique typically requires 10^{11}–10^{12} radioactive probe atoms and with regard to its sensitivity it is similar to MS if performed in the 'source mode' (as compared to the 'absorber mode' which requires a substantially larger number of non-radioactive probe atoms). The signal amplitude in a PAC experiment does not depend on the Debye–Waller-factor as it is the case for MS (like for all scattering techniques). Both properties qualify PAC as well suited technique for the investigation of crystallites and grain boundaries in the nc materials [7].

2.1 Analytical technique, PAC spectroscopy

By using radioactive probe atoms, information about local lattice surroundings is obtained on an atomic scale *via* measuring the hfi, *i.e.* the local electric field gradient and the magnetic field at the site of the probe nucleus. Electric field gradients (efg) are produced by the lattice surrounding about the probe atom with lower than cubic symmetry and by defects in the immediate vicinity of the probe atom. In this work, the wurtzite lattice structure of the nc ZnO host lattice gives rise to a characteristic efg at the ^{111}In probe atoms that are incorporated on substitutional sites of the Zn sublattice (Sect. 3.2). In nc Ni, the GB will produce distinct efg in spite of the fact that there is no efg in the unperturbed crystal lattice of Ni due to the cubic symmetry of the fcc lattice (Sect. 5.2). Magnetic fields are produced either by an external field or as internal field due to the magnetic properties of the host lattice. In the case of diamagnetic nc InP, an applied magnetic field is used in order to establish the fraction of probe atoms on regular In sites in a cubic, defect free lattice environment (Sect. 3.3). In nc NiO, nc Ni, and nc NiCu alloys internal magnetic fields are observed if the respective nc materials form the antiferromagnetic lattice of NiO (Sect. 4.2) or the ferromagnetic lattice of Ni (Sects. 5.2 and 5.3).

The local efg and magnetic fields are measured at the nuclear site of the radioactive isotope ^{111}In/^{111}Cd by PAC. Choosing V_{zz} as the largest component of the traceless efg tensor ($V_{xx} + V_{yy} + V_{zz} = 0$) and defining an asymmetry parameter $\eta = (V_{xx} - V_{yy})/V_{zz}$ with $|V_{xx}| \leq |V_{yy}|$, the efg tensor is characterised by the coupling constant $\nu_Q = eQV_{zz}/h$ and η. Here, Q is the quadrupole moment of the isomeric 245 keV level of the ^{111}Cd nucleus with spin $I = 5/2$ used for measuring the hfi with the efg. The magnetic field $\boldsymbol{B}$ is measured *via* the interaction with the magnetic moment $\boldsymbol{\mu}$ of the same isomeric state yielding the Larmor frequency $\hbar\omega_L = (\boldsymbol{\mu}\boldsymbol{B})/I$.

The hfi is detected in a delayed coincidence experiment by recording the γ–γ cascade, which is emitted after the radioactive decay of the probe ^{111}In to ^{111}Cd. The interaction causes a modulation of the exponential decay function of the intermediate ^{111}Cd level of the γ–γ cascade. In case of an electric quadrupole interaction, the experimental modulation function $R(t)$ is governed by a frequency triplet yielding

$$R(t) = A_2 \left(s_0 + \sum_{n=1}^{3} s_n \cos(\omega_n t) \right) , \tag{1}$$

where A_2 is the anisotropy coefficient of the γ–γ cascade. For the frequencies the relation $\omega_1 + \omega_2 = \omega_3$ holds and the magnitude of the frequency ω_1 is determined by the strength of the local efg, which usually is expressed by the quadrupole coupling constant $\nu_Q \sim \omega_1 \sim V_{zz}$. The ratio ω_2/ω_1 is determined by the asymmetry parameter η of the efg tensor. In case of a magnetic dipole interaction caused by an internal magnetic field, the modulation is governed by a frequency doublet yielding

$$R(t) = A_2 \left(s_0 + \sum_{n=1}^{2} s_n \cos(n\omega_L t) \right) , \tag{2}$$

where the Larmor frequency ω_L is proportional to the strength of the local magnetic field. In case of an external magnetic field, a single frequency is observed corresponding to $2 \cdot \omega_L$. If several, non-equivalent lattice sites are occupied by the probe atoms, the modulation function $R(t)$ is composed of different fractions f_i

$$R(t) = \sum_i f_i R_i(t) , \quad \text{with} \quad \sum_i f_i = 1 . \tag{3}$$

Each fraction of probe atoms is characterized by its own set of frequencies, which serve as a unique 'fingerprint' of the corresponding local structure. More details about PAC and its applications can be found in [8–10].

The characterization by PAC is supplemented by standard analytical techniques used in materials sciences such as X-ray diffraction (XRD), transmission electron microscopy (TEM), UV/VIS absorption spectroscopy (UV/VIS), photoluminescence spectroscopy (PL), and extended X-ray absorption fine structure spectroscopy (EXAFS). It should be noted that, with the exception of EXAFS, these techniques give information about the long range order, whereas hfi and EXAFS supply information on the short range order. In particular, the efg because of its r^{-3} dependence supplies information about the electric charge distribution of the nearest-neighbour lattice sites.

2.2 Sample preparation

Whereas the detailed description of sample preparation will be given in context with the specific materials addressed, there are some aspects that are common to experiments that investigate nc materials with help of radioactive isotopes. In case of PAC, similar to MS, the number of suitable radioactive isotopes is rather limited because of the different nuclear properties that are simultaneously required for the experiment [8]. The number of suitable probes is increased to some extent if experiments are performed close to the place of production of the respective isotopes for the lifetime of the parent isotope can be much shorter, being as short as 1 h, than the lifetimes required if the experiment is performed at the home laboratory. Such a possibility exists *e.g.* at the ISOLDE facility at CERN [11, 12]. As already mentioned above, here, the parent/daughter combination $^{111}In/^{111}Cd$ will be used with a parent half life of 2.8 days. The limited number of probes is the main reason for the limited number of investigations of nc materials performed up to now. For PAC experiments like for MS experiments in source mode, each sample to be investigated has to be freshly doped with the respective radioactive isotope what represents some technical burden. But, MS if performed in absorber mode offers some relief for in this case the stable daughter isotopes is used (*e.g.* ^{57}Fe in the case of the parent/daughter combination $^{57}Co/^{57}Fe$). This mode can be employed if the respective element is part of the sample under study or if the required higher concentration of probe atoms is not a critical parameter. Working with stable isotopes in the nc material explains the larger number of MS investigations performed in 'absorber mode' as compared to 'source mode'.

Doping of nc materials can be performed (i) during growth, (ii) by diffusion, or (iii) by implantation. The latter procedure (iii) represents the most defined and chemically clean way of introducing the radioactive dopant after the growth process. But, the additional heat treatment required for annealing of the implantation induced lattice damage excludes this procedure in case of nc materials because of the possible increase of the average size distribution of the nanocrystals. For this reason, procedures (i) and (ii) will be met in the following experiments whereby the diffusion step performed after growth offers the important advantage of placing radioactive probes preferentially into the region of GB because of the respective higher diffusion coefficients (see Sect. 5.2). In particular, procedure (i) requires the technical knowledge for growth of the nc materials, which for the experiments discussed in this paper has been provided by the close cooperation in the Sonderforschungsbereich 277 [13].

As known from literature, the properties of nc materials are closely connected to the way of synthesis [14]. The method of choice used for producing nc materials for the present investigations was the method of electrodeposition and the samples were doped according to procedure (i) or (ii). This technique produces porosity free material [15] and offers the possibility of easily introducing radioactive isotopes during the growth process [16]. Typical volume concentrations of the dopants introduced are in the range of 1 ppb if exclusively radioactive isotopes are introduced. This concentration can be easily increased to any desired value by adding stable isotopes of the same element, as will be illustrated in the case of In doping of ZnO (see Sect. 3.2).

3. Doping of semiconductors

3.1 Comparison with literature

In nanometer scale semiconductors, many optical and electrical properties exhibit strong size dependencies [17]. The doping of nc semiconductors, which is the basis for the successful application of these materials (*e.g.* for solar cells [18] or spintronics [19]), still poses a severe problem [20, 21]. Most effort for solving this problem has been focused on II–VI semiconductor nanocrystals that were doped with impurities, such as Mn, Cu, or rare earth-elements, such as Tb (see Ref. [21] and references therein). The doping with impurities that introduce shallow donor or acceptor levels has been reported by only a few groups [22–24]. In contrast to PAC, there exist only a few investigations performed by MS [25]. In case of doping with transition metal impurities, the study of the local hfi by ESR represents an elegant way if the doping concentration is sufficiently high [26]. Similarly, EXAFS can provide local information about the surrounding of the dopants as discussed below.

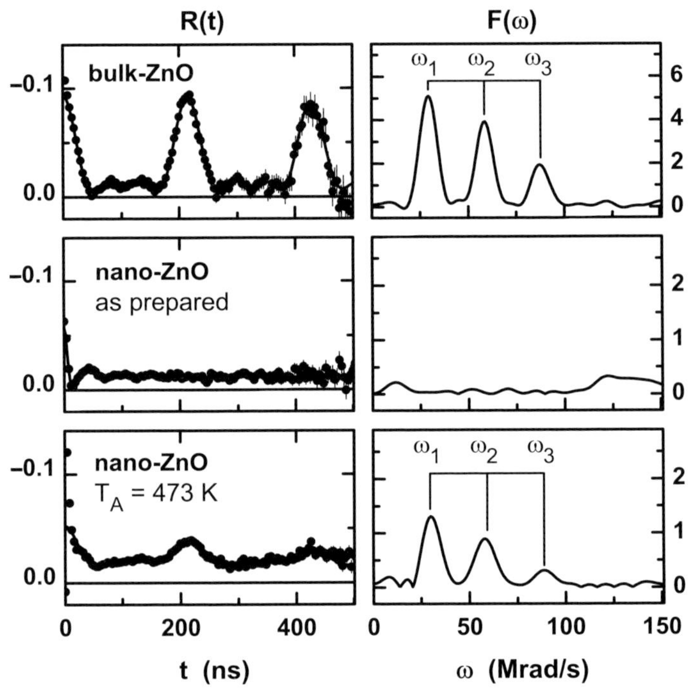

Fig. 1. PAC spectra $R(t)$ along with their Fourier transforms $F(\omega)$ measured at the radioactive dopant $^{111}In/^{111}Cd$ in different ZnO samples.

3.2 Doping of ZnO

Doping of the II–VI nc semiconductor ZnO was performed by introducing In atoms during the growth process of the nanocrystals. Prerequisite for the formation of an electrically active shallow donor state is the incorporation of the dopant In on substitutional Zn sites (In_{Zn}). Because of the non-cubic lattice structure of ZnO, the incorporation of the In dopants on Zn sites can easily be identified by detecting the well-known efg that is caused by the hexagonal ZnO lattice at the Zn sites [27]. The corresponding PAC spectrum $R(t)$ along with its Fourier transform $F(\omega)$ for bulk ZnO is shown in the top panel of Fig. 1. It shows the quadrupole interaction characterized by $\nu_Q = 31.2(1)$ MHz, $\eta = 0$ to be in agreement with the literature. As mentioned above, the data show the existence of one efg characterized by three frequencies ω_1, ω_2, and $\omega_3 = \omega_1 + \omega_2$. The value $\eta = 0$ for the asymmetry parameter is deduced from the ratio $\omega_2/\omega_1 = 2$, and the coupling constant is related to the basic frequency ω_1 by $\nu_Q = (10/3\pi) \cdot \omega_1$. Structural and optical properties of these nanocrystals were determined by the supplementary techniques XRD, TEM, UV/VIS, PL, and EXAFS.

3.2.1 Sample preparation

The nc ZnO was produced by the EDOC method (electrochemical deposition under oxidizing conditions) [28, 29]. The electrolysis process uses graphite as inert electrodes and takes place in a solution of $ZnNO_3$ and 0.1 M tetrabutylammonium-bromide (TBA-Br) dissolved in 2-propanol. For the electrolysis process a current density of $5–10\,\mathrm{mA/cm^2}$ was applied at a temperature of about 303 K. The nc ZnO crystals were doped *in situ* with ^{111}In atoms by adding radioactive $^{111}InCl_3$ solution to the electrolyte. By measuring the radioactivity of the produced nc ZnO, the relative concentration of ^{111}In was estimated to 1 ppb. Typically, 100 mg nc ZnO was produced during a single run. During this procedure the clusters were capped with TBA, which works as a stabilizer of the colloidal nanocrystals. In order to increase the dopant concentration, a mixture of *stable* $InCl_3$ and *radioactive* $^{111}InCl_3$ was added to the electrolyte. In this way, it was possible to vary the relative concentrations of In atoms in the range between 10^{-6} to 10^{-3}, which was verified by chemical analysis. For the hydrothermal treatment, which turned out to deliver the best annealing results (see below), the ZnO nanocrystals were dissolved in 2-propanol and the solution, encapsulated into a pressure vessel, was heated at temperatures between 373 K and 523 K. At the maximum temperature of 523 K, the pressure in the vessel was about 8.2 MPa.

3.2.2 Experimental results

PAC

Figure 1 shows two PAC time spectra $R(t)$ and their Fourier transforms $F(\omega)$ detected at ^{111}In doped nc ZnO in the 'as prepared' state and after hydrothermal annealing at 473 K. Without annealing (middle panels) and for temperatures up to 423 K, the PAC spectra exhibit a broad frequency distribution. This shows that the ^{111}In atoms are incorporated in non-unique crystalline environments characterized by a distribution of efg without a unique frequency. After annealing at 473 K (bottom panels), an efg characterized by $\nu_Q = 31(2)\,\mathrm{MHz}$ and $\eta = 0.2$ is observed and about 12% of all ^{111}In atoms are located in this environment. Increasing the annealing temperature to 523 K, this fraction increases to 25%. The coupling constant ν_Q of the observed efg matches very well that of the efg of ^{111}In on undisturbed Zn sites in bulk ZnO, mentioned above (see also top panel in Fig. 1). The occurrence of the efg proves that after this heat treatment a ZnO lattice is formed in the nc ZnO sample and that the ^{111}In probes are located on Zn sites. The difference in the asymmetry parameter η of the observed efg indicates a distortion of the ZnO lattice in the environment of the ^{111}In atoms. This distortion might be caused by the incorporation of ^{111}In atoms near the surface of the crystals or by internal strain present in the nanocrystals. Lattice distortions in nanocrystals are also known from EXAFS experiments [30] that will be discussed below. Figure 2 shows the fractions of In atoms incorporated in the Zn sublattice for the different annealing tem-

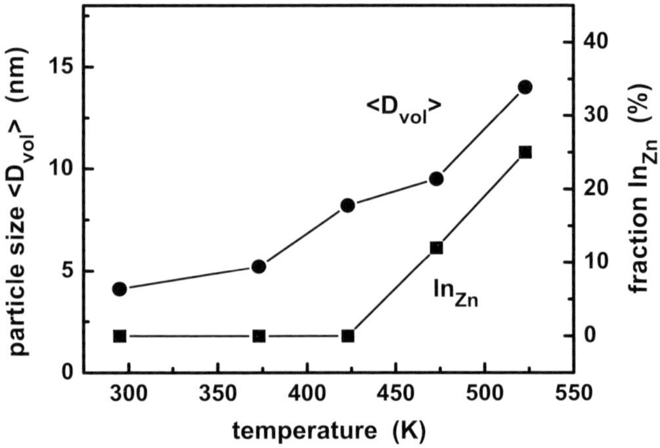

Fig. 2. Right axis: Fraction of In atoms on Zn sites (In_{Zn}) in nc ZnO observed during hydrothermal annealing at different temperatures. Left axis: Volume averaged particle size $\langle D_{vol} \rangle$ as determined by XRD.

peratures. The incorporation of In on substitutional Zn sites is the prerequisite of a successful doping process and should be achieved at a temperature as low as possible in order to avoid the growth of the nc ZnO crystals. It turned out that hydrothermal annealing realizes the incorporation of In on Zn sites at a significantly lower temperature than annealing in vacuum or under oxygen atmosphere [28].

In order to investigate the response of the ZnO host lattice to the hydrothermal treatment, this material was analyzed by XRD, TEM, UV/VIS, PL, and EXAFS. For this purpose, a sample of nc ZnO was divided into five parts in order to compare samples 'as prepared' and after hydrothermal treatments at 373 K, 423 K, 473 K, and 523 K, respectively. Because of lacking sensitivity, the samples analyzed by XRD and TEM were not doped with In atoms.

XRD and TEM

The XRD measurements were carried out using the Cu K_α radiation of a Siemens D 500 diffractometer in Θ–2Θ geometry. The spectra plotted in Fig. 3 exclusively show the diffraction pattern of the hexagonal ZnO lattice. The line shape of the Bragg peaks was analyzed by the method of Warren/Averbach [31] in order to extract size, size distribution, and microstrain of the nanocrystals [32]. After preparation, the nc ZnO crystals show a log-normal distribution with a median diameter $D_0 = 3.96$ nm, a geometrical standard deviation $\sigma = 1.11$, a volume-weighted mean grain size of $\langle D \rangle_{vol} = 4.1$ nm, and a microstrain of $\langle \varepsilon^2 \rangle = 1.50\%$. The corresponding size distribution is shown in Fig. 4 (solid line, top panel). With increasing annealing temperature the grain size and size distribution increases and, at an

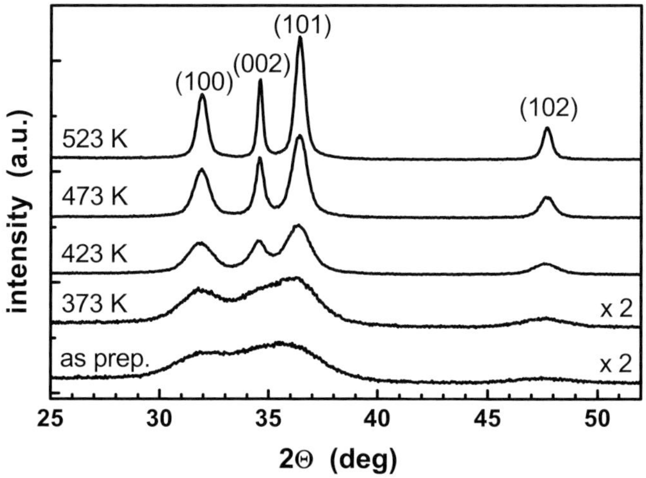

Fig. 3. XRD spectra of nc ZnO in the 'as prepared' state and after hydrothermal treatment at different temperatures.

annealing temperature of 473 K, reaches the values $D_0 = 7.47$ nm and $\sigma = 1.5$, corresponding to $\langle D \rangle_{vol} = 9.5$ nm (Fig. 4, bottom panel). Also, the microstrain disappears at 473 K what coincides with the onset of the incorporation of In dopants on regular Zn sites as observed by PAC (Fig. 2). The TEM images (JEM 2010 electron microscope; Fig. 5) show after preparation and subsequent hydrothermal treatments at 373 K (not shown) and 423 K nanocrystals of spherical shape. The sample annealed at 473 K still consists of nanocrystals of mainly spherical shape but some rod-shaped crystals are visible, too. After annealing at 523 K, the rod shaped crystals prevail with a length up to 200 nm and a diameter of about 15 nm. For the 'as prepared' state and the sample annealed at 473 K, Fig. 4 shows histograms of the corresponding grain size distributions determined by TEM, which both match well the log-normal distributions of the XRD analysis (solid lines).

UV/VIS and PL

The UV/VIS spectra were measured at nc ZnO dissolved in 2-propanol using an UNICAM UV 500 spectrometer. The absorption spectra in Fig. 6 exhibit a significant blue shift ΔE, which is assigned to a confinement effect expected to occur in nc semiconductors. With increasing annealing temperature, the spectra show a shift of the band gap of nc ZnO to smaller energies corresponding to the increase of the size of the nanocrystals as observed by XRD and TEM, and at a particle size of 17 nm, the material exhibits the band gap of bulk ZnO (3.37 eV). At the same time, with increasing annealing temperature up to 473 K, the total absorption in the UV range increases. Obviously, an increasing fraction of nanocrystals is involved in the absorption process, which

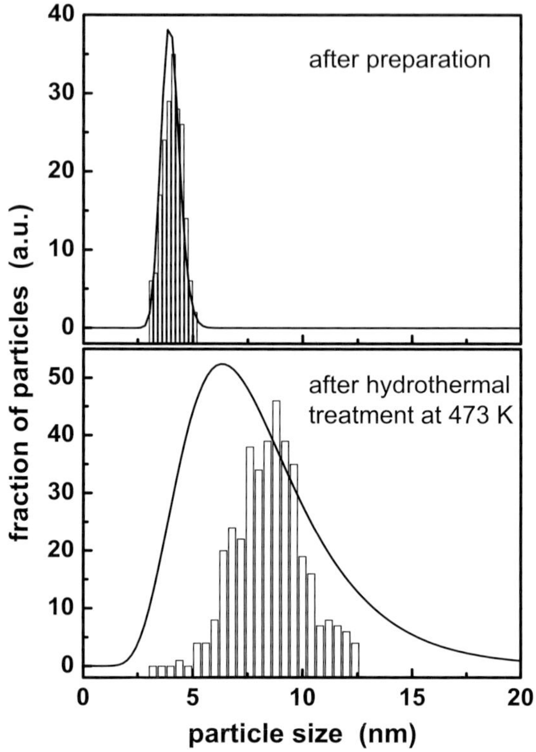

Fig. 4. Comparison of particle size distributions of nc ZnO measured by XRD (solid line) and TEM (histogram) after preparation of nc ZnO and after thermal treatment at 473 K.

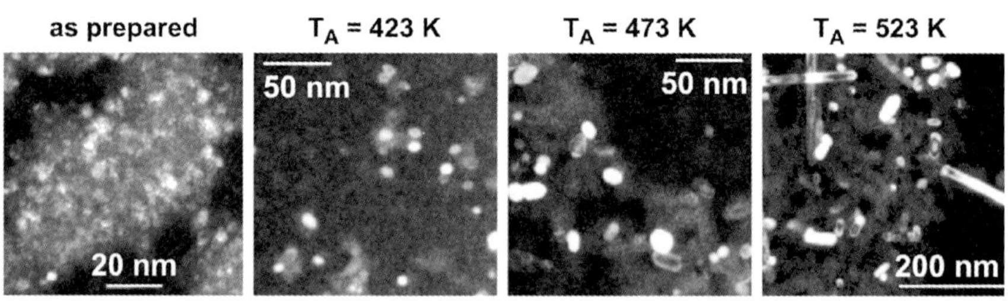

Fig. 5. TEM pictures of nc ZnO in the 'as prepared' state and after thermal treatment at different temperatures.

is caused by an ordering of the lattice inside of the nanocrystals. The same results were obtained in case of hydrothermally treated nc ZnO that was doped with ^{111}In. The change of the band gap energy ΔE, determined by the first maxima of the absorption spectra in Fig. 6, as a function of the particle size,

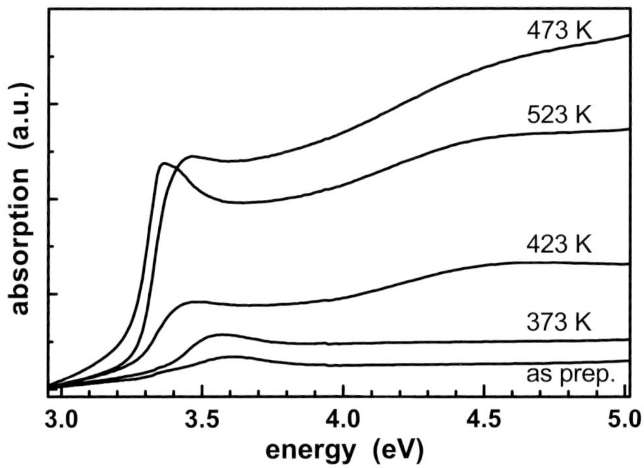

Fig. 6. UV/VIS absorption spectra of nc ZnO treated at different temperatures. The spectra were recorded at 295 K.

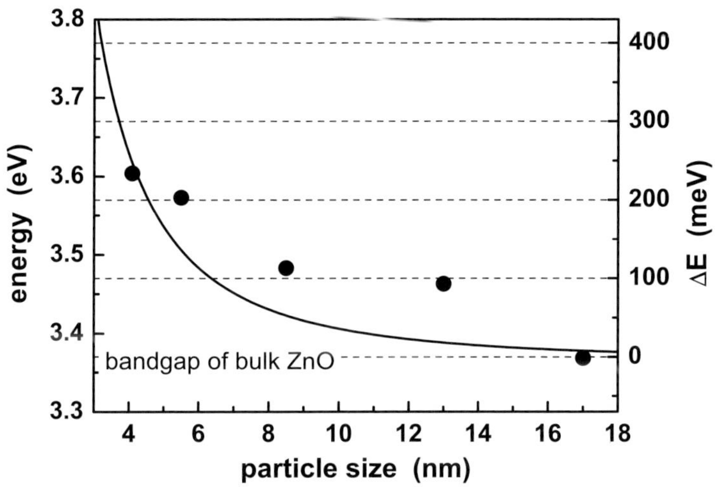

Fig. 7. Band gap energy E and change of band gap energy ΔE at 295 K plotted as a function of particle size; the data are deduced from the UV/VIS spectra in Fig. 6. The right axis shows the increase of the band gap energy with respect to that of bulk ZnO (3.37 eV). The solid line is a theoretical fit to the data (see text).

determined by XRD, is shown in Fig. 7. The solid line is a theoretical function that takes into account the size dependence of the band gap energy (quantum size effect) [33, 34].

The PL spectra were recorded at 1.6 K, using the 325 nm line of a He-Cd laser for excitation of the ZnO nanocrystals. The luminescence was analyzed

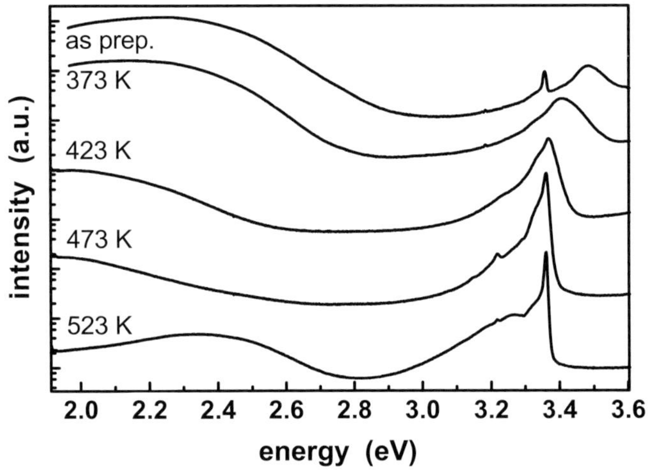

Fig. 8. PL spectra of nc ZnO treated at different temperatures. The spectra were recorded at 1.6 K. For clarity a vertical offset is added.

by a 0.5 m grating monochromator and detected by a CCD camera. After preparation, in the UV range the PL spectrum in Fig. 8 shows weak signals from a band–band recombination at 3.50 eV and a sharp line at 3.36 eV, which is close to the energy range where excitons bound to neutral donors are observed in bulk ZnO [35]. Thus, this line, which becomes more prominent at higher annealing temperatures, is assigned to a bound exciton, indicating the existence of donor like defects in the nominally undoped nc ZnO. That is consistent with undoped bulk ZnO which is always n-type. In the visible range of this spectrum, a broad deep emission band centred at 2.25 eV is present. For bulk ZnO, PL bands in this energy region are mainly caused by intrinsic structural defects, like the green band centered at 2.4 eV, which is reported to be caused by zinc vacancies [36]. After hydrothermal treatment at 373 K and 423 K, due to grain growth the positions of the band-band recombination and the deep emission band shift to lower energies and, at the same time, the intensity of the deep emission band decreases. After annealing at 473 and 523 K, the line at 3.36 eV becomes sharper and the structure of the deep emission band changes. The assignment of this line to a donor bound exciton is confirmed by measurements of the line intensity as a function of sample temperature for three different In concentrations (Fig. 9). The plot of these data *versus* $1/T$ yields an activation energy that agrees well with the activation energy of 24 meV assigned to the donor In in bulk ZnO [37].

EXAFS

The local structure of In and Zn atoms in nc ZnO (capped with TBA; hydrothermally annealed at 473 K) doped with a relative concentration of 10^{-3} In atoms

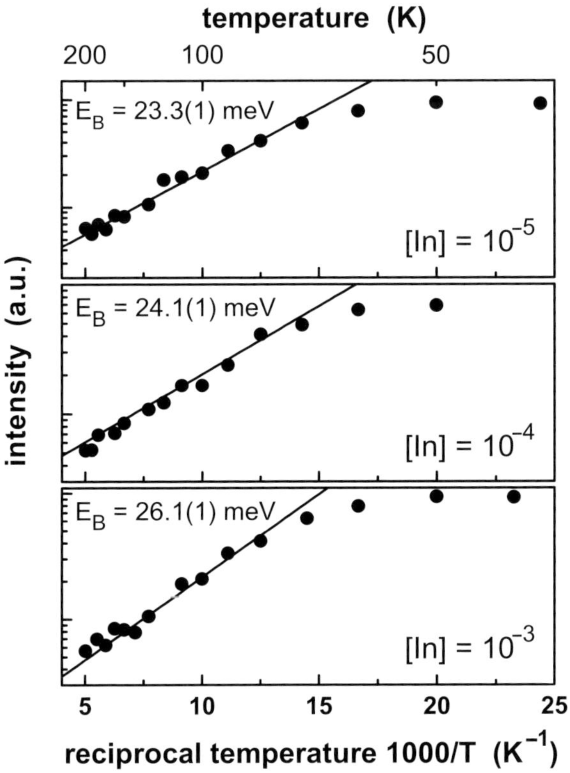

Fig. 9. Arrhenius plot of the intensity of the donor bound exciton observed in nc ZnO doped with ^{111}In and stable In at different relative concentrations.

was investigated by EXAFS measurements at a temperature of 20 K in fluorescence mode at the X1 beam line of HASYLAB at DESY. The EXAFS spectra plotted in Fig. 10 [38] show the K-edge absorption $k^2\chi(\mathrm{k})$ at the Zn host atoms in nc ZnO along with its Fourier transforms (panel b). The comparison with the bulk spectrum (Fig. 10 panel a) yields that the data can be fitted up to the fourth nearest-neighbour shell with the atomic distances known from bulk ZnO [35]. Of particular interest are the data in Fig. 10c showing the K-edge absorption $k^2\chi(\mathrm{k})$ at the dopant In along with its Fourier transform. Here, the first O shell ($O_{1\mathrm{th}}$) is clearly visible whereas the atoms of the second Zn shell ($Zn_{2\mathrm{nd}}$) and higher shells are almost invisible. An analysis of the first O shell about the In atoms yields an increased radial distance of about 2.17(5) Å (bulk: Zn-O_{NN}: 1.96(1) Å), which is caused by the larger covalent radius of In compared to Zn. The average number of atoms in the O shell is determined to be 3.3 ± 1.4. Since the number of O atoms in the first shell in bulk ZnO is 4, whereas this number would be 6 in bulk In_2O_3, it is suggested that the In atoms are incorporated on

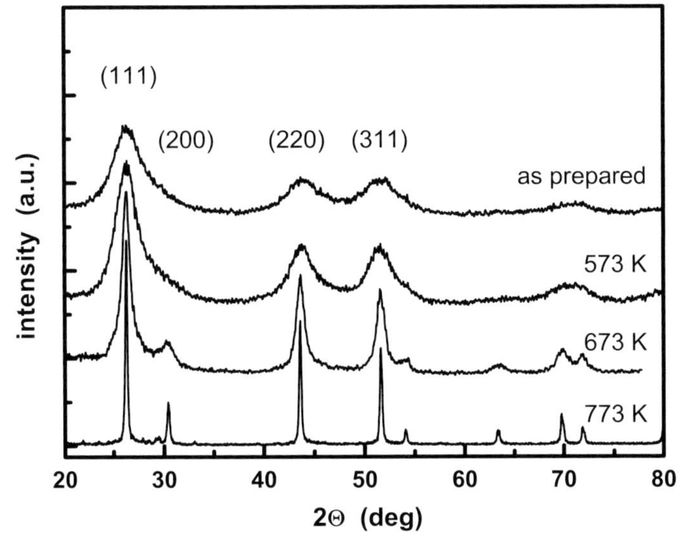

Fig. 11. XRD spectra of nc InP in the 'as prepared' state and after thermal annealing at different temperatures.

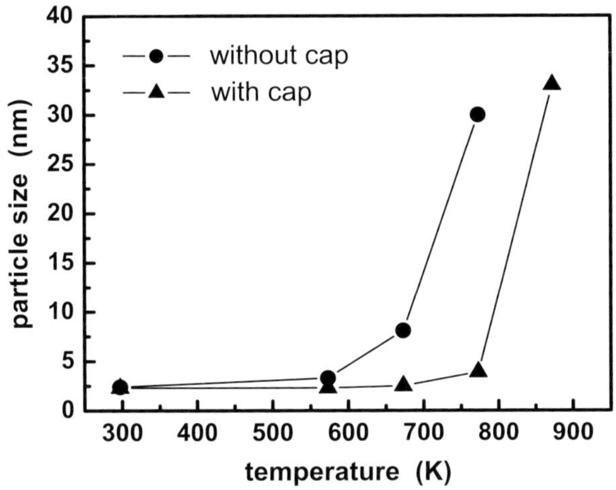

Fig. 12. Particle size of nc InP as a function of annealing temperature.

residing on In sites in a relaxed, defect free InP lattice vanish. In order to determine this unperturbed fraction f_c of dopant atoms quantitatively, an external magnetic field of 1.0 T is applied causing a unique Larmor spin precession ω_L of the corresponding ^{111}Cd nuclei, which in a PAC spectrum is visible by a modulation corresponding to a frequency of $2 \cdot \omega_L = 31.2$ Mrad/s (see *e.g.*

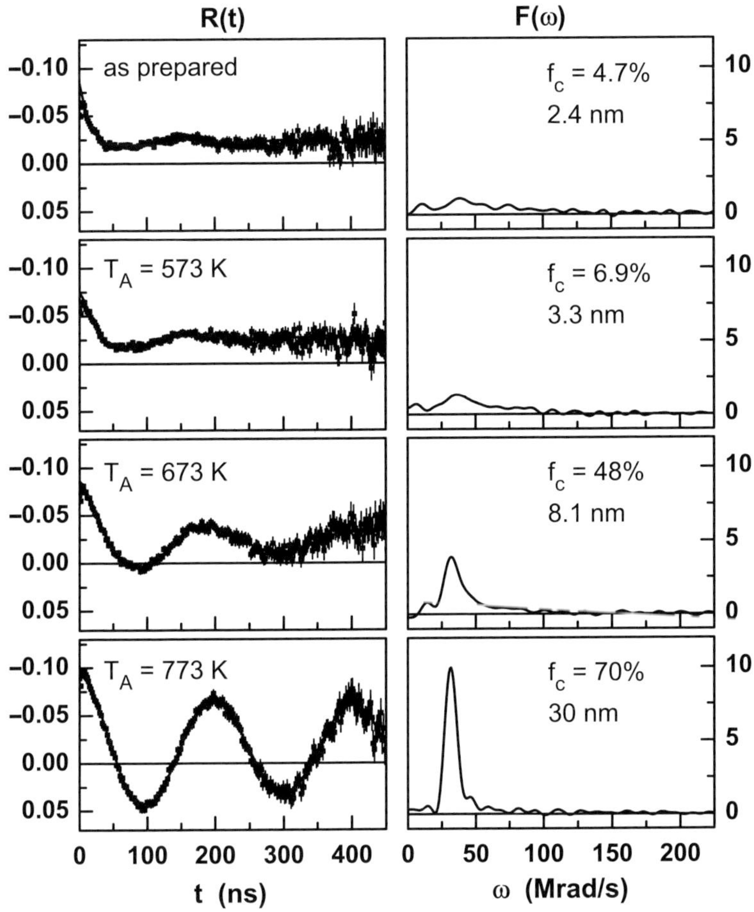

Fig. 13. PAC spectra $R(t)$ along with their Fourier transforms $F(\omega)$ measured at $^{111}In/^{111}Cd$ in differently treated InP samples (without cap). The respective cubic fractions f_c and the corresponding sizes of the crystals are indicated in the panels on the right.

bottom spectrum in Fig. 13). For the case of uncapped InP, the PAC spectra in Fig. 13 clearly indicate that there is no significant fraction f_c of ^{111}Cd dopants associated with this frequency below an annealing temperature of 673 K. Since InP does not give rise for any internal magnetic interaction, the strong damping visible in the PAC time spectra can only be caused by a distribution of strong efg due to lattice distortions, defects and surface sites making the interaction with the applied external magnetic field unresolvable [42]. Since the probe atoms are already located at a substitutional In site due to the doping procedure and the amount of the ^{111}Cd atoms without a unique Larmor precession can not be explained quantitatively by a statistical occupation of the surface

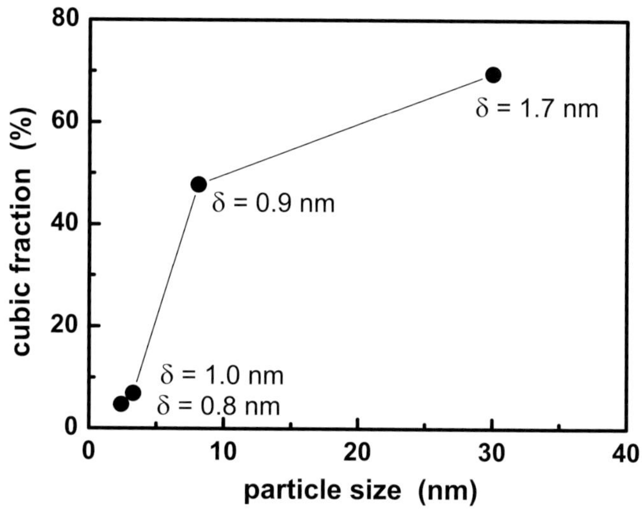

Fig. 14. Cubic fraction of probe atoms f_c as a function of InP particle size. Additionally, the thickness δ of the outer shell containing the distorted part of the InP lattice is given (see text).

sites of the crystals by the dopant atoms, distortions of the lattice of the crystals and defects have to be present throughout the crystal. Annealing at 673 K increases the cubic fraction to $f_c = 48\%$, but, at the same time, the particle size is increased to 8.1 nm.

The PAC data are consistent with theoretical calculations [43] and EXAFS experiments [30], showing the presence of distortions in the InP lattice, whereby the corresponding volume fraction decreases with increasing temperature. The occurrence of a temperature dependent unperturbed fraction f_c associated to an undistorted cubic lattice and a perturbed fraction $(1 - f_c)$ associated to a distorted lattice can be explained in a 'core–shell' model. Here, it is assumed that the nc InP particles consist of a cubic inner core surrounded by a distorted shell of thickness δ. Using the fraction f_c measured by PAC, the decrease of the thickness δ as a function of particle size (*i.e.* annealing temperature) is indicated in Fig. 14.

4. Size dependence of Néel temperature

4.1 Comparison with literature

The small volume of the crystallites and their interaction with the GB are expected to induce changes in the magnetic properties of nc ferromagnetic materials. Thus, a reduction of the magnitude of the saturation magnetization or a lowering of the ferromagnetic transition temperature T_C can be expected. As an example, 6 nm Fe is reported to have a saturation magnetization reduced to about 60% of that of polycrystalline Fe [1], and, in nc Gd, which is produced by inert gas condensation and subsequent compaction, a reduction of the Curie temperature by about 20 K is reported [44]. The shift of the Curie temperature $\Delta T_C = T_C(\text{bulk}) - T_C(d)$ observed for a particle of size d was measured by calorimetry and by ac magnetometry. Whereas there are several corresponding investigations in thin films, in 3D nanocrystals there are no experimental data available, up to now. Here, we report first results on the shift of the Néel temperature ΔT_N for nc antiferromagnetic NiO. By Rubinson *et al.* it is reported that the Néel temperature of NiO is reduced from 523 K for bulk NiO to 300 K for 6 nm sized nc NiO [45]. However, there is no information given for the size dependence of the Néel temperature.

4.2 Néel temperature of NiO

4.2.1 Sample preparation

Using PAC, the local magnetic field of the antiferromagnet NiO can be determined. Magnetic properties of antiferromagnetic materials can be determined in an elegant way, provided a preparation route can be found for introducing suitable radioactive probe atoms for the measurement of the magnetic hfi. With help of the EDOC method, NiO was prepared by using a 1 mm thick sheet of Ni metal as sacrificial anode. The electrolysis was performed in a water free solvent of 2-propanol using as conducting salt tetrabutylammoniumbromide (TBA). For doping nc NiO with the radioactive probe atom ^{111}In needed for the PAC experiments, this isotope was diffused into the sacrificial anode before electrolysis. At the carbon glass cathode air was blown through the solvent for oxidizing the precipitating Ni particles. After this procedure nano-$Ni(OH)_2$ particles doped with ^{111}In atoms are formed. The conversion to nc NiO is performed by a subsequent thermal heat treatment. Fig. 15 shows a sequence of XRD spectra obtained after annealing the sample at different temperatures. The spectrum obtained at 523 K still shows the coexistence of NiO and $Ni(OH)_2$. Thus, it is obvious that temperatures higher than 523 K are required in order to obtain a substantial fraction of NiO in the sample. The XRD spectra are also used for determining the resulting particle sizes effected by the heat treatment at the temperature indicated.

4.2.2 Experimental results and discussion

For bulk antiferromagnetic NiO, the corresponding Larmor frequency measured at the probe $^{111}\text{In}/^{111}\text{Cd}$ is well known, being $\omega_L = 242(3)$ Mrad/s at 300 K [46]. For a nc NiO sample annealed at 673 K, corresponding to a particle size of $d = 12$ nm, Fig. 16 shows a sequence of PAC spectra taken at different sample temperatures T_m varying between 18 K and 500 K. The fact that (i) the frequency doublet is observed and (ii) the magnitude of the measured Larmor frequency $\omega_L = 244(2)$ Mrad/s (for $d = 26$ nm) corresponds to the value

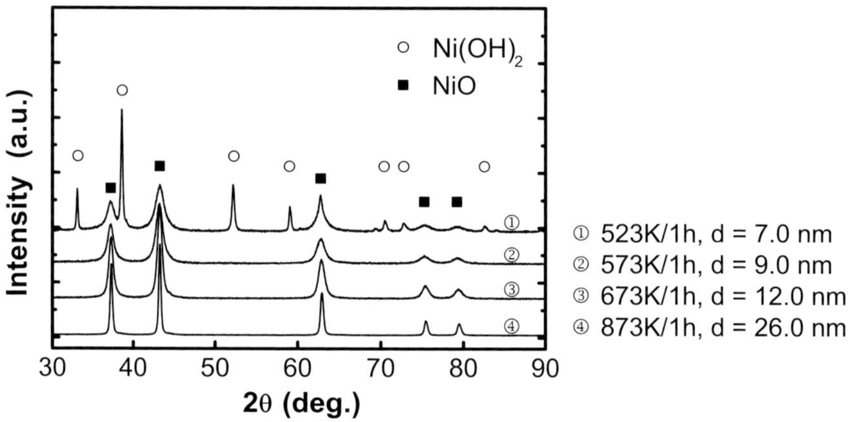

Fig. 15. XRD spectra of NiO after thermal treatment at different temperatures. The coexistence of NiO and $Ni(OH)_2$ after annealing at 523 K is clearly visible.

of bulk NiO, directly shows (i) that there is an internal magnetic field present in the sample and (ii) that the ^{111}In probe atoms are located on substitutional Ni sites of a NiO crystal. As expected, the signal of the magnetic interaction vanishes above the Néel temperature. The Néel temperature belonging to the crystal size d is determined by extrapolating the temperature dependent Larmor frequency to zero value.

The negative shift of the Néel temperature $-\Delta T_N$ as a function of the particle size is presented as a log–log plot in Fig. 17. The linear behaviour of the data points is governed by a power law what is expected for the magnetic transition temperature on the particle size. Using the relation

$$T_N(\text{bulk}) - T_N(d) = \Delta T_N \sim d^{-\lambda}$$

the value $\lambda = 3.1(1)$ for the exponent yields the best fit to the experimental data (see straight line in Fig. 17). This value is significantly larger than $\lambda = 1$ reported for nc ferromagnetic Gd [44]. Standard finite-size scaling theory relates λ to the critical exponent $\nu = 1/\lambda$ and predicts a value of $\lambda = 1.5$ [47].

It is noted that an effective exponent of $\lambda = 3$ is reported if corrections to the finite-size scaling are taken into account [48]. This value would be also in agreement with λ reported for experiments performed on antiferromagnetic CoO films [49]. Finally, taking into account the single experimental value reported by Rubinson *et al.* [45] for NiO, which was measured by ESR, it turns out that this value (see open square in Fig. 17) perfectly fits to the values obtained by the PAC measurements.

Regarding the limited number of investigations of the size effect on the behaviour of magnetism in nc magnetic materials, one reason is discussed in [44]: It is pointed out that grain growth has to be avoided during the study of the in-

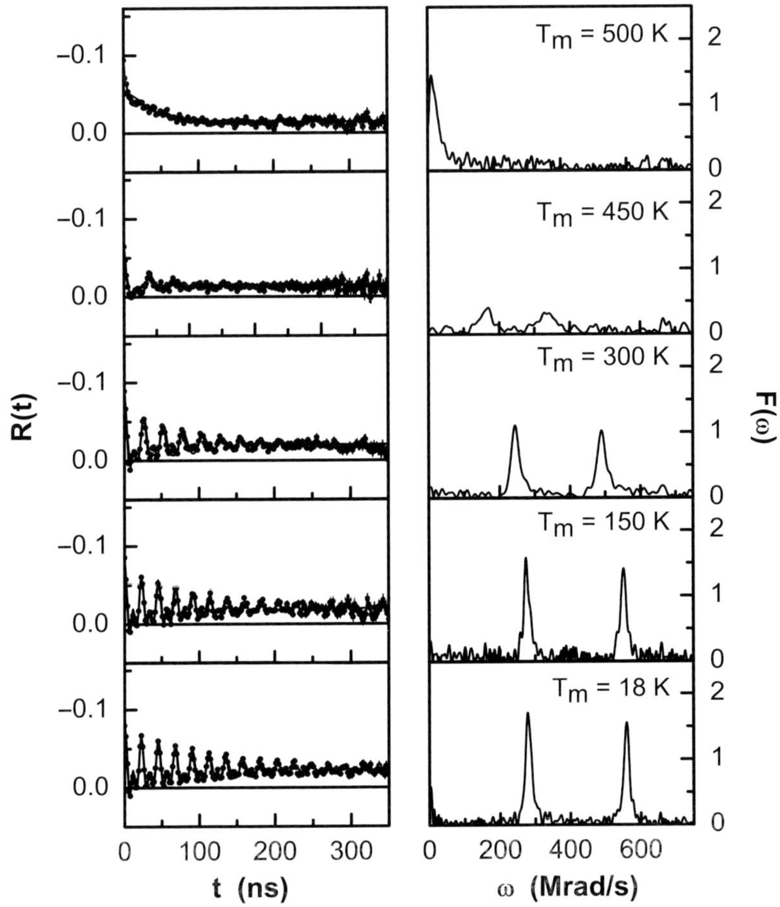

Fig. 16. PAC spectra of nc NiO annealed at 673 K (resulting particle size is 12 nm). The signal of the anti-ferromagnetic interaction vanishes between 450 and 500 K in this case.

fluence of d on the transition temperature. This is only possible as long as the temperature is kept below 20% of the melting point T_{melt}. In case of nc Gd with $T_C = 290$ K and $T_{melt} = 1570$ K and similarly in the present case of NiO with $T_N = 520$ K and $T_{melt} = 2250$ K this condition is met.

Finally, it is remarked that the investigation of the magnetic properties *via* the magnetic hfi yields a well defined measurement of the behaviour of the internal magnetic field if the probe atoms are exclusively situated in the crystallites. Other techniques, which measure integral properties of nc samples, observe an average of different contributions arising from both crystallites and GB. Moreover, in case of antiferromagnetic materials many experimental techniques are not sensitive to the magnetic behaviour of the sample.

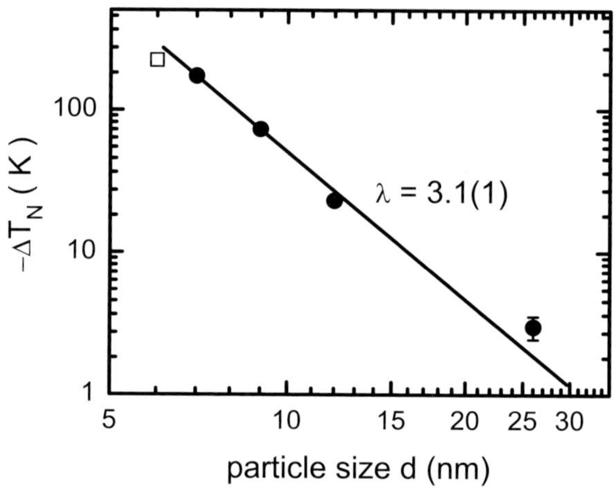

Fig. 17. Shift of the Néel temperature $-\Delta T_N$ as a function of particle size in nc NiO. Open square: Value obtained by ESR (Rubinson *et al.* [45]).

5. Grain boundaries and alloys

5.1 Comparison with literature

Although the properties of nc materials are determined to a large extent by the microscopic properties of the GB, a comprehensive picture of the structure of GB is still lacking [1]. The reason for this lack of information is found in the fact that only few experimental techniques are sensitive to the microscopic properties of GB in nc materials. Experimental methods, such as EXAFS [50], positron annihilation [51], and high resolution electron microscopy (HREM) [52, 53] were used for the investigation of GB structures in poly- and nanocrystalline materials, but the information about GB structures obtained so far is still rather limited. In addition, there are very few theoretical calculations and computer simulations [54], whereby most of these studies are focused on ordered GB structures. Currently, the most detailed experimental information about GB structures is supplied by HREM studies on bicrystals and tricrystals [55, 56], where only one or two GB are present, respectively. It is not obvious to what extent the experimental knowledge obtained by the HREM studies on bicrystals and tricrystals can be transferred to GB that exist in poly- or nanocrystalline materials. Likewise, other experimental techniques, in particular diffraction techniques that have been very successful in characterizing the structural properties of materials, often meet severe difficulties if applied to the investigation of GB structures in poly- and nanocrystals: Because of the small fraction of atoms located in GB, which is due to the thickness of GB typically in the order of 0.5–1 nm, the intensity of the signals of diffraction experiments often remain below the detection limit.

In this situation, the use of radioactive isotopes along with the locally sensitive hfi represents an important tool for characterizing GB. In contrast to PAC, there is a large number of investigations performed by MS. An extensive review is published by Campbell and Kaczmarek in addition to several reports on MS investigation on nc materials prepared by mechanical attrition [57–59]. In many cases, MS is performed in 'absorption mode' and the probe atom used is ^{57}Fe. In case of nc W the sensitivity to GB was improved by diffusing ^{57}Fe atoms into the sample at low temperatures (0.36–0.5 T_{melt}), which enhances the number of probe atoms inside of the GB [59]. An alternative way for enhancing the sensitivity to GB is achieved by placing the radioactive probe ^{57}Co in the GB of the *d*-transition metals Cr, Ta, and W with help of segregation effects [60]. These investigations deliver clear evidence for new signals that do not belong to the crystallites. But, in most cases only broad distributions of efg are observed, which yield no specific information so that their interpretation is difficult. In particular, there are no unique signals reported that belong to a distinct efg associated with a GB. For the PAC study discussed below, radioactive probe atoms are diffused into nc Ni at low temperatures (0.36–0.5 T_{melt}). The results on GB in Ni are published in more detail in [61] and reveal the existence of several distinct efg pointing to the existence of ordered GB. Similar results are obtained for nc Cu and Co [61].

The properties of nc alloys additionally depend on the presence of precipitates. For many experimental techniques it is difficult to detect precipitates giving information about the homogeneity of nc alloys. In case of nc alloys containing a magnetic component like Ni, precipitates of the magnetic component can be easily detected by looking for the Ni related magnetic hyperfine interaction as will be shown in the case of PAC investigations on nc NiCu.

5.2 Ordered grain boundaries in Ni

5.2.1 Sample preparation

Nanocrystalline samples of Ni were prepared by pulsed electrodeposition (PED) [62]. The corresponding preparation parameters are given in Table 1. The Ni samples were doped *ex-situ* with ^{111}In by diffusion at 623 K (30 min). The incorporation of the ^{111}In atoms on substitutional Ni sites was investigated as a function of the temperature of a pre-annealing step (60 min), *i.e.* annealing before ^{111}In diffusion, and post-annealing (30 min) after ^{111}In diffusion. The annealing steps were performed under vacuum and the PAC spectra were taken at 295 K after each annealing step.

5.2.2 Experimental results and discussion

After doping nc Ni *in-situ* with ^{111}In during the PED process, the magnetic hyperfine interaction known from ferromagnetic Ni is observed giving rise to the frequency doublet of ω_L and $2 \cdot \omega_L$ with $\omega_L = 97.6$ Mrad/s [63]. The agreement of the frequency observed in nc Ni with the bulk value of ferromagnetic

Table 1. Composition of the electrolyte, temperature, current density, on-and off-times of the current pulses, used for the deposition of nc Ni and $Ni_{0.5}Cu_{0.5}$ alloys. The grain size was determined by X-ray diffraction in case of Ni and by atomic force microscopy in case of Ni-Cu.

samples	grain size	electrolyte	preparation parameters		
	[nm]		T (°C)	j (A/cm^2)	$t_{on}(ms)/t_{off}(ms)$
nc Ni	19	$NiSO_4$ (40 g/l), K,Na-tartrate (120 g/l), NH_4Cl (40 g/l)	36.5	0.13	1/49
nc $Ni_{0.5}Cu_{0.5}$	30	$CuSO_4$(6 g/l), $NiSO_4$ (80 g/l) Na-citrate (100 g/l), saccharin(2 g/l) (optional)	15 . . . 70	0.1 . . . 0.3	1 . . . 2/48 . . . 49

Ni shows that the nanocrystallites possess the same lattice as large Ni crystals. There are no additional signals observable that might result from probe atoms located in GB. Since by *in-situ* doping the ^{111}In atoms are distributed homogeneously over the entire sample, the fraction of probe atoms in GB is too small for becoming visible in the PAC spectrum. A predominant doping of the GB of nc Ni was obtained *ex-situ* by diffusion at low temperatures ($T_{diff} < 0.5\,T_m$) after the preparation of the sample. After diffusion of ^{111}In at 623 K and subsequent annealing at different temperatures, a frequency distribution is observed in all PAC spectra of the Ni samples, which might correspond to disordered

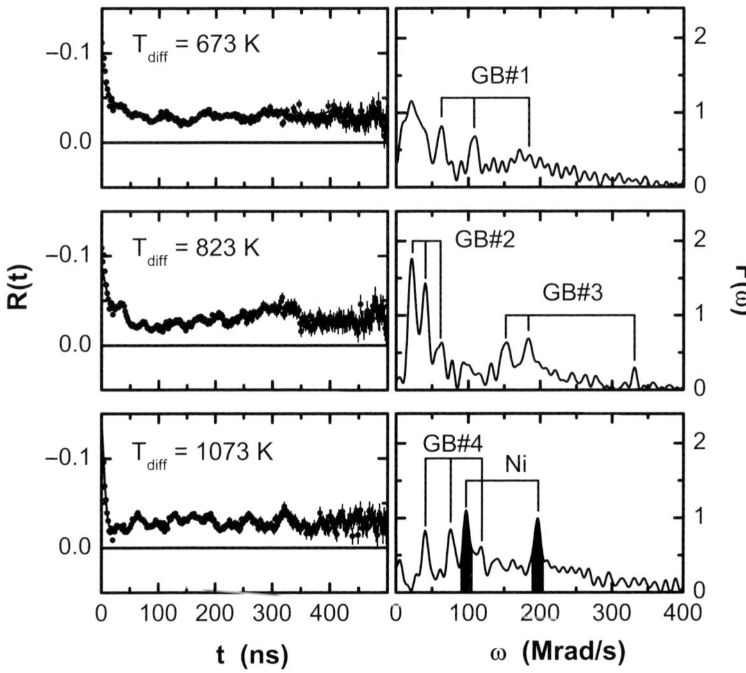

Fig. 18. PAC spectra and their Fourier transforms showing the presence of ordered structures in the GB of nc Ni *via* the occurrence of four unique efg (see Table 2). At 1073 K, the occurrence of the frequency doublet (black lines) indicates the incorporation of ^{111}In into the ferromagnetic Ni crystallites.

Table 2. Signals of unique efg observed in GB of nc Ni.

Signals	ν_Q (MHz)	η	$\Delta\nu_Q$ (MHz)	f (%)
GB#1	58(1)	0.38(2)	0	3–5
GB#2	21(1)	0.33(2)	6	10–15
GB#3	106(1)	0.77(2)	0	3–5
GB#4	41(1)	0.21(3)	0	3–7

GB structures. But, in addition, four distinct frequency triplets are observable (see Fig. 18 and Table 2). Since a hfi caused by an internal magnetic field gives rise to a frequency doublet and not by a triplet, these frequencies have to be assigned to four distinct efg. The corresponding values of the different efg are colleted in Table 2. The high experimental significance of these efg is well documented by the data in Fig. 18, showing the occurrence of a single efg (GB#1) at 673 K, two efg (GB#2 and GB#3) at 823 K, and of a forth efg (GB#4) at 1073 K. Etching of a sample after diffusion with ^{111}In effects no change of the measured PAC spectra, showing that lattice sites on the surface of the sample are not responsible for the detected efg. There are no defects known from the literature for bulk Ni that might explain the efg measured here [42].

Since most of the probe atoms should be placed in GB, it is concluded that the four observed unique efg arise from different ordered structures in the GB. The existence of ordered structures in GB was predicted, *e.g.* for Ag, on the basis of theoretical studies [64] and have been observed experimentally by transmission electron microscopy *e.g.* in case of a NiO bi-crystal [65]. However, there are no reports about direct experimental evidences of ordered structures in GB of polycrystalline samples. Since the PAC data obtained below 1073 K yield no frequencies that belong to magnetic hfi, it is concluded that in the GB of the ferromagnetic nc Ni no or only a weak magnetic field (< 0.5 T) is present. Because of the occurrence of well defined efg characterizing the different environments of the In atoms in GB (note that the frequency distribution $\Delta\nu_Q$ shown in Table 2 is always zero or close to zero) a conversion between different ordered GB structures upon annealing can easily be pursued by PAC as is described in [61].

After the diffusion anneal at 1073 K, again the magnetic frequency doublet belonging to ferromagnetic Ni is observable. The occurrence of this signal shows that a fraction of the In atoms now are incorporated into the Ni crystallites as is discussed in [66].

5.3 Ni precipitates in NiCu alloys

5.3.1 Sample preparation

Like Ni, the NiCu alloys were prepared by PED as described in Sect. 5.2.1 and Table 1. In contrast to the nc Ni samples, the nc NiCu samples were doped *in-situ* with ^{111}In by adding $^{111}InCl_3$ to the electrolyte. The formation of Ni precipitates was investigated as a function of the temperature of the electrolyte, the current density during deposition, and the optional addition of saccharin. The XRD experiments were performed at a Siemens D 5000 θ–θ spectrometer using the K_α radiation of Cu. In this way, the lattice constant of the NiCu alloys was determined yielding the composition of these samples. Additionally, the composition was determined by energy dispersive X-ray spectroscopy (EDX).

5.3.2 Experimental results and discussion

The composition of the Ni-Cu sample discussed here was determined by EDX to 50 at% Ni and 50 at% Cu. In Fig. 19b, the XRD spectrum recorded directly after the deposition by PED shows the fcc structure of Ni-Cu alloys and as an inset a blow-up of the $\langle 111 \rangle$-diffraction peak. The position of this peak at a scattering angle, that is located between those known for pure Cu and pure Ni, is in agreement with the composition determined by EDX. The diffraction peak is significantly broadened compared to a diffraction peak of a perfect crystal

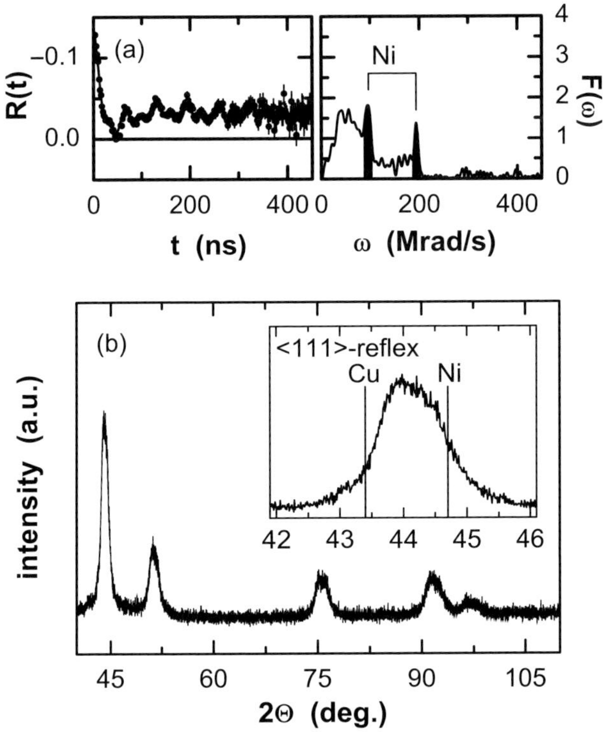

Fig. 19. Investigation of a nc $Ni_{50}Cu_{50}$ alloy by (a) PAC and (b) XRD. In the lower panel, the inset shows a blow-up of the $\langle 111 \rangle$ Bragg peak of the XRD spectrum.

and exhibits an asymmetric shape. In Fig. 19a, the PAC spectrum of the same sample shows a distribution of frequencies, centred at about $\bar{\omega} = 50$ Mrad/s. In addition, the sharp Larmor frequency $\omega_L = 97.6$ Mrad/s of ferromagnetic Ni shown by the frequency doublet in the Fourier transform $F(\omega)$ is visible. Obviously, the Ni-Cu sample is not homogenous after deposition and contains crystallites of pure Ni, in which $f_{Ni} = 20\%$ of the probe atoms are located. After annealing the sample at 473 K the fraction of ^{111}In in pure Ni crystallites increases slightly to 23% before it starts to decrease at 673 K. At a temperature of 873 K, finally, the Ni crystallites are no longer observed (*i.e.* $f_{Ni} < 2\%$) and only the frequency distribution about $\bar{\omega} = 50$ Mrad/s is present. Such a frequency distribution is consistent with a disordered Ni-Cu solid solution and is caused by the different arrangements of Ni and Cu atoms in the neighbourhood of the ^{111}In probe atoms.

A signature of the Ni precipitates obviously is, if at all, not easily visible in the XRD spectrum of Fig. 19b. Only the asymmetric shape of the diffraction peak might be an indication for the presence of Ni precipitates. It should be noted that the determination of the grain size by XRD is not reliable in this case

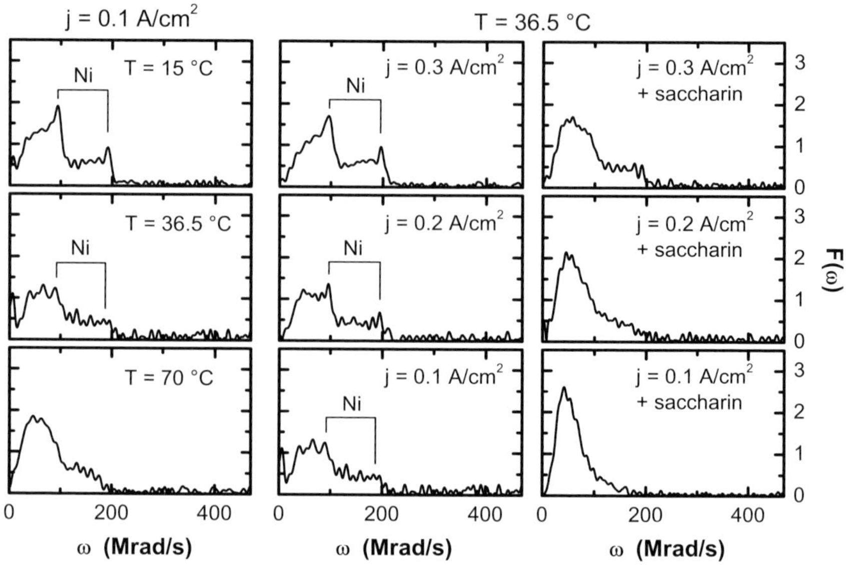

Fig. 20. Fourier transforms F(ω) of PAC spectra measured in different $Ni_{0.50}Cu_{0.50}$ alloys, which have been deposited *via* PED under different conditions (temperature, current density, content of saccharin).

since the shape of the diffraction peaks is affected by the presence of Ni crystallites in the Ni-Cu samples. In contrast to the XRD spectrum, the PAC data in Fig. 19a unambiguously prove the presence of Ni crystallites in the nc Ni-Cu alloy *via* the occurrence of the frequency doublet known for the magnetic hfi of ferromagnetic Ni.

Due to the high sensitivity of PAC to the Ni precipitates, the influence of the temperature of the electrolyte, the current density, and the addition of saccharin to the electrolyte on the homogeneity of nc NiCu alloys was studied. The PAC spectra obtained at various experimental conditions are shown as Fourier transforms $F(\omega)$ in Fig. 20. The temperature was varied between 15 °C and 70 °C keeping the current density constant at 0.1 A/cm^2 (left column). The current density was varied between 0.1 and 0.3 A/cm^2 keeping the temperature constant at 36.5 °C, whereby one series of samples was deposited without (centre column) and a second one with addition of saccharin (right column) to the electrolyte. The fraction of Ni crystallites, which directly is visible in the PAC data, is reduced by increasing the temperature at a constant current density (left column) and by decreasing the current density at constant temperature (centre and right columns). Obviously, the Ni-Cu alloy becomes more homogeneous at lower current densities and at higher temperatures. These conditions during deposition, however, are known to be correlated with an increasing grain size [62]. After adding saccharin to the electrolyte (2 g/l), no Ni crystallites are observed by PAC at all current densities. Only a frequency distribution as it is expected for a disordered $Ni_{0.50}Cu_{0.50}$ solid solution is visible (right column in Fig. 20). The addition of saccharin is obviously advantageous for producing homogeneous Ni-Cu alloys and, at the same time, is known to reduce the grain size during deposition [16, 62].

6. Conclusions

The potential of using radioactive probe atoms for the characterization of nc materials has been demonstrated. Although, the use of radioactive isotopes is connected with some special requirements, in particular a facility for handling radioactive isotopes and the need of equipment for producing the nc material in house, these special expenses are more than compensated by the advantages in getting precise and specific results as outlined below.

In particular, about the two constituents of nc materials, the crystallites and the GB, unique information is obtained by measuring unique signals of the hfi at the site of the radioactive probe atoms. Information obtained on the basis of the detected hfi in many cases is difficult to obtain by other experimental methods, if at all. The following two aspects were discussed in this context:

- By using the magnetic or electric hfi information about the short range ordering in the vicinity of the probe atom is obtained, in particular in case of the electric hfi, where the sensitivity of the efg decreases with r^{-3}.
- By using radioactive isotopes and the local character of the hfi, crystallites and GB can be investigated separately. In addition, the possibility of working with very low concentrations of probe atoms was of particular advantage in case of GB, the incorporation of dopants into nc semiconductors, and the detection of precipitates in nc alloys.

  Both aspects have been illustrated by addressing different areas being of interest for the characterization of nc materials:
- The controlled introduction of dopants into nc semiconductors was studied for nc ZnO and nc InP. It turned out that the as grown crystallites always require some additional heat treatment in order to provide a local environment of sufficient quality about the impurity atoms, which is required for acting as electrically active dopants.
- The influence of the crystal size on the magnetic transition temperature in antiferromagnetic nc NiO was observed by measuring the local magnetic field at the site of the probe atoms that are localized in the NiO crystals. In this way, information on the magnetic properties of the antiferromagnetic material was accessible and the influence of the crystal size on the shift of the Néel temperature was separated from a possible influence caused by the GB.
- In Ni, the structure of GB was investigated. The occurrence of distinct efg observed at the site of probe atoms that are localized inside the GB gives

47. V. Privman (Ed.), *Finite Size Scaling and Numerical Simulation of Statistical Systems*. World Scientific, Singapore (1990).
48. M. Henkel, S. Andrieu, P. Bauer, and M. Piecuch, Phys. Rev. Lett. **88** (1998) 4783.
49. T. Ambrose and C. Chien, J. Appl. Phys. **79** (1996) 5920.
50. T. Haubold, R. Birringer, B. Lengeler, and H. Gleiter, Phys, Lett. A **135** (1989) 461.
51. D. Segers, S. Van Petegem, J. F. Löffler, H. Van Swygenhoven, W. Wagner, and C. Dauwe, Nanostruct. Mater. **12** (1999) 1059.
52. Z. Horita, D. J. Smith, M. Nemoto, R. Z. Valiev, and T. G. Langdon, J. Mater. Res. **13** (1998) 446.
53. S. Ranganathan, R. Divakar, and V. S. Raghunathan, Scripta Mater. **44** (2001) 1169.
54. H. Van Swygehoven, D. Farkas, and A. Caro, Phys. Rev. B **62** (2000) 831.
55. I. Kaur, Y. Mishin, and W. Gust, *Fundamentals of Grain and Interphase Boundary Diffusion*. 3rd ed., John Wiley & Sons ltd., Chichester (1995) p. 388.
56. K. L. Merkle, J. F. Reddy, C. L. Willy, and D. J. Smith, Phys. Rev. Lett. **59** (1987) 2887.
57. S. J. Campbell and W. A. Kaczmarek, *Mössbauer Spectroscopy Applied to Magnetism and Materials Science*. Vol. 2, G. J. Long and F. Grandjean (Eds.), Plenum Press, New York, (1966) p. 273.
58. B. Fultz and H. N. Frase, Hyperfine Interact. **130** (2000) 81.
59. G. Rixecker, Hyperfine Interact. **130** (2000) 127.
60. S. M. Klotsman, V. N. Kaigorodov, M. I. Kurkin, and V. V. Dyakin, Interface Sci. **8** (2000) 323.
61. Z. Guan, H. Wolf, X. Li, Th. Agne, and Th. Wichert, to be published.
62. H. Natter, M. Schmelzer, and R. Hempelmann, J. Mater. Res. **13** (1998) 1186.
63. St. Lauer, Z. Guan, H. Wolf, H. Natter, M. Schmelzer, R. Hempelmann, and Th. Wichert, Nanostruct. Mater. **12** (1999) 955.
64. Q. Ma, C. L. Liu, J. B. Adams, and R. W. Balluffi, Acta Metall. Mater. **41** (1993) 143.
65. R. W. Cahn, P. Hassen, and E. J. Kramer (Eds.), Materials Science and Technology, Vol. 1, *Structure of Solids*. VCH Publisher Inc., Weinheim (1993).
66. H. Wolf, Z. Guan, X. Li, and Th. Wichert, Hyperfine Interact. **136/137** (2001) 281.

Theoretical Studies of Structural, Energetic, and Electronic Properties of Clusters

By Michael Springborg[1,*], Yi Dong[1], Valeri G. Grigoryan[1], Violina Tevekeliyska[1], Denitsa Alamanova[1], Elisaveta Kasabova[1], Sudip Roy[2], Jan-Ole Joswig[3], and Abu Md. Asaduzzaman[4]

[1] Physical and Theoretical Chemistry, University of Saarland, D-66123 Saarbrücken, Germany
[2] Inorganic and Physical Chemistry, Technical University Darmstadt, D-64287 Darmstadt, Germany
[3] Physical Chemistry, Technical University Dresden, D-01062 Dresden, Germany
[4] LRRS, University of Bourgogne, F-21078 Dijon Cedex, France

Clusters / Theoretical Calculations / Structure / Electronic Properties

Size in combination with low symmetry makes theoretical studies of the properties of clusters a challenge. This is in particular the case when the studies also shall identify the structures of the lowest total energy. We discuss here various methods for calculating the structural, energetic, and electronic properties of nanoparticles, emphasizing that the computational method always should be chosen carefully according to the scientific questions that shall be addressed. Therefore, different approximate methods for calculating the total energy of a given structure are discussed, including the embedded-atom method and a parameterized density-functional method. Moreover, different approaches for choosing/determining the structures are presented, including an *Aufbau/Abbau* method and genetic algorithms. In order to illustrate the approaches we present results from calculations on metallic and semiconducting nanoparticles as well as on nanostructured HAlO.

1. Introduction

Theoretical studies of materials are an important part of science that contributes with information that often complements experimental studies in a very constructive way. However, such studies rely heavily on some kind of information on the possible structure of the system of interest. Thus, if the system of interest contains just slightly more than 10 atoms, one usually assumes that at least

* Corresponding author. E-mail: m.springborg@mx.uni-saarland.de

Z. Phys. Chem. **222** (2008) 387–405

Table 1. Calculated number of (second line) inequivalent total-energy minima and (third line) inequivalent saddle points for Lennard–Jones clusters as a function of the number of atoms N in the cluster. From [3].

N	7	8	9	10	11	12	13
Minima	4	8	21	64	152	464	1328
Saddle points	12	40	152	584	911	2803	8453

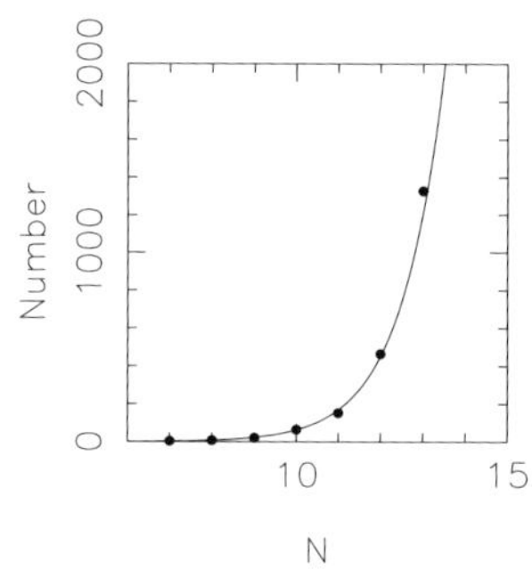

Fig. 2. A schematic representation of number of inequivalent local total-energy minima from the results of Table 1 (the black circles) together with a fit with an exponential (the full curve).

and not quantum particles, and there is no variational principle that can be used in systematically getting arbitrarily close to the true ground-state structure, even when assuming that the total energy can be calculated accurately for any structure.

Instead, most often one has to change the structure 'by hand' until one feels confident that the structure of the lowest total energy has been identified. But it shall be stressed that there is absolutely no approach that with absolute certainty can guarantee that precisely that structure has been found. Trying to prove so would require complete knowledge about the total-energy surface as a function of all internal degrees of freedom. This information is beyond what ever will be available, even if the computers keep on becoming more and more powerful, and the programs more and more efficient.

That this information will never become available, can be seen by, *e.g.*, considering the largest possible computer, *i.e.*, the complete universe. Lloyd [4] estimated that the complete universe, viewed as an enormous quantum computer that has been operating since the Big Bang, could have performed of the order of 10^{120} binary operations. Although this number is very large, it is also finite. Therefore, it means that considering, *e.g.*, a system of 42 atoms, assuming that a total-energy calculation for this system and a given structure is just one single binary operation, and considering just 10 different values for each internal structural degree of freedom, exactly 10^{120} calculations are needed for obtaining this, relatively crude, total-energy hyper-surface. Thus, the calculations would have required the use of the complete universe since the Big Bang!

In total, it is obvious that for any but the absolutely smallest systems it is not possible to explore anything but very limited parts of the total-energy hyper-surface, and, moreover, that it is never possible to be absolutely sure that the true global total-energy minimum has been found. Any total-energy minimum may provide a total energy that is close to that of the global minimum, although the structures of the two may be markedly different, simply due to the very large number of local total-energy minima. It is also obvious that any attempt to identify the global total-energy minimum has to be based on some kind of qualified search in the multi-dimensional structure space.

It is one of the purposes of this presentation to discuss some of the approaches that we are using in searching for the global total-energy minimum in the structure space. In Sect. 2 we shall discuss the methods we are using in calculating the total energy for a given structure as well as the approaches we are applying in determining the structure. Various applications, including metallic and semiconducting clusters, are discussed in Sect. 3 and a brief summary is presented in Sect. 4.

2. Methods

As discussed in Sect. 1, theoretical studies of the properties of nanoparticles require an efficient identification of the structures that shall be studied as well as efficient methods for the calculation of the properties of the chosen structures. Depending on the system and the properties of interest, different methods for the two aspects may be chosen. We shall here discuss the different methods that we are using in our studies.

2.1 Total-energy methods

The chemical bonding in extended, condensed-phase systems can be of very different types. In metallic systems the electrons are considered delocalized and packing effects are often the main driving force for the structure. On the other hand, for many semiconductors and insulators, the bonds between the atoms are partly covalent and partly ionic, so that directional bonding as well as electron transfers are important, *i.e.*, a precise description of the electronic degrees of freedom is important. Yet other types of systems include, *e.g.*, molecular crystals with weak interactions between the molecules, but here we are concerned with the first two types of systems.

One approach that we have been using for clusters of metal atoms is based on the embedded-atom method (EAM) of Daw, Baskes, and Foiles [5–8]. According to this method the total energy of the system is written as a sum over atomic energies,

$$E_{\text{tot}} = \sum_{i=1}^{N} E_i \,, \tag{1}$$

with

$$E_i = F_i\left(\rho_i^h\right) + \frac{1}{2}\sum_{i \neq j=1}^{N} \phi_{ij}(r_{ij}) \,. \tag{2}$$

The first term is the so-called embedding energy, which is obtained by considering each atom as an impurity embedded into a host provided by the rest of the atoms. The second term describes electron–electron interactions and is represented in terms of short-ranged pair potentials. The local density at site i is assumed being a superposition of atomic electron densities,

$$\rho_i^h = \sum_{j\,(\neq i)} \rho_j^a(r_{ij}) \,, \tag{3}$$

where $\rho_j^a(r_{ij})$ is the spherically averaged atomic electron density provided by atom j at the distance r_{ij}. Notice that compared to non-interacting, neutral atoms, E_{tot} is negative for stable structures. In our studies we use either the form of Daw, Baskes, and Foiles [5–7] or the more recent one of Voter and Chen [9, 10]. Both forms provide expressions for the different functions entering the potential, but in the latter also information for the diatomic molecule has been used in determining the values of the parameters entering the potential.

Another simplified description of the cohesion of metallic systems is provided by the Gupta potential [11]. Then, the total energy is written in terms of repulsive and attractive many-body terms,

$$E_{\text{tot}} = \sum_{i=1}^{N} [V^r(i) - V^m(i)] \,, \tag{4}$$

where

$$V^r(i) = \sum_{j=1\,(\neq i)}^{N} A(a,b) \exp\left[-p(a,b)\left(\frac{r_{ij}}{r_0(a,b)} - 1\right)\right] \tag{5}$$

and

$$V^m(i) = \left\{\sum_{j=1\,(\neq i)}^{N} \zeta^2(a,b) \exp\left[-2q(a,b)\left(\frac{r_{ij}}{r_0(a,b)} - 1\right)\right]\right\}^{\frac{1}{2}} . \tag{6}$$

The summations run over all (N) atoms of the system of interest. Furthermore, r_{ij} is the distance between atoms i and j, and A, r_0, ζ, p, and q are fitted to experimental values of the cohesive energy, lattice parameters, and elastic constants for the crystal at 0 K.

It is clear that none of the methods above includes any explicit description of electronic degrees of freedom. This makes them computationally efficient and allows, therefore, for many total-energy calculations also for larger systems. This issue is essential for our studies of the structural properties of metal nanoparticles. On the other hand, the lack of an explicit description of electronic degrees of freedom makes them inadequate when attempting to study semiconductor or insulator nanoparticles. For those we use instead a parameterized density-functional tight-binding (DFTB) method of Seifert *et al.* [12–14].

This method is based on the density-functional theory of Hohenberg and Kohn [15] in the formulation of Kohn and Sham [16]. Any Kohn–Sham orbital ψ_i of the system of interest is expanded in terms of atom-centered localized basis functions, χ_{jm}, where $\mathbf{R}_j$ is the position of the jth atom, and m characterizes the radial and angular dependencies of the basis function,

$$\psi_i(\mathbf{r}) = \sum_{jm} c_{ijm} \chi_{jm}(\mathbf{r} - \mathbf{R}_j) \,. \tag{7}$$

The Kohn–Sham single-particle operator is approximated as

$$\hat{h}_{\text{eff}} = \hat{t} + V_{\text{eff}}(\mathbf{r}) = \hat{t} + \sum_j V_j^0(|\mathbf{r} - \mathbf{R}_j|) \,, \tag{8}$$

with $\hat{t} = -\frac{\hbar^2}{2m}\nabla^2$ being the kinetic-energy operator and $V_{\text{eff}}(\mathbf{r})$ being the effective Kohn–Sham potential which is approximated as a simple superposition of the potentials of the neutral atoms. Furthermore, we assume that the matrix elements $\langle \chi_{j_1 m_1} | V_j^0 | \chi_{j_2 m_2} \rangle$ vanish unless $j_1 = j$ or $j_2 = j$. All non-vanishing matrix elements are calculated accurately.

From the Kohn–Sham eigenvalues of the isolated atoms and of the system of interest we calculate the total energy (relative to that of the isolated atoms) according to

$$E_{\text{tot}} \simeq \sum_i^{\text{occ}} \epsilon_i - \sum_j \sum_m \epsilon_{jm} + \frac{1}{2} \sum_{j \neq j'} U_{jj'}(|\mathbf{R}_j - \mathbf{R}_{j'}|) \,. \tag{9}$$

Here, ϵ_i is the energy of the ith orbital for the system of interest and ϵ_{jm} is the energy of the jth orbital for the isolated, mth atom. Moreover, U_{jk} is a short-range pair potential between atoms j and k that is adjusted so that results from parameter-free density-functional calculations on two-atomic systems as a function of the interatomic distance are accurately reproduced.

2.2 Structure-determination methods

Through the discussion in Sect. 1 it was argued that the determination of the structure of the global total-energy minimum becomes increasingly prohibitive as the size of the system grows. Ultimately, it is not possible to identify that structure with absolute certainty for just intermediate-sized systems, and one has to apply methods that build on chemical intuition and simultaneously attempt to be as unbiased as is felt desirable. To which extent one combines the keeping of good building blocks and randomness in the construction of reasonable structures, is a matter of a subjective choice that also depends on the scientific question that is being addressed.

Therefore, in our studies we have used more different approaches in identifying structures that shall be studied. The simplest approach is based on assuming that the structure of interest resembles a fragment of the corresponding infinite crystalline material. We have considered such structures by cutting out a spherical part of the crystal. Depending on the position of the center of the sphere, different structures may be obtained. Each of these initial structures is subsequently relaxed to the closest structure of a local total-energy minimum. Due to this approach, it is rarely possible to obtain sets of structures that vary continuously over a whole size range but instead only certain, discrete sizes are obtained.

Besides this highly biased approach we also apply two different unbiased methods. One of those, the *Aufbau/Abbau* method was developed earlier by us [17, 18] and consists of the following steps:

1. We assume that we have obtained the structures of the global total-energy minima for the clusters with up to N atoms.
2. In addition, we consider clusters with $N+K$ atoms with $K \simeq 5-10$. For each of those sizes (*i.e.*, N and $N+K$) we randomly generate and relax a large set of structures, from which the one of the lowest total energy is selected.
3. This leaves us with two 'source' clusters with N and $N+K$ atoms. For the cluster with N atoms we add at a random position a single atom and the structure is relaxed. This is repeated many hundred times. In parallel, we consider the $N+K$ structures of $N+K-1$ atoms that can be obtained by removing one of the atoms from the cluster with $N+K$ atoms and relaxing the structure. The structure of the lowest total energy is kept.
4. The procedure of step 3 is repeated $K-1$ times until we have obtained two sequences of cluster structures with N, $N+1$, $\cdots$, $N+K-1$, and $N+K$ atoms. From these two sets the structures of the lowest total energies are chosen.
5. The steps 3 and 4 are repeated with the new sets of structures, until no lower total energies are found. Then it is assumed that the global total-energy minima for this size range have been identified and we proceed to the next set of sizes.

As an alternative we also apply the so-called genetic algorithms. The genetic algorithms are based on the principles of natural evolution and are, therefore, also called evolutionary algorithms [19, 20] and have been found to provide an efficient tool for global geometry optimizations. Our versions of the genetic algorithms are as follows.

For a given value of N we construct a set of M different structures. This set could, *e.g.*, be a set of completely randomly generated structures or, alternatively, be constructed by adding a single atom at a random position to the cluster of $N-1$ atoms. Each of these M structures is relaxed to its nearest total-energy minimum. A new set of structures is obtained through cutting and pasting. Either we cut each of the M structures randomly into two parts that are interchanged. Or we split the set of M clusters into pairs and for the members of each pair cut them into two parts that are interchanged between the two members. Each new structures is let relax to the nearest total-energy minimum. Out of the total set of $2M$ structures, the M ones of the lowest total energy are kept as the next generation. This procedure is repeated until the lowest total energy is unchanged for a large number of generations.

3. Applications

In this section we shall briefly review the results of our studies from the last few years. We have used the methods discussed in the previous section. The main objective of this section is to give an impression of the possibilities and limitations of our approaches, rather than presenting a detailed discussion of any of the systems. For the latter, the interested reader is referred to the original literature, mentioned in the subsections below.

3.1 Metal clusters with one type of atoms

First, we shall consider clusters of metals containing just one type of atoms. In one set of studies we have studied the energetics and structures of clusters of Ni, Cu, Ag, and Au using the EAM or Gupta potentials in combination with the *Aufbau/Abbau* structure-optimization method [18, 21–24]. We shall here discuss how the size of the particularly stable clusters depends on the metal. To this end we define the stability function

$$\Delta_2 E(N) = E_{\text{tot}}(N+1) + E_{\text{tot}}(N-1) - 2E_{\text{tot}}(N) \qquad (10)$$

which is has positive (negative) peaks for particularly stable (unstable) clusters. Here, $E_{\text{tot}}(N)$ is the total energy of the cluster of N atoms.

This function is shown in Fig. 3 for the four different metals. The results for the Ni, Cu, and Au clusters were obtained using the EAM, whereas we used the Gupta potential for the Ag clusters.

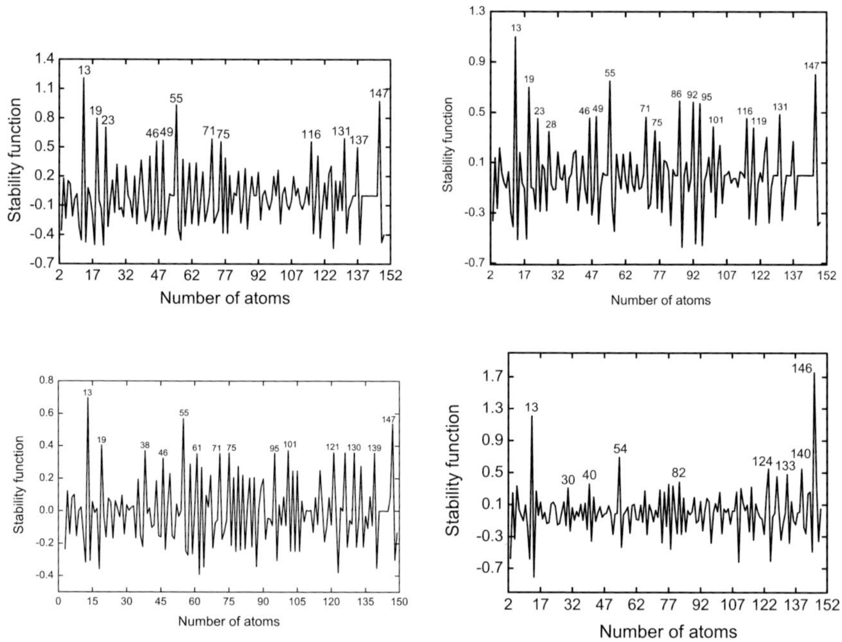

Fig. 3. Stability function (in eV) for (top, left) Ni, (top, right) Cu, (bottom, left) Ag, and (bottom, right) Au.

The figure shows first of all that the stability function for Au is overall different from that of the other metals. Indeed, in many cases we find the same values of N for the particularly stable clusters for Ni, Cu, and Ag, whereas other values are found for Au. Nevertheless, it is also clear that no two metals are absolutely identical. This means that simple rules based, *e.g.*, on packing arguments are not able to produce the correct structures in the general case, but material-specific descriptions of the interatomic interactions have to be taken into account.

A further result of our analysis is that the particularly stable clusters according to Eq. (10) in many cases also are clusters for which the energetically next-lying isomer of the same size is found at a particularly high energy. Thus, in many cases particularly stable clusters are unique in all respects.

Analysing the resulting structures in detail it turned out that the Ni and Cu clusters have an overall icosahedral packing. On the other hand, for Ag clusters our results suggest a decahedral packing, but a detailed comparison with experimental results indicates that also for Ag clusters an icosahedral packing should be preferred. In fact, also EAM calculations for smaller Ag clusters suggest an icosahedral packing. Thus, this result does point to some of the limitations related to using simplified descriptions of the interatomic interactions for metal clusters when the systems become sufficiently small.

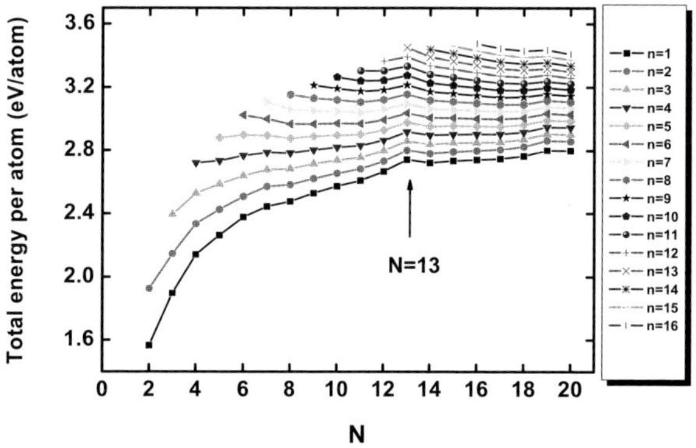

Fig. 4. The total binding energy per atom for bimetallic Ni_nCu_{N-n} clusters as a function of N for different values of n.

A much more difficult case is that of Au clusters. The results of our calculations with the EAM method showed that many clusters have a low symmetry although a fairly compact structure. On the other hand, gold clusters are among the most intensively studied clusters (see, *e.g.*, Ref. [25]) and it has been found that Au_N clusters are planar up to around $N = 15$. This is not found with the EAM calculations. Therefore, we decided to study gold clusters using the DFTB method in combination with genetic algorithms for structure optimization for N up to 58 [26–28]. Indeed we found that clusters for the smallest values of N were planar, although the threshold to three-dimensional structures was found for a too small value of N ($N = 7$). This may be ascribed to the approximations inherent in the DFTB method. In this study we found, furthermore, that for N up to around 20 the clusters formed hollow cages, and the atoms of all clusters had a fairly low coordination. In total, gold appears as an element that can form stable, low-dimensional structures, making it different from many other metals.

3.2 Metal clusters with two types of atoms

In a recent study we have examined the properties of bimetallic Ni_nCu_{N-n} clusters using the EAM method in combination with genetic algorithms. For a given N and n, the fact that by interchanging a Cu and a Ni atom, a new structure occurs (a so-called homotype) leads to an enormous increase in the computational demands when attempting to optimize the structure using an unbiased approach.

A key result is shown in Fig. 4, where the binding energy per atom, $-E_{\text{tot}}/N$, is depicted. In the figure the value $N = 13$ is marked which corresponds to particularly stable clusters. It is remarkable that all structures for a given N turned out to be very similar, and for $N = 13$ the structures are close to that of an icosahedron. Furthermore, it was a general finding that the Ni atoms were those at the center of the cluster and the Cu atoms those at the surface. The larger surface energy of Ni than of Cu as well as the smaller size of Ni than of Cu can explain this finding of core/shell-like structures.

3.3 Na clusters

In order to study further the role of electronic orbitals we considered clusters of one of the most simple metals, *i.e.*, Na_N clusters. These have been the prototype for cluster studies for about a quarter of a century (see, *e.g.*, Ref. [25]), and the fact that the valence orbitals of these clusters are formed mainly by Na $3s$ functions has led to the consensus that directional bonding does not exist in these systems. This is further supported by the early finding [29] that experimental abundance spectra could be well understood by considering a spherical jellium model. It was found that the sizes of the particularly stable clusters could be understood as being a consequence of the filling of orbitals for the valence electrons that are moving in a spherical potential provided by the spherical, smeared out distribution of the core electrons and the nuclei.

A first step beyond the spherical-jellium model is provided by the EAM potential, and a further improvement by the DFTB method. Therefore, we studied Na_N clusters using these two approximate descriptions of the interatomic interactions together with the *Aufbau/Abbau* method and the genetic algorithms for the structure determination [30].

The stability function as obtained with the two approaches is shown in Fig. 5. According to the spherical-jellium model (and experiment), particularly stable clusters are found for $N = 2, 8, 18, 20, \cdots$. The DFTB method, that does include electronic degrees of freedom, finds peaks at those values, although not particularly pronounced. One reason is that the interatomic interactions have a significant covalent character. Furthermore, essentially none of the clusters has a roughly spherical shape according to the calculations. Thus, the earlier findings that experiment and the spherical-jellium model give the same sequence of particularly stable clusters, does not necessarily have to imply that all clusters are spherical. The results of the DFTB calculations also show a clear even-odd oscillatory pattern that can be ascribed to the filling of electronic orbitals.

On the other hand, with the EAM method electronic effects are only very indirectly included. Instead, packing effects are those that first of all dictate particular stability. Therefore, the EAM calculations do not give the same sequence of particularly stable clusters as found in the spherical-jellium or DFTB calculations. Thus, electronic effects and not packing effects are the important ones for Na_N clusters. Interestingly, it has been found [31] that for Na nanowires the stability of the thinnest ones are dictated by electronic effects, but for slightly thicker ones there is a crossover after which point packing effects are the most important ones.

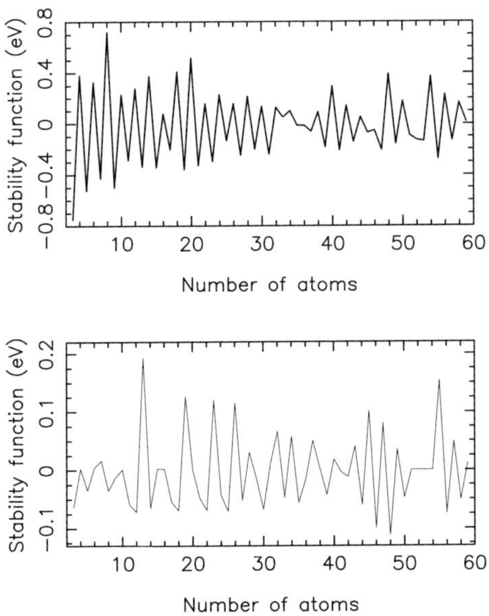

Fig. 5. The stability function (in eV) for Na_N clusters as function of the size of the clusters and obtained with (upper panel) the DFTB method and (lower panel) the EAM method.

3.4 Semiconductor clusters

Whereas the central issue of the preceding subsections was that of determining the structural properties of the clusters, for semiconductor nanoparticles the electronic properties are those that have received the most attention. When the spatial extension of the clusters becomes comparable with that of an exciton, the optical properties of the semiconductor will depend markedly on the size of the material.

For AB semiconductors an additional, relevant (and interesting) issue is that the macroscopic, crystalline material may exist in (at least) two structures that are energetically very close, *i.e.*, the zincblende and the wurtzite structure [32, 33]. Accordingly, it is likely that for the finite-sized clusters, the energetic ordering of those two structures may change. In fact, both structures have been observed experimentally for nanoparticles of the same material (see, *e.g.*, Ref. [34]).

In order to obtain further information into the electronic properties and the relative total energy per atom pair of the two structures for finite systems, we have studied a larger range of II–VI and III–V semiconductor nanoparticles [35–43]. We shall here discuss briefly some of the main results.

Since the electronic properties and the relative stability of wurtzite- and zincblende-derived clusters are the central issues, we have used the DFTB method that includes electronic degrees of freedom and not attempted to op-

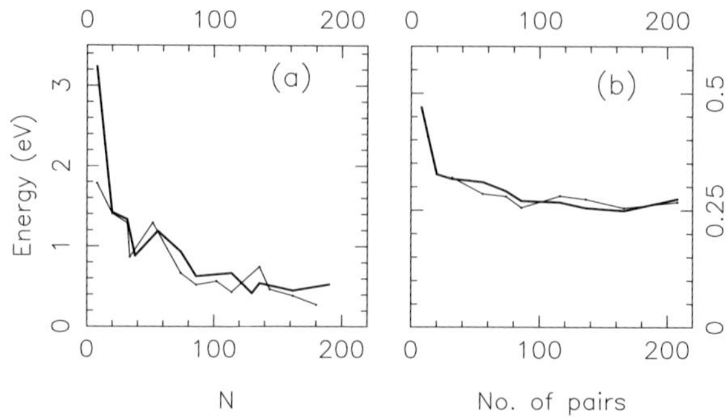

Fig. 6. The variation in total energy per atom pair for stoichiometric (a) CdS and (b) CdSe clusters for (thick curves) zincblende- and (thin curves) wurtzite-derived clusters.

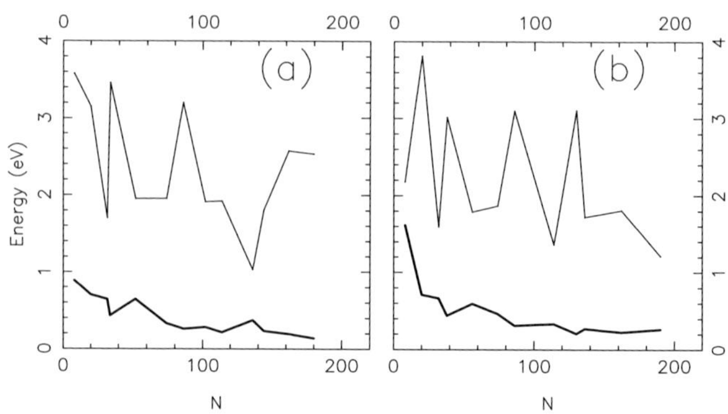

Fig. 7. Variation in the total energy per atom pair (thick curves) and band gap (thin curves) for CdS clusters with (a) zincblende- and (b) wurtzite-derived structures as a function of the total number of atoms in the cluster, N.

timize the structure completely. Instead, starting from the infinite, periodic crystals we have cut out a spherical part that subsequently was allowed to relax to the closest local total-energy minimum. In most cases we considered stoichiometric clusters that were obtained by placing the center of the above-mentioned sphere on the midpoint of a nearest-neighbour bond.

In Fig. 6 we show representative examples for the relative total energy of clusters derived from the two different crystal structures. It is clear that the relative total energy depends critically both on size and on material of the clusters. Moreover, although the total energy does show an overall decreasing behaviour as a function of size of the cluster, it is also structured.

In order to understand this aspect further we show in Fig. 7 the variation of the total energy and the energy gap between the highest occupied (HOMO) and lowest unoccupied (LUMO) molecular orbital for a single example, *i.e.*, CdS clusters. Similar results are obtained for the other systems. The figure demonstrates a clear correlation between the two quantities: high stability coincides with a high energy gap, which may be interpreted in terms of the so called hard-and-soft-acids-and-bases (HSAB) principle [44]. Moreover, for the less stable structures, *i.e.*, the ones with a small HOMO–LUMO gap, at least one of the HOMO and LUMO orbitals is localized to the surface region. Furthermore, because of the surface states, less stable clusters are particularly reactive.

These results have two important implications. First, in experimental studies it is often assumed that one obtains a more or less narrow size distribution that is very regular as a function of size. Moreover, that the size of the clusters can be estimated through the energy of the optical transition that is a monotonous function of size. However, our results give that the total energy is structured as a function of size, suggesting that the size distribution also will be structured, and also that the energy for optical transitions is less regular as a function of cluster size.

Secondly, since the stability of the clusters is strongly dependent on the presence/absence of low-coordinated atoms on the surface, it should be possible to stabilize the clusters through properly chosen surfactants or by modifying the surface in other ways [45]. For our stoichiometric AB clusters, the outermost atoms are formed to equal parts of A and of B atoms. Considering non-stoichiometric clusters constructed, *e.g.*, as a spherical cut-out of the crystal with the center at an atomic position, the surface will contain exclusively one type of atoms. In one study, on InP clusters [37], we considered this case and did indeed observe a strong dependence of the properties on the surface, although the properties did not only depend on the type of the outermost atoms. A further study [41] demonstrated that surfactants could modify the electronic properties of the clusters, leading in most, but not all, cases to a larger HOMO–LUMO gap. Alternatively, one may obtain more stable structures by considering the so-called core/shell nanoparticles, which are systems containing a core of one material coated by a shell of another. We studied such systems with the two materials being either CdS and CdSe or Si and Ge [40, 43] and found, besides a saturation of dangling bonds on the core part of the nanoparticle, that new properties may emerge from such systems. This includes that in some few cases the HOMO and LUMO are localized to the two different materials, making a charge separation upon optical excitation possible. Furthermore, the more stable systems were obtained when the shell material was the one with the lowest surface energy.

3.5 Nanostructured HAlO

As a last example we discuss results for nanostructured HAlO. Experimentally, little was known about it except that it is stoichiometric, and, in addition, that it can be used as a substrate for organized structures of organic materials. It turned out that theoretical studies could give useful information that ultimately could extend and support the experimental information about it.

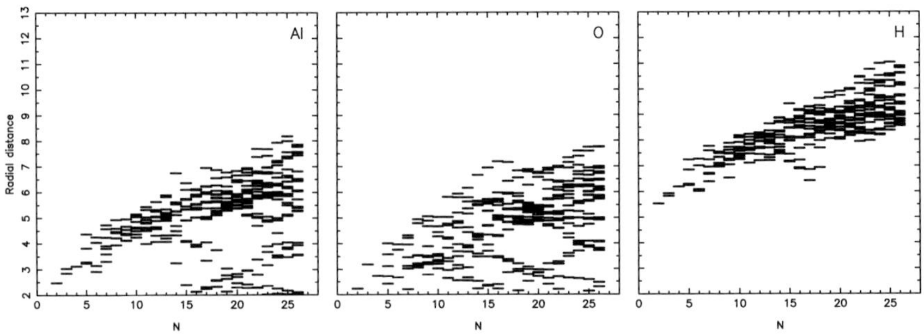

Fig. 8. The radial distances (in a.u.) for Al, O, and H atoms, separately, as a function of the size of the cluster n for $(HAlO)_n$ clusters. In each panel, a small horizontal line shows that at least one atom of the corresponding type has that distance to the center of the cluster for a given value of n.

Using the DFTB method in combination with genetic algorithms for the structure optimization we determined the structures for isolated $(HAlO)_n$ clusters with n up to 26 [46]. We shall here summarize the main findings.

In Fig. 8 we show the so-called radial distances for the atoms of the different clusters. For a given $(HAlO)_n$ cluster the radial distance $\mathbf{r}_i$ for each of the $3n$ atoms is defined as

$$r_i = |\mathbf{R}_i - \mathbf{R}_0| \,, \tag{11}$$

where $\mathbf{R}_0$ is the geometrical center of the cluster,

$$\mathbf{R}_0 = \frac{1}{3n} \sum_{i=1}^{3n} \mathbf{R}_i \,. \tag{12}$$

Figure 8 shows that the aluminum and oxygen atoms are those with the smallest radial distances, whereas all hydrogen atoms have larger radial distances, implying that the hydrogen atoms are placed on the surface of an Al–O core. Analysing the structures further gave that the H atoms preferred to be bonded to Al atoms (*i.e.*, not to O atoms), and that the Al–O core contained only heteroatomic interatomic bonds. Through additional calculations for the smallest values of n using a parameter-free, density-functional method, these general conclusions were confirmed.

The results above have some interesting consequences. First, whereas the number of hydrogen atoms grows linearly with n, the area of the surface of a roughly spherical cluster grows with $n^{2/3}$, so that the available area per hydrogen atom decreases as a function of n. Thus, there will be an upper limit above which the structural motifs will have to change. One possibility for larger structures is to form Al–O layers with additional H atoms attached to the Al atoms. Indeed, such structures were also found to be stable, offering an explanation for why both types of structures are produced experimentally (see Ref. [46]).

Second, the macroscopic material may contain such clusters in close contact. Our theoretical calculations on the interaction between two clusters showed that, in this case, the hydrogen atoms will move away from the region between the two clusters, whereas the Al and O atoms will try to fill the space between them.

It may be possible to remove hydrogen, thereby obtaining a material consisting of only Al and O. Again, our theoretical calculations showed that stoichiometric AlO clusters contain only heteroatomic bonds. Moreover, the clusters become in general more compact than the cores of the HAlO clusters with the same number of Al and O atoms. Finally, although it may be suggested, we did not find any indications of a segregation into Al_2O_3 and Al.

4. Conclusions

The purpose of the present work has been to discuss the problems related to theoretical studies of the properties of nanoparticles with more than just a couple of 10s of atoms. Due to the size and (possibly) low symmetry of the systems of interest, it is important to apply methods that are sufficiently efficient and sufficiently accurate when calculating the properties (including the total energy) for a given structure. These methods have to be combined with specially developed approaches for the determination of the structures of interest. Here, one may either study specifically constructed structures or, alternatively, attempt to determine the structure of the lowest total energy.

In the present work we applied different approximate methods for the calculation of the total energy of a given structure, of which some methods explicitly include electronic degrees of freedom (which becomes particularly important for systems with directional, covalent bonds) and others did not (which is a reasonable approximation for some metallic systems). Moreover, we studied systems for which we optimized the structure using unbiased approaches as well as others for which the scientific questions could be addressed by considering specially constructed structures.

The systems of our interest ranged from metallic nanoparticles, via semiconducting ones, to nanostructured HAlO. Due to the spatial limitations of this overview it was not possible to go into details, but our maybe most important, general conclusion is that, for each material, separately, the size dependence of the system properties is unique.

Acknowledgement

We gratefully acknowledge *Fonds der Chemischen Industrie* for the very generous support. This work was supported by the SFB 277 of the University of Saarland and by the German Research Council (DFG) through projects Sp439/11 and Sp439/14.

Photoemission Investigation of the $\overline{L}$-Gap Surface States on Clean and Rare Gas-Covered Noble Metal (111)-Surfaces

By Stefan Hüfner[1,*], Friedrich Reinert[2], Stefan Schmidt[1], Georg Nicolay[1], and Frank Forster[2]

[1] Universität des Saarlandes, 7.2 Experimentalphysik, Postfach 151150, D-66041 Saarbrücken, Germany
[2] Universität Würzburg, EP II, Physikalisches Institut, Am Hubland, D-97074 Würzburg, Germany

Photoemission Spectroscopy / Shockley-Type Surface States / Adsorbates / Physisorption / Noble Metal Surfaces

High resolution photoemission spectroscopy is used to investigate the linewidth and dispersion of the $\overline{L}$-gap surface states on (111)-surfaces of noble metals. The effect of a coverage of rare gases is demonstrated.

1. Introduction

Photoemission spectroscopy (PES) has been established as one of the most important methods to study the electronic structure of molecules, solids and surfaces [1, 2]. Furthermore, PES has widespread practical implications in various fields [3] like surface chemistry or material science, and has significantly contributed to the understanding of fundamental principles in solid state physics.

In one of the four famous publications in 1905 – Einstein's *annus mirabilis* – he introduced [4] the concept of the *photon* and deduced the relation between the photon energy $h\nu$ and the maximum kinetic energy $E_{\text{kin}}^{\text{max}}$ of the emitted electrons, the fundamental photoelectric equation:

$$E_{\text{kin}}^{\text{max}} = h\nu - \Phi_0 \,. \tag{1}$$

* Corresponding author. E-mail: huefner@mx.uni-saarland.de

Z. Phys. Chem. **222** (2008) 407–431

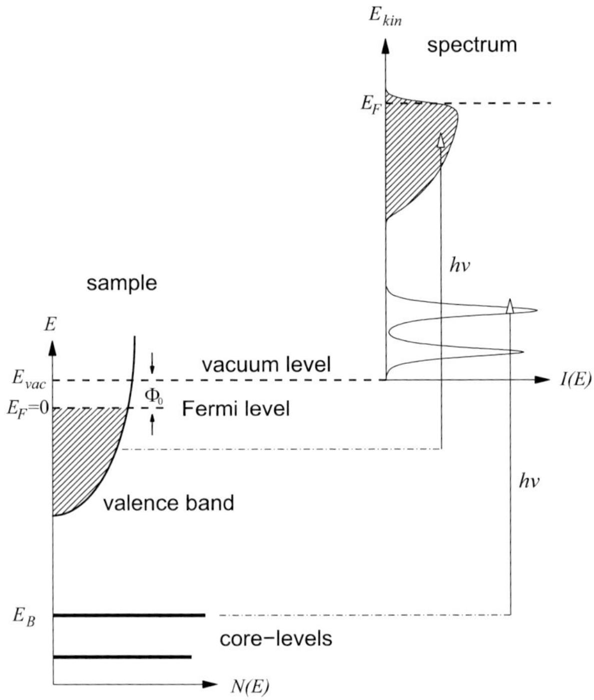

Fig. 1. Schematic view of the photoemission process in the single particle picture. Electrons with binding energy E_{B} can be excited into free-electron states above the vacuum level E_{vac} by photons with energy $h\nu > E_{\mathrm{B}} + \Phi_0$. The photoelectron distribution $I(E_{\mathrm{kin}})$ can be measured by the analyser and is – in first order – an image of the occupied density of electronic states $N(E_{\mathrm{B}})$ of the sample.

Φ_0 is a characteristic constant of the sample surface and is known as the *work function* today.

At that time, the maximum kinetic energy of the photoelectrons could be determined under vacuum conditions by the retarding-field technique. However, it took several years until Einstein's formula was experimentally confirmed by observing a linear dependence between kinetic energy and $h\nu$ (a review about the history of photoemission can be found in Ref. [5]).

The fundamental principle of the photoemission process is sketched in Fig. 1 [6]. This simplified picture shows the attractiveness of photoemission spectroscopy, because in that view, which will be refined later, the properties of the photoelectrons basically reflect the electronic eigen states of the investigated system. Basically one distinguishes between ultraviolet photoemission (UPS), mainly for the (angular resolved) investigation of valence band states (ARUPS), and X-ray photoemission spectroscopy (XPS), providing the investigation of core-level states at higher binding energies.

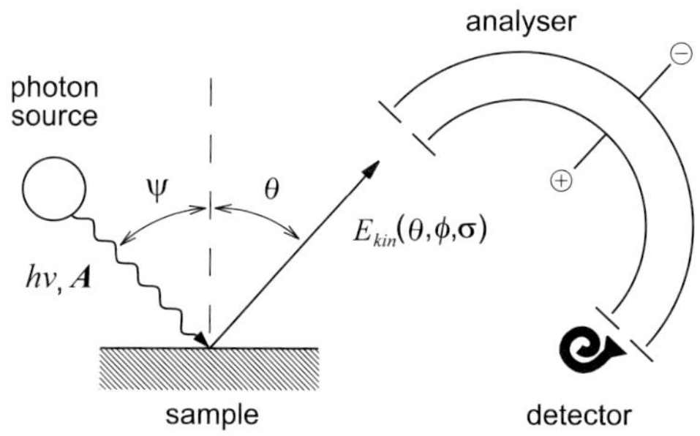

Fig. 2. Principle of a modern photoemission spectrometer. Monochromatic photons with energy $h\nu$ and polarisation (**A** is the vector potential of the electromagnetic field) are produced by a light source, *e.g.* an Al K_α X-ray anode for XPS or a helium discharge lamp for UPS, and hit the sample surface under an angle Ψ with respect to the surface normal. The kinetic energy $E_{\rm kin}$ of the photoelectrons can be determined by use of electrostatic analysers (usually by an additional retarding field) as a function of the experimental parameters, *e.g.* emission angle (θ, ϕ), the electron spin orientation $\boldsymbol{\sigma}$, or the photon energy or polarisation. The whole setup is evacuated to ultra high vacuum (UHV, typically $p \lesssim 10^{-10}$ mbar).

2. Experimental considerations

Today, a photoemission experiment for spectroscopy is basically performed in the same way as more than hundred years ago (see Fig. 2). Photons from a monochromatised light source, which can be a laboratory source – for vacuum ultraviolet (VUV) or soft X-ray radiation – or a synchrotron device, are directed on a sample and the photoelectrons, liberated by the photoelectric effect, are analysed with respect to emission angle and kinetic energy by an electrostatic analyser.

In the laboratory, gas discharge lamps and soft X-ray sources have been used for the photoexcitation with photon energies ranging from 11.8 eV (Ar I) to 1486.6 eV (Al K_α). The same energy range is covered by synchrotron radiation, which has become increasingly important because it allows measurements that can not be performed with usual VUV or X-ray sources in the laboratory. The main difference compared to laboratory sources is that the photon energy can be selected by use of a monochromator from a continuous spectrum over a wide energy range. Other important advantages of the synchrotron light are *e.g.* very high intensity and brightness, variable polarisation, small photon spots, or the possibility of time resolution. However, for many applications the use of the – comparatively simple and cheap – laboratory sources is still advantageous.

Using standard laboratory sources, the kinetic energies of the produced photoelectrons in the usable spectrum lie in the energy range of the order of 10 eV up to ≈ 1.5 keV (*cf.* Eq. (1)). The escape depth of photoelectrons in this energy range is rather small, typically of the order of a few Å. Therefore, a characteristic property of PES experiments is its high *surface sensitivity*. The consequently small information depth of PES can be a drawback if one is interested in studying bulk properties of solids. However it seems that for many materials investigated so far, the observed electron spectra reflect predominantly the bulk properties.

In this contribution advantage will be taken of the high surface sensitivity of PES by investigating prototype surface state systems. These are the surface states in the $\overline{L}$-gap of the (111)-surface of the noble metals. Data will be presented on the electronic structure of these surface states on the clean surfaces and on those covered with rare gases. The systems of noble metals surfaces covered with rare gases are the simplest possible adsorbate systems.

3. Surface states on clean (111)-surfaces of noble metals

3.1 Introduction

Since the first spectroscopic investigation of the $\overline{L}$-gap surface states on the (111) surface of noble metals [7, 8] there has been a large effort to study in detail the dispersion relations, temperature dependence, the interaction with adsorbates, and the line-shape of these paradigmatic electronic systems. These surface states are prototype quasi-two-dimensional electron states that – as shown by Shockley [9] – appear in a projected energy gap of the bulk bands, because of the termination of the infinite crystal by the surface.

Just recently the internal lifetime of the Ag(111) surface state was determined independently by use of a scanning tunneling microscope (STM) and high-resolution photoemission spectroscopy (PES) at the band minimum [10, 11]. Such experimental and theoretical [12] investigations of the lifetime of surface states are important for the understanding of many processes that take place at the surface of solids, like *e.g.* chemical reactions and catalytic processes [13]. In addition, the lifetime of electronic states in solids is not very well understood, although it is closely related to fundamental solid state properties like superconductivity, magnetism or transport properties.

In principle, there are three experimental methods to determine the lifetime of electronic states: tunneling spectroscopy [10] using the tip of a STM, high-resolution photoemission spectroscopy [11] and time resolved pump-probe experiments [14–17] using two or more photon excitations. On the example of published photoemission spectra of the the Ag(111) surface state from the last 25 years, Fig. 3 demonstrates the technological development in photoemission spectroscopy, which consists mainly in an improvement of the energy resolution by nearly two orders of magnitude. In addition there are significant improvements concerning the angular resolution, the detection efficiency, and not last the UHV conditions.

Another comparatively new experimental approach for the investigation of surface states is the spectroscopy with a scanning tunneling microscope by

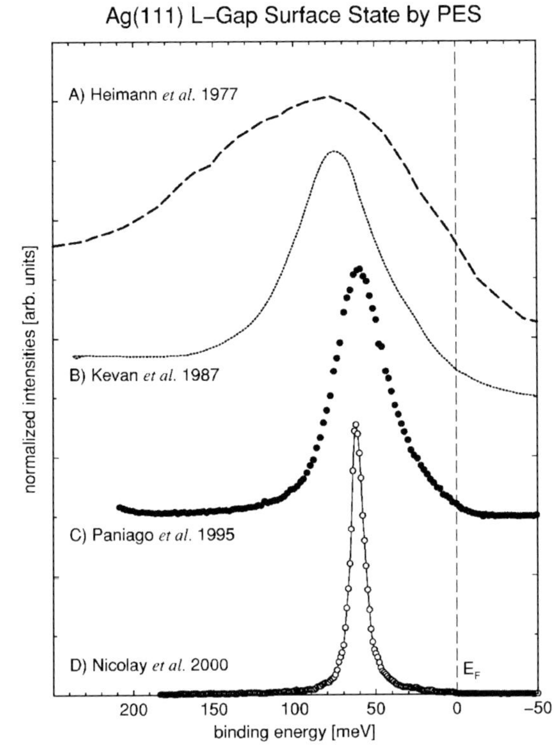

Fig. 3. Technological development in PES since the first observation of the Ag(111) surface state in photoemission spectra: A) from Ref. [8] measured at room temperature (RT) with Ar I ($h\nu = 11.83$ eV), angular integrated; B) from Ref. [30] at RT with $h\nu = 13$ eV, $\Delta E \approx 60$ meV and $\Delta\theta = 1°$; C) from Ref. [21] at $T = 56$ K with Ar I, $\Delta E = 21$ meV and $\Delta\theta = 0.9°$; D) present data at $T = 30$ K with He I ($h\nu = 21.23$ eV), $\Delta E = 3.5$ meV and $\Delta\theta = \pm 0.15°$.

measuring the current–voltage characteristics at a particular position of the surface (scanning tunneling spectroscopy: STS) [18]. From the width of the step-like onset in the $\mathrm{d}I/\mathrm{d}V$ curve one can determine the lifetime width of the surface state at the band minimum. New experiments by STS [19] have shown significantly larger lifetimes than the classic high-resolution photoemission spectra of the last 5 years [20–25]. In addition to the lifetime measurements, a detailed analysis of the voltage dependence of the Friedel oscillations at point and line defects on the surface allows to extract Fermi surfaces and dispersion relations by Fourier transform methods and appropriate model fits [26–28].

In contrast to the photoemission spectroscopy (PES) experiment, the tunneling experiment has the advantage that it investigates the microscopic topology on an atomic scale before the spectroscopic measurements are performed [29]. Therefore it is possible to choose an almost perfect, defect-free area on the single crystal, which is typically of the size of $\gtrsim 3000\,\mathrm{nm}^2$. On the other hand, photoemission with the discharge lamp integrates over a comparatively large sample area of about 1 mm^2, with all its steps and defects and other imperfections. This principle difference between the two methods was believed to be the reason for the observed discrepancy in the lifetime width of the noble metal surface states [10, 19].

In the particular case of the Au(111) surface LaShell *et al.* [25] observed by photoemission – with high angular and energy resolution – a characteristic splitting of the surface state and explained this splitting as due to a spin–orbit coupling (SOC), that breaks the spin degeneracy in the system. This explanation was corroborated by recent tight-binding model calculations [31]. The authors did rule out several other possible explanations, especially those related to the herringbone reconstruction [32, 33] ($23 \times \sqrt{3}$), which has been considered in the literature as responsible for influencing the STS results [34]. The observed splitting leads to two peaks in the EDCs with an energy separation proportional to the in-plane wave vector $k = |\mathbf{k}_{||}|$ reaching values of about $\Delta E \gtrsim 110$ meV close to k_F. The experimental values for the Fermi vectors given in Ref. [25] are 0.153 $Å^{-1}$ and 0.176 $Å^{-1}$ at room temperature. A dependence of the k_F-values on the measured k-space direction ($\bar{\Gamma}$–$\bar{M}$ or $\bar{\Gamma}$–$\bar{K}$) or on the photon energy could not be detected. However, the physical origin of the observed splitting is still discussed, in particular because STS and PES show quantitatively and qualitatively different results for the Au(111) surface state and an equivalent splitting for the other noble metal surface states could not be observed by PES.

3.2 Lifetime width at normal emission

Figure 4 shows the photoemission spectra of the three investigated surface states on a common energy scale, measured at normal emission (equivalent to the center of the surface Brillouin zone $\bar{\Gamma}$). Here the peak dispersion has its minimum and the lines appear with the smallest experimental line-width. The peak positions at normal emission for the Au(111), Cu(111), and Ag(111) surface states are 484 meV, 435 meV, and 63 meV, respectively.

To extract the intrinsic line-width of the surface states we used a simple model for the description of the experimental photoemission data at normal emission. In the case of Ag and Cu the spectra can be described by a single Lorentzian with a full width at half maximum Γ (FWHM), broadened by the finite experimental resolution, which we approximated by a Gaussian with fixed FWHM $\Delta E = 3.5$ meV. Depending on the individual sample surface the spectra showed a slight asymmetry, which can be seen more pronounced in the spectra B) and C) in Fig. 3. For our data at normal emission we can rule out that this asymmetry is due to the finite angular resolution of the spectrometer, confirmed by respective data modelations (see below). The asymmetry is most probably due to a stepped surface with additional defects, and differs slightly from preparation to preparation [35]. In our model we describe this in-

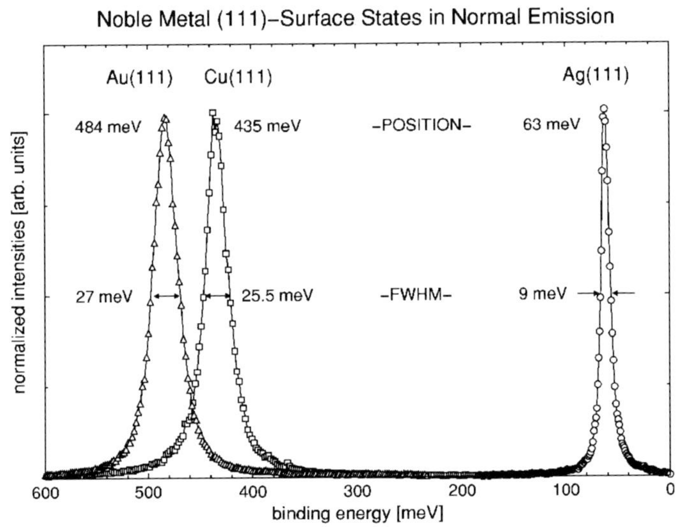

Fig. 4. L-gap surface states at normal emission ($\bar{\Gamma}$-point) and $T = 30$ K for Au, Cu, and Ag on a common energy scale, measured with He I_α radiation ($h\nu = 21.23$ eV). Note that the experimental peak positions and line-widths (FWHM) differ slightly from the fitted values given in Table 1.

trinsic asymmetry by a line-shape consisting of two independent Lorentzian halfs with FWHM $\Gamma^<$ for the high binding energy side and $\Gamma^>$ for the low binding energy side towards the Fermi energy. This leads to an asymmetry parameter $A_0 = (\Gamma^< - \Gamma^>)/(\Gamma^< + \Gamma^>)$ [35]. To compare our data with the STM results or other values from the literature we considered only the high binding energy side $\Gamma^<$ for the determination of the intrinsic lifetime width (see *e.g.* Ref. [21]). In summary the model curve for the normal emission fit consists of four independent parameters, *i.e.* the binding energy E_0^{B}, the intrinsic lifetime width $\Gamma = \Gamma^<$, the asymmetry parameter A_0 and the energy resolution $\Delta E = 3.5$ meV, which was constant for all spectra analyzed here.

Figure 5 shows the result of a least-squares fit with this model for the Ag(111) surface state. The sheer measured line-width (open circles) amounts to FWHM = 9 meV, including experimental broadening and asymmetry. The resulting model parameters from the least-squares fit of this spectrum are: binding energy $E_0^{\mathrm{B}} = 63.3$ meV, intrinsic Lorentzian FWHM $\Gamma = 6.2$ meV, asymmetry parameter $A_0 = -0.09$. The analysis of the data on different surfaces showed slight variations of the individual parameters, especially for the asymmetry parameter A_0. A small asymmetry was usually correlated with a high binding energy. However, within a comparatively small scatter the results could be repeatedly reproduced.

The Cu(111) surface state was analyzed analogously (see Fig. 6), the result for the intrinsic line-width is given in Table 1, the other fit parameters are given

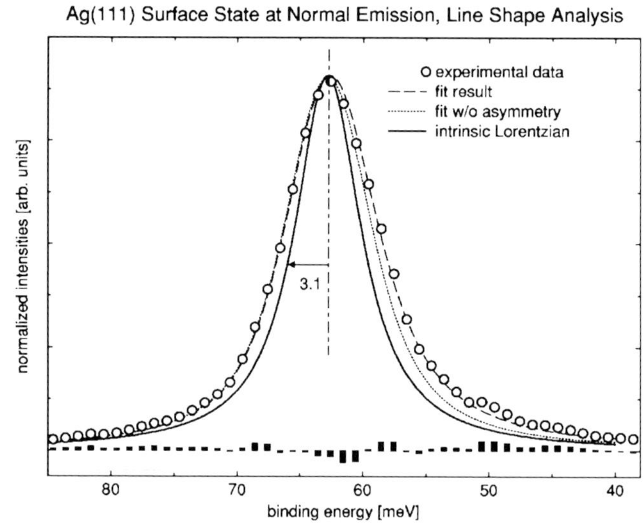

Fig. 5. Least-squares analysis of the L-gap surface state of Ag(111) at normal emission measured with He I_α radiation at $T = 30$ K. The experimental data (circles) were fitted by an asymmetric Lorentzian, convoluted with the resolution function (Gaussian). The least-squares result is plotted by a dashed line, the intrinsic Lorentzian (FWHM = 6.2 meV) by a solid line. The dotted curve is the fit result without the asymmetry ($A_0 = 0$, see text). The residue of the least-squares fit is given by black bars.

in the figure captions. For both Ag and Cu there is no influence of the finite angular resolution to the line-shape of the spectra at the $\bar{\Gamma}$ point, not even for the asymmetric tailing to lower binding energies.

The understanding of the Au(111) state is more complex: although the normal emission spectrum gives the most narrow line, the binding energy does not correspond to the band minimum, because the line consists of the contributions from the two (probably spin–orbit) split parabolas shifted in opposite k direction from the normal emission (see 10). However, the analysis of the narrowest experimentally observed line (see Fig. 7) in the way described above yields a value of $\Gamma = 25 \pm 1$ meV for Au(111), and accordingly a surface state lifetime of approximately $\tau = 27$ fs. Because of the particular splitting of the Au(111) surface state dispersion, the finite angular resolution leads – in contrast to Ag and Cu – to a slight additional broadening at $\bar{\Gamma}$; at angles $0.5° \lesssim \theta \lesssim 5°$ off normal emission this broadening transforms to a split peak in the EDCs. We included the dispersion of the two parabolas in the line-shape fit by a convolution in **k**-space with a semi-elliptic angular window function. The size of the window is given by the angular resolution of $\Delta\theta = \pm 0.15°$. As anticipated from the dispersion relations in Fig. 10, the respective fit gives a slightly larger binding energy of the band minimum and a reduced intrinsic line-width of $\Gamma = 21 \pm 1$ meV.

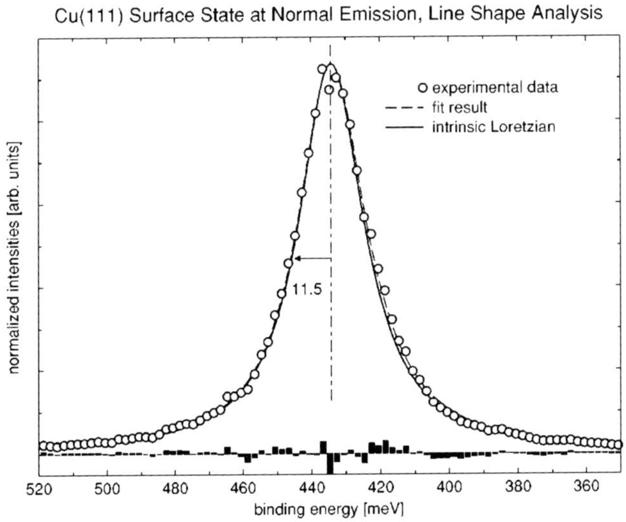

Fig. 6. Least-squares analysis of the L-gap surface state of Cu(111) at normal emission measured with He I_α radiation at $T = 30$ K. The experimental data (circles) were fitted by an asymmetric Lorentzian, convoluted with the resolution function (Gaussian). The least-squares result is plotted by a dashed line, the intrinsic Lorentzian (FWHM= 23.0 meV) by a solid line. Fit parameters for this spectrum: $E_0^B = 434.6$ meV, $A_0 = -0.018$, $\Gamma = 23.0$ meV.

Table 1. Comparison of experimental and theoretical results for the natural line-width of the investigated surface states. All Γ values are given in meV. The theoretical and the STM results are taken from Ref. [19], the PES values are taken from the least-squares fit at normal emission. The lifetime $\tau = \hbar/\Gamma$ was calculated from the PES line-width. The value of the PES result on Au(111) is the fit result including the existence of the split dispersion (see text).

	Γ_{STS}	Γ_{theory}	Γ_{PES}	τ [fs]
Ag	6	7.2	6 ± 0.5	110
Cu	24	21.7	23 ± 1	29
Au	18	18.9	21 ± 1	31

Table 1 gives a comparison of the intrinsic lifetime widths Γ of the three surface states at the band minimum, determined by photoemission as described above, by STS, and by new theoretical calculations [19]. The very close agreement between the experimental results is quite striking, in view of the very different sampling areas. However, this indicates that there is either an additional broadening mechanism in the STS data, which has not been revealed and is of the same order of magnitude as the unavoidable defect scattering in PES [36], or that the effect of surface imperfection is less important for the data

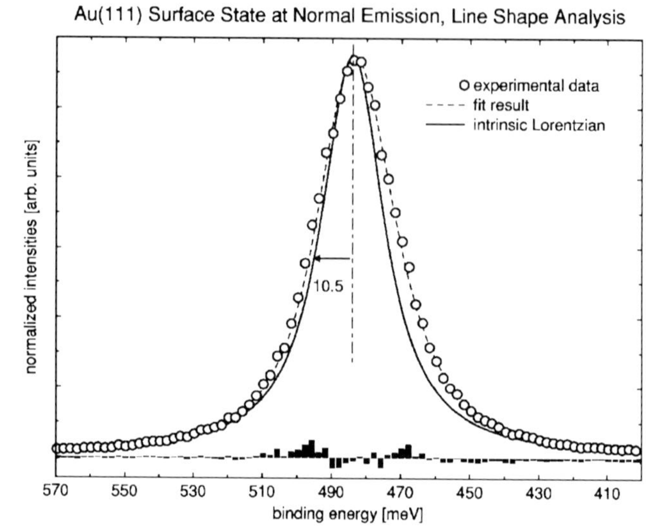

Fig. 7. Least-squares analysis of the L-gap surface state of Au(111) at normal emission measured with He I_α radiation at $T = 30$ K. The experimental data (circles) were fitted by an asymmetric Lorentzian, convoluted with the resolution function (Gaussian in energy space) and the angular broadening (semi-elliptic in k-space). The least-squares result is plotted by a dashed line, the intrinsic Lorentzian (FWHM = 21 meV) by a solid line. Fit parameters for this spectrum: $E_0^{\mathrm{B}} = 487.3$ meV, $A_0 = -0.035$, $\Gamma = 21$ meV, $\Delta\theta = \pm 0.15°$.

presented here than anticipated. In this context we emphasize that we reduced the time between surface preparation and experiment as far as possible in our experimental setup. The influence of surface quality clearly deserves further investigation.

It should be mentioned that the STS data of the Au(111) surface state are characteristically different from the others, because the $\mathrm{d}I/\mathrm{d}V$ spectra – although averaged over several tip positions – show a high-frequency distortion, probably due to the herringbone reconstruction [19]. This makes the interpretation of the STS data on Au(111) less straightforward than the data on Cu and Ag.

3.3 Dispersion relation: an MDC analysis

The left-hand panels of Figs. 8–10 show the measured dispersion relation of the three surface states as a grey-scale plot. The surface states can clearly be detected as parabolic curves with an opening at the top, positioned inside the gap of the projected band structure of the bulk (only faint in the chosen intensity scale). The respective two-dimensional data sets – *i.e.* intensity *vs.* angle and energy – have been measured with one or two shot experiments for Ag and Cu/Au, respectively. The slight asymmetry in intensity is due to the asymmetry

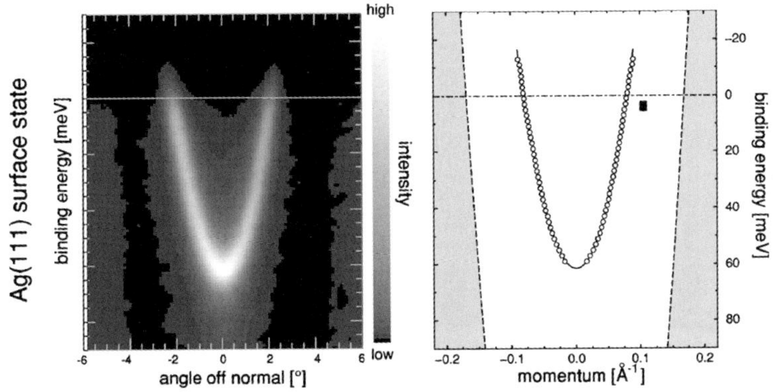

Fig. 8. Dispersion of the Ag(111) surface state. Left panel: grey-scale plot from parallel detection directly from the spectrometer; right panel: dispersion of MDC maxima (circles) and least-squares fit result (solid line). The shaded area symbolizes the $\overline{L}$-gap of the 111 surface. The crossing of the Fermi level at $\pm 2.17°$ is equivalent to Fermi vectors at $k_F = \pm 0.08$ Å^{-1}. The black box indicates the experimental uncertainties due to finite angle and energy resolution.

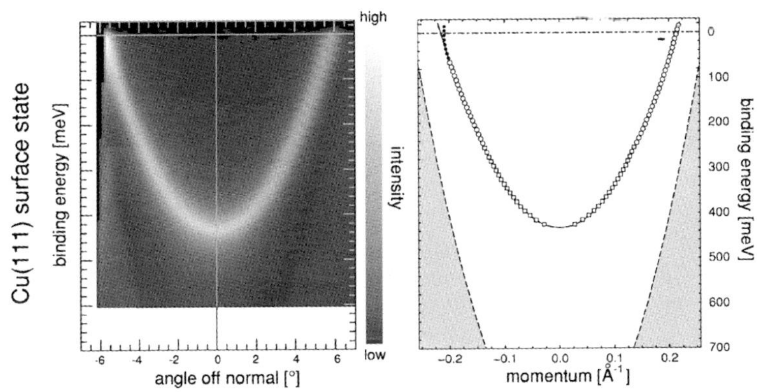

Fig. 9. Dispersion of the Cu(111) surface state (*cf.* Fig. 8). The crossing of the Fermi level at $\pm 5.85°$ is equivalent to Fermi vectors at $k_F = \pm 0.202$ Å^{-1}. Here, the left panel is in a logarithmic intensity scale. The bending of the data points on the left (negative) branch of the parabola are due to distortions at the detector edge. The complete dispersion is assembled from two data sets at low (circles) and high binding energies (squares).

of the experimental setup: the VUV light comes from the right hand side and – according to the characteristics of the photoemission process – favors the emission of the parabola branch left of the normal emission. The data sets for Cu and Au were assembled from two successively measured sets at low (circles) and high binding energy (squares), because not the whole energy range of the surface state dispersion could be detected simultaneously. Due to some non-

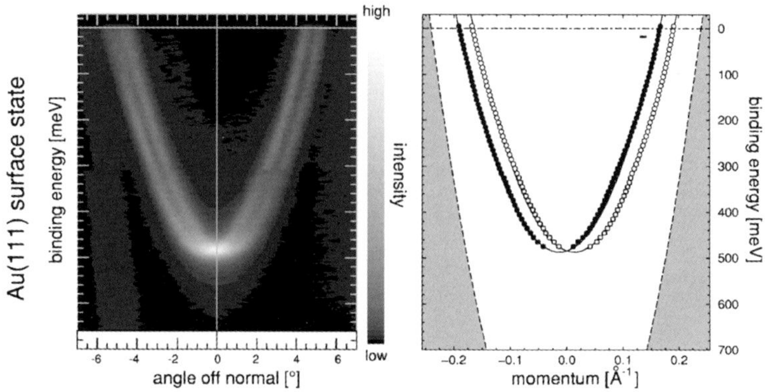

Fig. 10. Dispersion of the Au(111) surface state (*cf.* Fig. 8). The splitting in k is clearly visible on the grey scale plot on the left side. The crossing of the Fermi level at $\pm 4.55 \pm 0.1^\circ$ and $\pm 5.2 \pm 0.1^\circ$ is equivalent to Fermi vectors at $k_F = \pm 0.167\,\text{Å}^{-1}$ and $k_F = \pm 0.192\,\text{Å}^{-1}$, respectively. The complete dispersion is assembled from two data sets at low (circles) and high binding energies (squares).

linearities in the parallel detection of the angles, the matching of both regions is not perfect and appears as slight kinks in the branches of the parabolas. However, this apparative deviation from the parabolic behavior is only marginal (*i.e.* within the given experimental errors) and was taken into consideration for the least-squares analysis.

The k-splitting of the Au(111) surface state can be seen directly from the grey-scale plot in Fig. 10 without any further data analysis. There is obviously no such splitting for Ag and Cu on the scale of our instrumental angular and energy resolution, which means clearly below 5 meV or $0.01\,\text{Å}^{-1}$. From the atomic $p_{3/2}$–$p_{1/2}$ spin–orbit splittings – which are 470 meV, 110 meV and 31 meV for Au, Ag, Cu, respectively [37] – one would expect close to E_F an estimated p-band splitting of 26 meV for Ag and and 7 meV for Cu [25]. Such a splitting should actually be resolved under our experimental conditions.

To analyze the dispersion of the surface states we used cuts in horizontal direction, which then describe the angular or momentum distribution at a certain energy. In analogy to the usual photoemission spectra or energy distribution curves (EDC) these are called MDC (momentum distribution curves). Thus, one is able to investigate strongly dispersing features without the influence of the broadened Fermi–Dirac distribution on the data, which results in a shift of the peak maximum positions and a change of line-shape in the EDCs at energies close to E_F. We analyzed the MDCs by least-squares fits with two (four in the case of Au(111)) Lorentzians, extending up to energies of about $15\,\text{meV} = 6k_BT$ above E_F [38]. The resulting positions are plotted as circles and squares in the right panels of Figs. 8–10. The maxima follow a parabolic dispersion, which can be translated to the physical parameters given in Table 2. These parameters are temperature dependent [21], *e.g.* the values measured by

Table 2. Summary of the (parabolic) dispersion fit results. The two parabolas of the Au(111) surface states are centered at ± 0.013 Å^{-1}.

	E_0^B [meV]	m^*/m_e	k_F [Å^{-1}]
Ag	63 ± 1	0.397	0.080
Au	487 ± 1	0.255	0.167/0.192
Cu	435 ± 1	0.412	0.215

LaShell *et al.* [25] have been extracted from data taken at room temperature and therefore show slight differences to ours. For the Au surface state we used two parabolas with identical parameters for band minimum E_0^B and curvature, shifted in k-direction symmetrically from the $\bar{\Gamma}$ point by $\pm\Delta k$.

The resulting parabolic dispersions of the peak maxima give the Fermi vector k_F and – from the curvature of the parabola – the effective electron mass $m^* = \hbar^2[\mathrm{d}^2E(k)/\mathrm{d}k^2]^{-1}$. The resulting values are summarized in Table 2. Note that the observed maximum binding energies E_0^B of the investigated surface states (*cf.* Table 2) are typically $\approx 4\%$ smaller than the STS values [19]. The two parabolas of the Au(111) surface state are shifted by $\Delta k = \pm 0.013$ Å^{-1} from the $\bar{\Gamma}$ point, equivalent to a linear dependence of the energy-splitting on k with $\Delta E(k) \approx 0.8$ [eV/Å] $\times\, k$.

A renormalized band dispersion close to E_F, as was observed for a Mo(110) surface state [39], is not resolvable in our data for any of the investigated surfaces (see Figs. 8–10). Furthermore, in contrast to observed methodical discrepancies in the determination of the Fermi vector on other compounds [40], several standard methods gave the same results for k_F in our case [41].

3.4 Fermi surface mappings – FSM

Finally we present in Fig. 11 the three two-dimensional Fermi surfaces (FSM) as measured by PES, which are in principle also determinable by Fourier transform STM [27]. The Fermi surfaces are centered at the $\bar{\Gamma}$ point, the slight asymmetries and deviations from the circular shape are due to problems with the spectrometer lens and the surface aging during the tilt series. The data of the FSMs have been taken after the high-resolution EDC measurements, on surfaces which show already slight degradation effects. Therefore the Fermi vector k_F, defining the volume of the Fermi surface, has already decreased by approximately 2% (for Cu/Au) up to 9% (for Ag), in comparison to freshly prepared surfaces. Again clearly observable is – in contrast to Ag and Cu – the split Fermi surface of the Au(111) surface state. The two individual Fermi surfaces form two closed concentric circles connected to the inner and outer branches of the $E(k)$ parabolas in Fig. 10, with the k_F values given in Table 2.

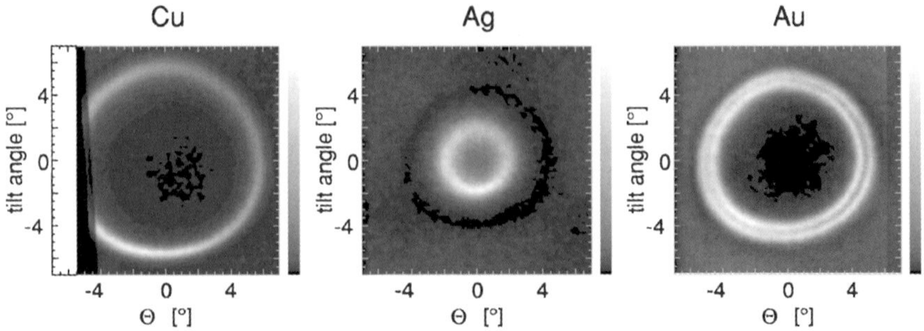

Fig. 11. Fermi surface maps (FSM) of the investigated noble metal surfaces states centered at $\bar{\Gamma}$. The data were taken after a surface exposition of about 0.5 h in a 3 h run. Θ labels the parallel angular detection of the analyzer, 'tilt angle' the rotational degree of freedom of the manipulator. The time order of the tilt goes from negative to positive angles.

3.5 Conclusion – surface states on clean (111)-surfaces of noble metals

In conclusion, we have presented a comprehensive high-resolution photoemission study of the $\overline{L}$-gap surface states on Ag(111), Cu(111), and Au(111), from that new values of the lifetime, binding energy, and dispersion can be established. The photoemission line-widths are in striking agreement with recent theoretical and spectroscopic results by scanning tunneling microscopy (STM), which gave significantly larger values for the lifetime than previous photoemission measurements. However, there remain qualitative discrepancies in the case of the Au(111) surface, in respect to the split Fermi surface and band dispersion observed by photoemission. An eventual clarification of this point requires further theoretical and experimental investigations on the noble metal surface states.

4. Surface states on (111)-surfaces of noble metals covered with rare gases

In order to study the simplest possible adsorbate systems the clean noble metal surfaces were covered with rare gases and then studied by PES. The adsorbates were applied slowly at a partial pressure of $p_{\mathrm{rg}} \approx 3 \times 10^{-8}$ mbar at low temperatures *in situ* at the measurement position, while the coverage was monitored in real-time by a comparison of the photoemission intensities of the surface state of covered and uncovered areas [42, 43]. The individual preparation temperature of each system must be high enough to provide sufficient mobility of the adsorbates to form well ordered films, but still below the desorption temperature of the first monolayer. The values are given in the respective figure captions. In addition, we checked the surface quality of the substrates and the coverage procedure by low energy electron diffraction (LEED) with the sample holder mounted on a closed-cycle refrigerator in the preparation chamber.

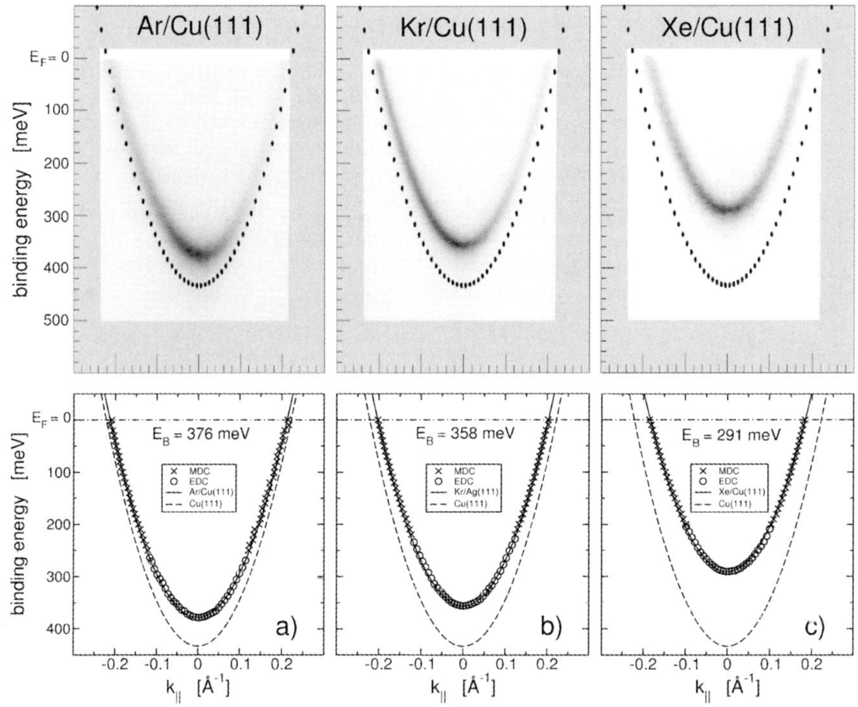

Fig. 12. Photoemission data on the Cu(111) surface covered with 1 ML of different rare gases. The grey-scale plot in the upper panel gives the photoemission intensity *vs.* binding energy and parallel momentum along the $\overline{\Gamma}$–$\overline{K}$ direction (dark is high intensity). The dispersion of the clean Cu(111) surface is indicated as a dashed line. The temperatures during preparation and measurement were kept constant at $T = 25$ K, 35 K, and 50 K for Ar, Kr, and Xe, respectively. Lower panel: analysis of the dispersions by a parabolic least-squares fit.

4.1 Rare gases on Cu(111)

The rare gas monolayers were prepared on a clean Cu(111) surface as described above. In the case of Ar and Kr, the rare gas atoms form an incommensurate structure with a nearest neighbor distance slightly expanded compared to 1×1. In contrast, Xe forms a commensurate $\sqrt{3} \times \sqrt{3}R30°$ super-lattice on Cu(111), with the Xe atoms located at on-top sites [45] (see below). This is the only commensurate super-structure among the adsorbate systems investigated here.

The influence of the rare gas monolayer to the surface state dispersion along $\overline{\Gamma}$–$\overline{K}$ is shown in Fig. 12. The grey-scale plot in the upper row represents the photoemission intensity I *vs.* the binding energy E_B (vertical axis) and the parallel momentum $k_{\|}$ (momentum along $\overline{\Gamma}$–$\overline{K}$; horizontal axis); as a reference the dispersion of the clean surface is indicated as a dashed line. The

Table 3. Summary of the photoemission results for maximum binding energy E_B^0, the effective band mass m^*/m_e, and the resulting Fermi vector k_F of the Shockley-state for different adsorbate–substrate combinations. The temperatures of the measurements on the monolayer systems are given in the figure captions of Figs. 12–14. The reference data on the clean surfaces were measured at $T = 30$ K and belong to experimental settings also used for the coverage experiments (there are small differences compared to the highest-resolution data in Table 2, obtained using different experimental parameters).

↓Ads. \ Subst.→		Cu(111)	Ag(111)	Au(111)
clean	E_B^0 [meV]	434 ± 2	62 ± 3	479 ± 2
	m^*/m_e	0.43 ± 0.01	0.42 ± 0.03	0.26 ± 0.01
	k_F [Å^{-1}]	0.222 ± 0.001	0.082 ± 0.001	$0.1690.193 \pm 0.001^*$
Ar	E_B^0 [meV]	376 ± 3	-1 ± 3	422 ± 3
	m^*/m_e	0.46 ± 0.03	0.46 ± 0.04	0.26 ± 0.03
	k_F [Å^{-1}]	0.046 ± 0.003	–	$0.158/0.184 \pm 0.001^*$
Kr	E_B^0 [meV]	358 ± 2	-8 ± 2	400 ± 3
	m^*/m_e	0.44 ± 0.02	0.44 ± 0.04	0.27 ± 0.02
	k_F [Å^{-1}]	0.204 ± 0.002	–	$0.152/0.182 \pm 0.001^*$
Xe	E_B^0 [meV]	291 ± 2	$(-52 \pm 2^\dagger)$	340 ± 2
	m^*/m_e	0.44 ± 0.02	$(0.42 \pm 0.06^\dagger)$	0.28 ± 0.02
	k_F [Å^{-1}]	0.183 ± 0.002	–	$0.140/0.173 \pm 0.001^*$

† values from STM, see Ref. [44],
∗ spin–orbit splitting

lower row displays the result of a least-squares fit of the peak dispersion by a parabola $E_B = E_B^0 - \hbar^2 k_\parallel^2 / 2m^*$, giving the effective mass m^* and the maximum binding energy E_B^0 at normal emission. The peak positions have either been extracted from the MDCs (close to the Fermi level) or from the EDCs (at high binding energies), depending on from which cut the position can be determined more precisely. Obviously the surface state of the covered surfaces is shifted to smaller binding energies (towards the Fermi level), whereas the curvature, *i.e.* the effective mass m^*/m_e, remains constant within the experimental errors (see Table 3). The relative shift increases from left to right. Thus, it is the smallest for Ar ($\approx$ 58 meV) and the largest for Xe ($\approx$ 143 meV).

In contrast to the photoemission data on the clean surface, the projected band gap could not be resolved in these experiments, presumably due to the increased impurity scattering on defects in the adsorbate layer and – in the case of the incommensurate Ar and Kr system – due to the increased background intensity because of the back-folded bulk states.

4.2 Rare gases on Ag(111)

In principle, the same dependence of the surface state on the rare gas coverage can be observed for the Ag(111) substrate. The main difference is that the band

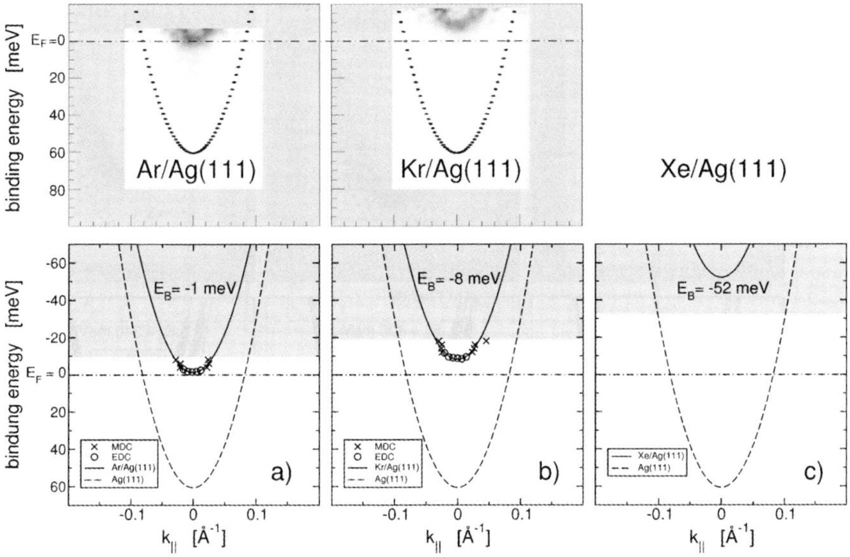

Fig. 13. Photoemission data on the Ag(111) surface covered with 1 ML of different rare gases. The grey-scale plot in the upper panel gives the photoemission intensity *vs.* binding energy and parallel momentum along the $\overline{\Gamma}$–$\overline{K}$ direction after the normalization procedure described in the text. The dispersion of the clean Ag(111) surface is indicated as a dashed line. The temperatures during preparation and measurement were kept constant at $T = 22$ K, 43 K for Ar and Kr, respectively. Lower panel: analysis of the dispersions by a parabolic least-squares fit. The surface state dispersion for Xe is out of the recoverable region of $5k_{\mathrm{B}}T_{\mathrm{des}}$ (grey area, T_{des} is the desorption temperature of the monolayer) and has been constructed from the results of STM measurements [44] for comparison ($T \approx 4$ K).

minimum for clean Ag(111) appears at a binding energy of only ≈ 63 meV. Therefore, the interaction with the rare gas shifts the surface state close to or even above the Fermi level. Above the Fermi level, the photoemission intensity decreases steeply and the spectral information is suppressed by the Fermi distribution . However, at finite temperatures the information about the electronic states above E_{F} can partly be restored by the application of a normalization method introduced by Thomas Greber *et al.* on Ni(111), provided that the energy resolution ΔE is high enough (*i.e.* comparable to the thermal broadening $4k_{\mathrm{B}}T$ or better), the noise is low and the excitation is free from satellites [38]. The accessible energy range for photoemission can be extended to $5k_{\mathrm{B}}T$ *above* E_{F}, just enough to recover the tail of the parabola for the data of the Ar and Kr monolayer system, respectively. In the case of the Xe monolayer, the surface state is shifted to 52 meV above the Fermi level, *i.e.* far beyond the limit of the accessible energy range at 32 meV, taking the maximum experimental temperature defined by the desorption temperature of 75 K [46]. For comparison we add for this system the results of an investigation by Hövel *et al.* [44], who measured the surface state spectroscopically with an STM, for which the

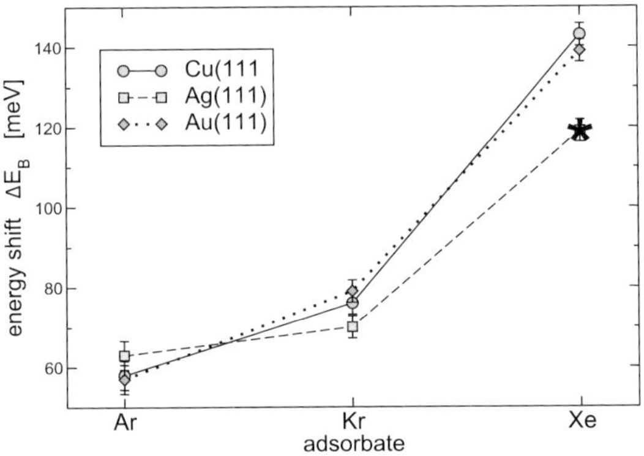

Fig. 16. Relative shift of the Shockley-state in comparison to the clean substrate surface at $T = 30$ K. The systems were measured at different temperatures (see captions Figs. 12–14), slightly influencing the measured binding energies. The data point for Xe on Ag(111) has been taken from an STM measurement from Hel *et al.* at $T = 5$ K [44].

can cause a current dependent shift of the Shockley-state binding energies, *e.g.* up to 15 meV to higher binding energies for clean Ag(111) [49].

The observed monotonic increase of the binding energy is not so obvious as it appears at the first glance. In the case of Cu(111), the rare gas with the lowest nobleness, *i.e.* Xe, has a different structural order than the other investigated gases. While Kr and Ar form an incommensurate overlayer, Xe adsorbs on Cu(111) as a $\sqrt{3} \times \sqrt{3}R30°$ superstructure (see Fig. 17). In this case, all Xe atoms adsorb on identical on-top sites [45], whereas for the incommensurate phase the adsorption site varies for different rare gas atoms. Although different sites might differ significantly in the local adsorption energy, the effect averages out in the influence on the binding energy shift of the Shockley-state.

As already demonstrated in Fig. 15, the same continuous behavior can be observed for the spin–orbit splitting (SOS) of the Shockley-state on Au(111). This indicates that the mechanisms responsible for the SOS are related to the electrostatic potentials at the surface that are also responsible for the binding energy shift. However, there is no obvious and simple explanation up to now, although slab-layer calculations by density functional theory methods are able to describe the change of both binding energy and SOS [50]. The work function change, that is typically of the order of 0.5 eV [51], plays only a minor role in this context because its contribution to the gradient of the potential is not sufficient to explain the observed magnitude of the SOS [31]. Another parameter is the increased surface corrugation [52], but a clear correspondence to the observed changes in the SOS is not discernible. Photoemission measurements on thin epitaxial Ag films on Au(111) show a *decrease* of the SOS [53, 54] with increasing film thickness. But in this case, there is a gradual change from the properties of the Au(111) to the ones of the Ag(111) surface, on which the

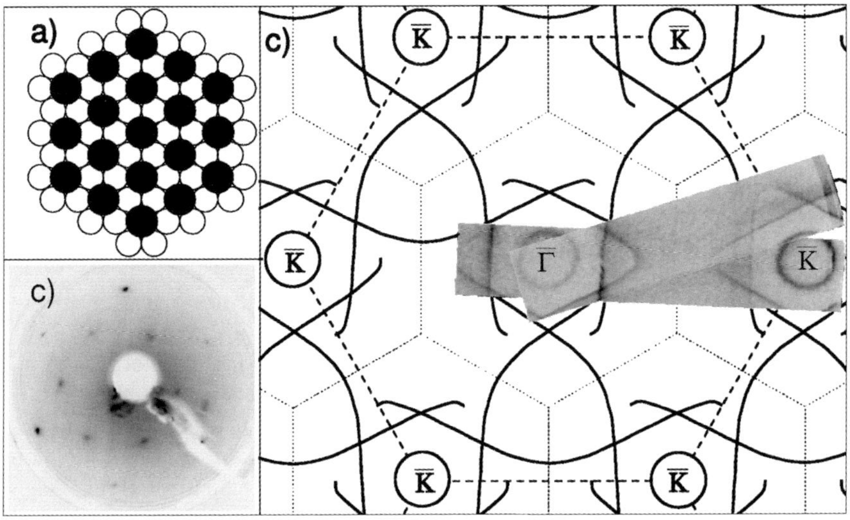

Fig. 17. Superstructure and Fermi surface map (FSM) of 1 ML Xe on Cu(111). The upper panel on the left a) shows the geometric structure of the Xe ad-atoms (black) on the substrate surface (white). This $\sqrt{3} \times \sqrt{3}R30°$ on top superstructure can be observed as additional spots in the form of the inner hexagon in the LEED pattern given in panel b) ($E_0 = 120$ eV). Panel c) displays the measured FSM as a grey-scale plot in addition to the back-folded Fermi surfaces from LDA calculations [60]. $\overline{K}$ and $\overline{\Gamma}$ mark the high symmetry points of the surface Brillouin zone (SBZ) of the clean Cu(111) surface.

SOS of the Shockley-state is one order of magnitude smaller than for clean Au(111) [55].

The binding energy shift is usually explained by the picture of the Pauli repulsion, where the repulsive short-range interaction is a consequence of the overlapping wavefunctions of the closed-shell of the rare-gas atoms with the wavefunction of the substrate, exponentially decaying outside the crystal. In this weak-interaction picture [56–59], the orthogonalization of the metal states to the nearly unperturbed atomic wavefunctions induces an oscillation of the metal wavefunctions in the region of the rare-gas core and the charge from the evanescent tail is driven back into the substrate. As a consequence, the binding energies of the Bloch electrons – including the Shockley-state electrons – are shifted towards the Fermi level, so that the net interaction potential can be described by the integral over the energy shifts of all involved states plus the attractive van der Waals potential. Since the Shockley-states are spatially confined to the surface region, they show the strongest influence of this effect. The electrons from the surface density of states, which is shifted through the Fermi level, are transferred into the Bloch-states of the bulk.

Possibly, the spatial modification of the Shockley-state wave function by the Pauli repulsion results in a shift of the electron density to regions where the net gradient of the potential is larger. Since the Rashba term α_R, giving the SOS of the Shockley-state (see. *e.g.* Ref. [31]), is proportional to the gradient of the potential, this would result in an increase of the observed splitting with an increasing perturbation of the Shockley-state wave function. However, for a quantitative understanding of the observed SOS increase of up to 30% in the case of Xe, further theoretical studies are necessary.

4.5 Commensurate phase and back-folding: Xe on Cu(111)

Finally we briefly discuss the $\sqrt{3} \times \sqrt{3}R30°$ superstructure of the Xe monolayer on Cu(111). The reduced hexagonal surface Brillouin zone (SBZ) of the adsorbate system appears as additional spots in the LEED pattern (panel b) in Fig. 17). In correspondence with the new SBZ, the Shockley-state can be observed at the back-folded $\overline{\Gamma}$ points, *i.e.* at the $\overline{K}$ points of the Cu(111) SBZ. In photoemission, such a back-folding of the surface states was observed for other systems, *e.g.* Xe on HOPG (highly oriented pyrolytic graphite) [61], Cu_3Au [62], the 'herringbone' reconstructed Au(111) surface [63], or Si(111)$\sqrt{21} \times \sqrt{21}$-(Ag,Au) [64]. Panel c) of Fig. 17 shows the Fermi surface map (FSM) of the Xe covered Cu(111) surface. The FSM gives the photoemission intensity *vs.* $\mathbf{k}_{\parallel} = (k_x, k_y)$ as a grey-scale plot over two overlapping measured stripes (high intensity appears dark). The vertical widths of the stripes is determined by the parallel detected angular range, the length represents a tilt angle range from $-64° \cdots + 16.5°$ with a step width of 0.5°. Note that the circular Fermi surface of the back-folded Shockley-state centered at $\overline{K}$ appears even more intense than the one of the 'original' at $\overline{\Gamma}$. Around the two measured Shockley-states at $\overline{\Gamma}$ and $\overline{K}$, there appears an additional intensity forming nearly straight and narrow lines. The origin of this additional intensity becomes clear when we compare the FSM with the back-folded Fermi surface map of the Cu(111) surface. As known from the He *I* measurements on Cu(111) published in the literature (see Ref. [2] and references therein), the cut through the belly-neck shaped bulk Fermi surface of Cu forms an intensity distribution with a three-fold rotational symmetry, with the necks appearing at the $\overline{M'}$ points of the SBZ. The FSM of the Xe covered Cu(111) surface is constructed by back-folding this bulk FSM (plus the circular Shockley-state FSMs centered at $\overline{\Gamma}$) at the new zone boundaries of the reduced SBZ. The result, given as solid lines in Fig. 17, nicely coincides with the observed intensity distribution. The overlapping bulk bands form a hexagon around the circular Fermi surfaces of the Shockley-state at normal emission ($\overline{\Gamma}$ point) and the former $\overline{K}$ points. Obviously, although the coupling between adsorbate and substrate is weak, the Xe-adlayer results in a perfect back-folding of the angular resolved photoemission intensity of the Cu(111) surface. Although the back-folding indicates that the Xe-monolayer is well ordered there must exist additional defects in comparison to the clean surfaces: the $\overline{L}$-gap, which could be observed as a minimum intensity ring around the Shockley-state Fermi surface of the clean surfaces [65], can not be resolved here. However, the influence of this additional defect scattering on the linewidth of the Shockley-state is small [42].

4.6 Conclusion – surface states on rare gas-covered (111)-surfaces of noble metals

In conclusion, we have presented a systematic photoemission study on Shockley-states on noble metal (111) surfaces covered with a rare-gas monolayer. Although the interaction between adsorbate and substrate is weak, the monolayer causes a significant binding energy shift and – in the case of Au(111) – an increase of the spin–orbit splitting of the Shockley-state as compared to the one of the clean surface. The observed effects can be qualitatively understood by the influence of the Pauli repulsion between the closed-shell atoms and the Shockley-state wave function out of the crystal. The high sensitivity to surface modifications make the Shockley-states an important tool for the understanding of the physics at interfaces, in particular for the interaction between adsorbates and solids. We hope that this work stimulates the investigation of the Shockley-states in first-principles calculation on adsorbate and thin-film systems.

5. Summary

A comprehensive PES study of the $\overline{L}$-gap surface states of the noble metals was presented. From this, new data on the lifetime, binding energy and dispersion could be obtained. The investigation of rare gas-covered noble metal surfaces showed significant adsorbate effects in particular on the binding energy of the surface states.

Acknowledgement

This work was supported by the Deutsche Forschungsgemeinschaft (grant nos. Hu 149-19 and SFB 277). We would like to thank Dr. Jörg Kliewer, Dr. Paul Steiner, Dr. Valeri G. Grigoryan and Prof. Michael Springborg for very helpful discussions.

References

1. M. Cardona and L. Ley (Eds.), *Photoemission in Solids I*. volume 26 of *Topics in Applied Physics*. Springer, Berlin, Heidelberg, New York (1978).
2. S. Hüfner, *Photoelectron Spectroscopy. Principles and Applications*. Springer, Berlin, Heidelberg, New York, 3rd edn. (2003).
3. F. Himpsel and P.-O. Nilsson (Eds.), *Focus on Photoemission and Electronic Structure*. Number 7 in New J. Phys. Deutsche Physikalische Gesellschaft & Institute of Physics (2005).
4. A. Einstein, Ann. Phys. **17** (1905) 132.

5. H. P. Bonzel and C. Kleint, Phys. Stat. Solidi. B **49** (1995) 107.
6. F. Reinert and S. Hüfner, New J. Phys. **7** (2005) 97. Focus Issue on 'Photoemission and Electronic Structure'.
7. P. O. Gartland and B. J. Slagsvold, Phys. Rev. B **12** (1975) 4047.
8. P. Heimann, H. Neddermeyer, and H. F. Roloff, J. Phys. C Condens. Matter **10** (1977) L17.
9. W. Shockley, Phys. Rev. **56** (1939) 317.
10. R. Berndt, W.-D. Schneider, and S. Crampin, Appl. Phys. A **69** (1999) 503.
11. G. Nicolay, F. Reinert, S. Schmidt, D. Ehm, P. Steiner, and S. Hüfner, Phys. Rev. B **62** (2000) 1631.
12. P. M. Echenique, J. M. Pitarke, E. V. Chulkov, and A. Rubio, Chem. Phys. **251** (2000) 1.
13. E. Bertel and M. Donath (Eds.), *Electronic Surface and Interface States on Metallic Systems*. World Scientific, Singapore (1994).
14. W. Steinmann, Appl. Phys. A **49** (1989) 365.
15. T. Hertel, E. Knoesel, M. Wolf, and G. Ertl, Phys. Rev. Lett. **76** (1996) 535.
16. J. Cao, Y. Gao, R. J. D. Miller, H. E. Elsayed-Ali, and D. A. Mantell, Phys. Rev. B **56** (1997) 1099.
17. W. Wallauer and T. Fauster, Surf. Sci. **374** (1997) 44.
18. J. Li, W.-D. Schneider, R. Berndt, O. R. Bryant, and S. Crampin, Phys. Rev. Lett. **81** (1998) 4464.
19. J. Kliewer, R. Berndt, E. V. Chulkov, V. M. Silkin, P. M. Echenique, and S. Crampin, Science **288** (2000) 1399.
20. B. A. McDougall, T. Balasubramanian, and E. Jensen, Phys. Rev. B **51** (1995) 13 891.
21. R. Paniago, R. Matzdorf, G. Meister, and A. Goldmann, Surf. Sci. **336** (1995) 113.
22. R. Paniago, R. Matzdorf, G. Meister, and A. Goldmann, Surf. Sci. **331–333** (1995) 1233.
23. R. Matzdorf and A. Goldmann, Surf. Sci. **359** (1996) 77.
24. R. Matzdorf, Appl. Phys. A **63** (1996) 549.
25. S. LaShell, B. A. McDougall, and E. Jensen, Phys. Rev. Lett. **77** (1996) 3419.
26. J. Li, W.-D. Schneider, and R. Berndt, Phys. Rev. B **56** (1997) 7656.
27. L. Petersen, P. T. Sprunger, P. Hofmann, E. Lægsgaard, B. G. Briner, M. Doering, H.-P. Rust, A. M. Bradshaw, F. Besenbacher, and E. W. Plummer, Phys. Rev. B **57** (1998) R6858.
28. L. Bürgi, L. Petersen, H. Brune, and K. Kern, Surf. Sci. **447** (2000) L157.
29. J. Li, W.-D. Schneider, S. Crampin, and R. Berndt, Surf. Sci. **422** (1999) 95.
30. S. D. Kevan and R. H. Gaylord, Phys. Rev. B **36** (1987) 5809.
31. L. Petersen and P. Hedegård, Surf. Sci. **459** (2000) 49.
32. C. Wöll, S. Chiang, R. J. Wilson, and P. H. Lippel, Phys. Rev. B **39** (1989) 7988.
33. L. Petersen, P. Laitenberger, E. Lægsgaard, and F. Besenbacher, Phys. Rev. B **58** (1998) 7361.
34. W. Chen, V. Madhavan, T. Jamneala, and M. F. Crommie, Phys. Rev. Lett. **80** (1998) 1469.
35. A. Beckmann, K. Meinel, C. Ammer, M. Heiler, and H. Neddermeyer, Surf. Sci. **375** (1997) L363.
36. F. Theilmann, R. Matzdorf, G. Meister, and A. Goldmann, Phys. Rev. B **56** (1997) 3632.
37. C. E. Moore, Natl. Bur. Stand. (U.S.) Circ. No. 467, Vols. II & III, U.S. GPO, Washington. D.C. (1949) (1952).
38. T. Greber, T. J. Kreutz, and J. Osterwalder, Phys. Rev. Lett. **79** (1997) 4465.
39. T. Valla, A. V. Fedorov, P. D. Johnson, and S. L. Hulbert, Phys. Rev. Lett. **83** (1999) 2085.

40. L. Kipp, K. Rossnagel, C. Solterbeck, T. Strasser, W. Schattke, and M. Skibowski, Phys. Rev. Lett. **83** (1999) 5551.
41. G. Nicolay, F. Reinert, S. Hüfner, and P. Blaha, Phys. Rev. B **65** (2002) 033407.
42. F. Forster, G. Nicolay, F. Reinert, D. Ehm, S. Schmidt, and S. Hüfner, Surf. Sci. **532–535** (2003) 160.
43. G. Nicolay, F. Reinert, F. Forster, D. Ehm, S. Schmidt, B. Eltner, and S. Hüfner, Surf. Sci. **543** (2003) 47.
44. H. Hövel, B. Grimm, and B. Reihl, Surf. Sci. **477** (2001) 43.
45. J. L. F. Da Silva, C. Stampfl, and M. Scheffler, Phys. Rev. Lett. **90** (2003) 66104.
46. K. D. Gibson and S. J. Sibener, J. Chem. Phys. **88** (1988) 7893.
47. J. Unguris, L. W. Bruch, E. R. Moog, and M. W. Webb, Surf. Sci. **109** (1981) 522.
48. G. Nicolay, *Untersuchungen zur elektronischen Struktur in situ-präparierter Edelmetalloberflächen*. PhD thesis, Universität des Saarlandes, Saarbrücken, Germany, October (2002).
49. L. Limot, T. Maroutian, P. Johansson, and R. Berndt, Phys. Rev. Lett. **91** (2003) 196801.
50. F. Forster, A. Bendounan, F. Reinert, V. G. Grigoryan, and M. Springborg, Surf. Sci., in press (published online 10. Oct. 2007).
51. M. Wolf, E. Knoesel, and T. Hertel, Phys. Rev. B **54** (1996) R5295.
52. J. Henk, A. Ernst, and P. Bruno, Phys. Rev. B **68** (2003) 165416-1.
53. H. Cercellier, Y. Fagot-Revurat, B. Kierren, D. Malterre, and F. Reinert, Surf. Sci. **566–568** (2004) 520.
54. H. Cercellier, C. Didiot, Y. Fagot-Revurat, B. Kierren, L. Moreau, D. Malterre, and F. Reinert, Phys. Rev. B **73** (2006) 195413.
55. F. Reinert, J. Phys. C Condens. Matter **15** (2003) S693.
56. L. W. Bruch, M. W. Cole, and E. Zaremba, *Physical Adsorption: Forces and Phenomena*. Oxford University Press, Oxford, New York (1997).
57. E. Bertel and N. Memmel, Appl. Phys. A **63** (1996) 523.
58. E. Zaremba and W. Kohn, Phys. Rev. B **15** (1977) 1769.
59. J. Harris and A. Liebsch, J. Phys. C Solid State Phys. **15** (1982) 2275.
60. P. Aebi, R. Fasel, D. Naumovic, J. Hayoz, T. Pillo, M. Bovet, R. G. Agostino, L. Patthey, L. Schlapbach, F. G. Gil, H. Berger, T. J. Krentz, and J. Osterwalder, Surf. Sci. **402–404** (1998) 614.
61. F. Patthey, M.-H. Schaffner, W.-D. Schneider, and B. Delley, Surf. Sci. **454–456** (2000) 483.
62. R. Courths, M. Lau, T. Scheunemann, H. Gollisch, and R. Feder, Phys. Rev. B **63** (2001) 135110-1.
63. F. Reinert and G. Nicolay, Appl. Phys. A **78** (2004) 817.
64. J. N. Crain, K. N. Altmann, C. Bromberger, and F. J. Himpsel, Phys. Rev. B **66** (2002) 205302-1.
65. F. Reinert, G. Nicolay, S. Schmidt, D. Ehm, and S. Hüfner, Phys. Rev. B **63** (2001) 115415.

Computer Simulations of Phase Transitions and Dynamics in Confined Systems

By Heiko Rieger[1,*], Raja Paul[2], Jae-Dong Noh[3], and Grégory Schehr[4]

[1] Universität des Saarlandes, Theoretische Physik, Pf 151150, D-66041 Saarbrücken, Germany
[2] BIOMS/IWR, Universität Heidelberg, Im Neuenheimer Feld, D-69120 Heidelberg, Germany
[3] Department of Physics, The University of Seoul, Seoul 130-743, South Korea
[4] Laboratoire de Physique Theorique d'Orsay, Univ. Paris-Sud 11, 91405 Paris Cedex, France

Monte-Carlo Simulations / Non-Equilibrium Dynamics / Nanopores / Orientational Transition / Disordered Systems

Results of theoretical studies of spatially confined systems *via* Monte Carlo simulations are reported. Three experimentally relevant systems are considered: 1) Ad- and desorption phenomena in nanoporous materials like Vycor glass; 2) Dipolar molecules on a graphite surface that undergo an orientational transition; 3) Non-equilibrium dynamics in two-dimensional disordered materials.

1. Introduction

Monte Carlo simulations are an important tool for the theoretical study of phase transitions and dynamics in confined system. The confinement can either be dimensionally, as in the case of surfaces and interfaces, or geometrically by small lateral sizes, as in the case of pores and micro- or nano-crystals. The advantage of computer simulations is that the system size in any spatial direction can easily be varied – a fact that is also systematically used in finite size scaling to study phase transitions of infinite systems.

A number of experimental setups studied in this SFB can be mapped onto statistical physics models that are perfectly suited for a theoretical investigation *via* computer simulations. One is ad- and desorption phenomena in nanoporous materials like Vycor glass, which can be described by grand canonical lattice gas models in special geometries. Another is dipolar molecules on

* Corresponding author. E-mail: h.rieger@mx.uni-saarland.de

Z. Phys. Chem. **222** (2008) 433–469

a graphite surface that undergo an orientational transition, which we argue to be in the same universality class as the transition(s) occurring in the 6-state clock model on the fully frustrated triangular lattice. For both cases we performed a detailed numerical study, which we report here. In addition we investigated intensively the non-equilibrium dynamics in several two-dimensional disordered materials, which we also review here.

2. Condensation phenomena in nano-pores

Adsorption and desorption isotherms of a gas condensed into nano-pores show hysteresis and have become a very useful tool for the classification of porous materials [1–3]. Due to the effect of surface tension the formation of meniscii inside the pore is delayed on the desorption branch, resulting in a non-vanishing hysteresis area of the sorption isotherms. A theoretical analysis and detailed description of the early research on this phenomenon is described in a review by Everett [4]. Recent investigation from a point of view of the stability of adsorbed multilayers [5] and other [6–8] analytical and numerical approaches together with density functional theory also proposed that the hysteresis phenomenon is an intrinsic property of the phase transition in a single idealized pore and arises from the existence of metastable states. Although real nano-porous materials consist of an interconnected network of simple pores [9, 10], quite recently it has been shown by numerical analysis [11] and experiments [12–14] that hysteresis can occur in isolated pores also. Moreover the shape of the hysteresis loops are influenced by the characteristic features of the pore geometry.

Here we seek insight into the non-equilibrium dynamics in general and the adsorption/desorption hysteresis in particular, within a SINGLE nano-pore – on the basis of a microscopic model. Using extensive Monte Carlo simulations, we characterize the sorption isotherms for different pore geometries and compare our results directly to the experimental observations. Moreover we get insight into the mechanism for the extremely slow dynamics in the confined geometry of a single pore during adsorption and desorption processes as well as during domain growth. We study the phase separation kinetics of a binary liquid in nano-pores at low temperatures. The demixing transition of a binary fluid within a solid gel has been investigated in [15]. Other works [16–18] showed that a binary liquid, unlike Ising-like complete phase separation, does not separate into two phases completely. Instead the adsorbed material forms many small domains far inside the coexistence region.

It has been suggested that this behavior is a consequence of a random pore geometry and can be understood on the basis of the random field Ising model (RFIM) [19–21]. However, Monette *et al.* [22] argued that this picture does not apply to Vycor, with much lower porosity than gels. These are the systems we are interested in and we focus mainly on isolated (single) pores with a homogeneous wall-particle potential.

A very slow segregation dynamics (as also predicted within the RFIM picture) takes place in single-pores without randomness because the confinement in a small pore slows down domain growth in certain regions of the wetting phase diagram [22] and as a result macroscopic phase separation is not observed on short time scales. Moreover, the late stage domain evolution, obtained from a two point correlation function $C(r, t)$ follows the same Lifshitz–Slyozov [24] growth law of $t^{1/3}$. Unlike earlier works [25] on domain evolution in porous networks, we concentrate on a single pore which is more applicable for systems with low porosity (Si mesopores [14]). Randomness in terms of an irregular pore structure or presence of impurity atoms in the pore wall, which is inherent in real systems, may nevertheless have a substantial effect on the domain growth. An average over many isolated pores corresponding to an assembly of non-interconnected (unlike Vycor) pores, as we consider it in this paper, was experimentally realized [14] by a B-doped Si wafer *via* electrochemical etching.

2.1 Simulation model

A standard model for a binary liquid mixtures is the Ising lattice-gas model with spin occupancy variables $\sigma_i = 0, 1$ governed by the Hamiltonian:

$$\mathcal{H} = -W_{\text{pp}} \sum_{\text{bulk}\langle i,j \rangle} \sigma_i \sigma_j - W_{\text{wp}} \sum_{i \text{ n.n. of wall}} \sigma_i \,, \tag{1}$$

where W_{pp} and W_{wp} are the particle–particle and wall–particle couplings, respectively. We denote an occupied site as a particle and an empty site as a vacancy. Experimental observations suggest that in case of a glass (Si) adsorbent, the pore wall has a very strong affinity towards the adsorbed gases, which implies $W_{\text{wp}} > W_{\text{pp}}$. Most of our simulations, unless otherwise specified, are performed at a fixed "wettability" $W_{\text{wp}}/W_{\text{pp}} = 1.5$. A change in this value does not qualitatively affect our results as long as $W_{\text{wp}} > W_{\text{pp}}$. We set $W_{\text{pp}} = 1$ which also defines the units in which energies and temperatures (devided by k_{B}) are measured). The geometry of the simple pore (see Fig. 1a) is chosen to be a rectangle of size $L \times h$ in two dimension and parallelepiped of size $L \times h \times h$ in three dimension with $h \ll L$. The unit in which lengths are measured is one lattice constant. An ink-bottle pore (see Fig. 1b) consists of two simple pores of unequal radii h_1 and h_2. Standard conserved order parameter dynamics (Kawasaki) is employed to study the diffusion and the domain growth kinetics in the nano pores. A spin σ_i at site i chosen randomly and exchanged with one of its nearest neighbor spin σ_j at site j, with Metropolis acceptance probability

$$P(\{\sigma\} \longrightarrow \{\sigma'\}) = \min\left[1, \frac{\exp(-\beta\mathcal{H}\{\sigma'\})}{\exp(-\beta\mathcal{H}\{\sigma\})}\right], \tag{2}$$

tion. As a result the the loop area for bigger pore is also increased as shown in Fig. 3 (ii)a and b and (iii)a and b. This features can be used to compare pores of different sizes.

Experimentally [14] one observes a two step decrease of the desorption isotherms for ink-bottle pores. Our data, as shown in Fig. 3 (v)b, for a both-end open pore in 3d of similar geometry, agrees with this experimental observation. The two step decrease of the desorption isotherm arises from the sequential emptying of the different sections (of diameter h_1 and h_2) of the pore. The heights of the two steps suggest that the wider section of the pore emptied earlier than the smaller section. However, the one end open ink bottle geometry Fig. 3 (iv)b does not show this two-step feature, in contrast to the experimental finding [14]. There the two-step feature for ink-bottle pores vanishes only for much smaller temperatures, when the adsorbent is already in the solidphase. Since the temperature is the same in Fig. 3 (iv)b and (v)b a possible reason for the discepancy we find could be that our system sizes and the ratio h_2/h_1 are too small to detect this feature. In 2d ink-bottle pores we do not observe a two step process at all, neither when only one end is open, Fig. 3 (iv)a, nor when both ends are open, Fig. 3 (v)a.

2.3 Evaporation

If a partially or completely filled pore is kept in vacuum, the density inside it decreases with time. This is what we call evaporation and investigate it for different pore geometries. In experiments [14] one measures the vapor pressure change inside a previously equilibrated pore subject to the pressure variation in the reservoir. However, we carry out the simulation in a slightly different way by keeping the initial pore density just above the desorption threshold.

To study the variation in pore density ρ_f as a function of time we fill pores completely and allow it to evaporate in an empty reservoir (vacuum). The change in ρ_f has been recorded as a function of time t. Decay of the pore density can be very well described by a stretched exponential law,

$$\rho_f(t) \sim \exp\left[-(t/\tau)^{\beta}\right] . \tag{3}$$

The simulations are carried out for pores with both simple and ink-bottle geometry. For the simple pore we study the evaporation at temperature $T = 0.4$, for systems with $L = 128$, 256, 400, 512 and $h = 7$ and finally average over 50 ensembles. The pore density ρ_f as a function of time t is plotted in Fig. 4a in a log-linear scale. It is noticed that a pure stretched exponential decay is found only above a certain value of the pore filling fraction as shown by continuous lines. These values of ρ_f are nothing but the ratio between the number of surface to bulk molecules and decreases as L becomes larger. The surface molecules are attached rather strongly ($W_{wp} > W_{pp}$) to the walls and slow down the evaporation rate for a while but finally drop off suddenly to zero. As the length of the system is increased, the particles deep

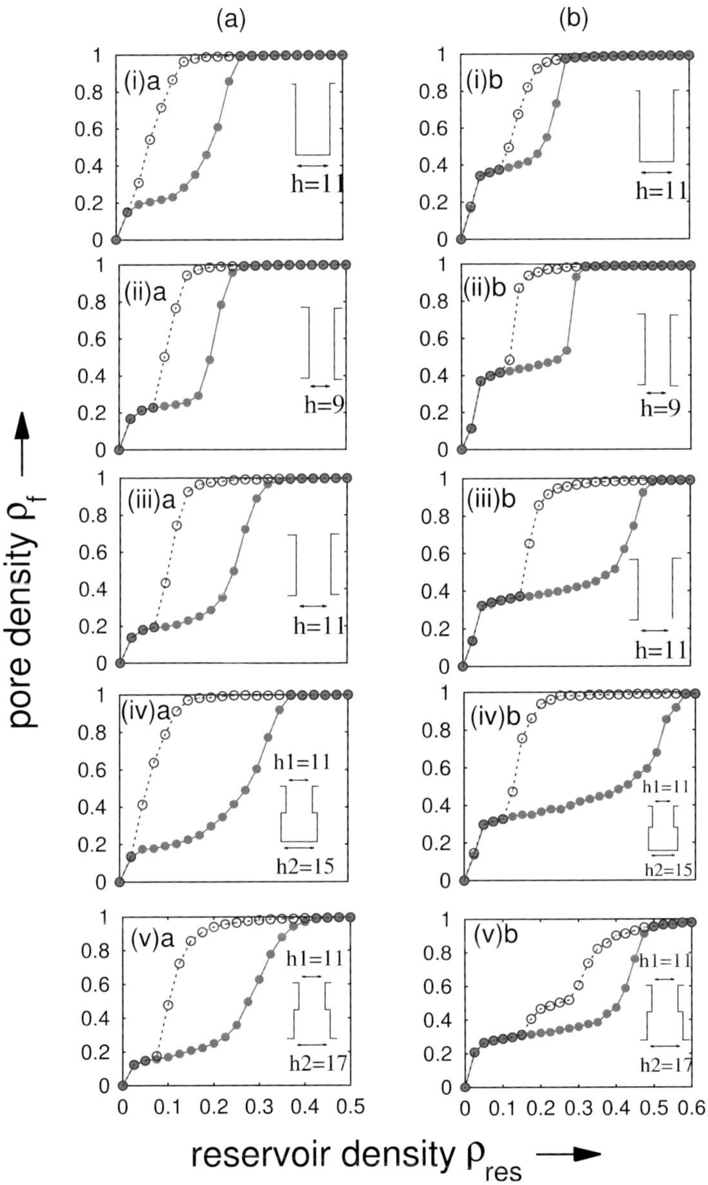

Fig. 3. Adsorption (filled circles) and desorption (empty circles) isotherms for nano pores. (a): Filling fraction ρ_f of the pore as a function of the reservoir density ρ_{res} in 2 dimension. The equilibration time t_w for each data point 2^{22}. (i) has one reservoir of length 10 attached to the open end. II, III and IV have one reservoir attached on each side. (b): Hysteresis isotherms in 3 dimension. Equilibration time $t_w = 2^{22}$, pore length $L = 50$ and reservoir length is 10. For each hysteresis isotherm data are averaged over 50 disorder samples.

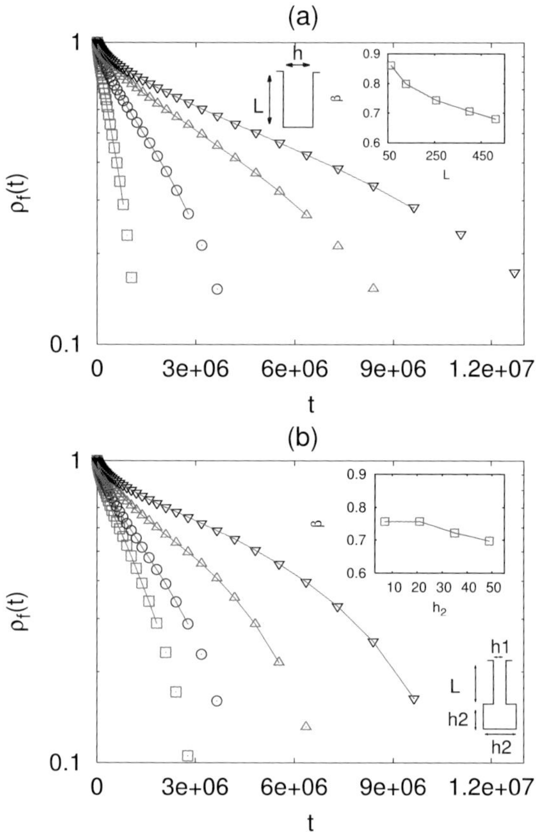

Fig. 4. (a): Evaporation from simple pores of diameter $h = 7$ and different lengths $L = 128$, 256, 400 and 512 from left to right at temperature $T = 0.4$. Data for the filling fraction $\rho_f(t)$ are shown in a log-linear scale – note that the decay is initially slower than exponential. The full line connects the data points that can very well be fitted to a stretched exponential with the exponent β given in the *inset*. Inset: The decay exponent β as a function of the pore length L. (b): $\rho_f(t)$ for ink-bottle pores of fixed length $L = 200$, tube diameter $h_1 = 7$ at $T = 0.4$ for different sack width $h_2 = 7$, 21, 25, 49 from left to right. The decay law is similar to the simple pores. Inset: β as a function of h_2.

inside the pore requires much more trial attempts to diffuse till the open end which effectively increases the evaporation time for longer pores. This leads to the decay exponent β to decrease with L, as shown in the *inset* of (a).

A similar study in case of ink-bottle pores was carried out with systems of fixed $L = 200$ and $h_1 = 7$ for different values of $h_2 = 7$, 21, 35, 49 as shown in Fig. 4b. The temperature is kept fixed at $T = 0.4$ and data are averaged over 50 ensembles. Like the previous case, here also, ρ_f shows a pure stretched exponential decay above a certain value, which becomes smaller as h_2 is increased. Moreover the effective length $(L+h_2)$ of the pore increase with h_2 results in the decay exponent β (*inset* of (b)) to drop off similarly to the simple pore. A further analysis (not shown) on both simple and ink-bottle pore shows that β is independent of the temperature.

2.4 Domain evolution and random walks

In this section we study the temporal evolution of domain structures inside pores starting with a random (high temperature) initial configuration of given density. Now we use periodic boundary conditions between the two ends of the pore and the total number of particle is conserved. Figure 5a, shows the domain structure-profile at different time steps when the system is quenched to $T = 0.3$ from $T = \infty$. Fig. 5b is the snap-shot of a cut along the pore axis. The black regions correspond to the occupied sites and are called "particle-domains", whereas empty white regions are termed "blobs".

In the initial stage, the growth is dominated by nucleation and spinodal decomposition. As soon as the domain- or the blob-size becomes comparable to the diameter of the pore, the above two mechanisms do not work any more. Transverse directional growth is completely stopped because of the pore wall and the horizontal movement of the blobs is slowed down by the presence of the particle-domains.

The snapshots in Fig. 5 confirm that the random motion of the blobs plays an important role in the late stage of growth. A closer look to Fig. 5b, which is horizontally the occupancy along the pore-axis and vertically time in log scale, shows in the late stage, the blobs move to and fro along the axis of the pore and penetrate through the particle-domains to coalesce with the neighbors. During this process it also transfers a holes (vacancies, white regions) from its surface to the neighboring blobs.

To elucidate the random motion, we study the time evolution of a single blob as shown in Fig. 5c. Initially it is a perfect rectangle of linear size (l_b) comparable to the pore diameter (h) and placed in the center of the pore. The temperature is kept fixed at $T = 0.3$. The system is allowed to evolve and the mean square deviation $\langle x^2(t) \rangle$ of the center of mass (COM) of the blob along the axis of the pore is measured at each time step. Since the boundary of the blob perpendicular to the wall, fluctuates very rapidly, the true COM may not lie on the geometrical axis of the pore. In this case we trace the actual COM and project it on to the axis. The size of the blob has to be chosen large enough to avoid disintegration of the main blob.

Since we start with an exactly rectangular blob, a true random-walk motion is not immediately observed. In the early stages, the blob walls, perpendicular to the pore axis roughens, leading to $\langle x^2(t) \rangle \sim t^{2/3}$ as in surface roughening described by the KPZ-equation [26]. At late stages $(t > 10^4)$, the blob performs

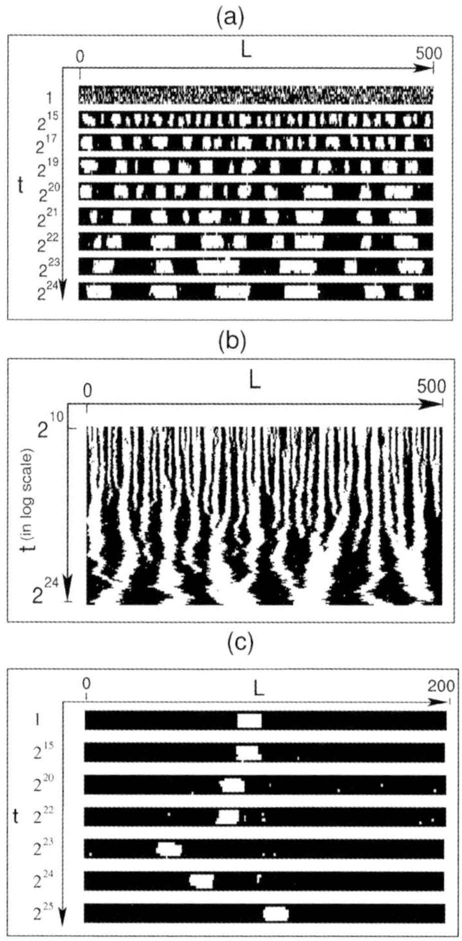

Fig. 5. (a): Domain evolution in nano-pores. The figures are produced by taking snap-shot for a system of length $L = 300$ and height $h = 7$ at temperature $T = 0.3$. (b): Axial snap-shot of the domains for the same system in Fig. a. A horizontal cut through the figure at any instant t will give the location and size of domains/blobs along the axis of the pore. The time is plotted in logarithmic scale. (c): Time evolution of a single rectangular blob of linear size $l_b = 13$ in a pore of dimension 200×7 at $T = 0.3$.

a random walk with a blob-size dependent diffusion constant $\mathcal{D}(l_b)$,

$$\langle x^2(t) \rangle = \mathcal{D}(l_b)\, t\,. \tag{4}$$

Fig. 6a, shows the mean square displacement $\langle x^2(t) \rangle$ of the COM of blobs with $l_b = 11, 15, 19, 23, 27, 31, 35, 39$ as a function of time t. For each values of l_b the simulation is carried out in a system with fixed $L = 200$, $h = 7$ at

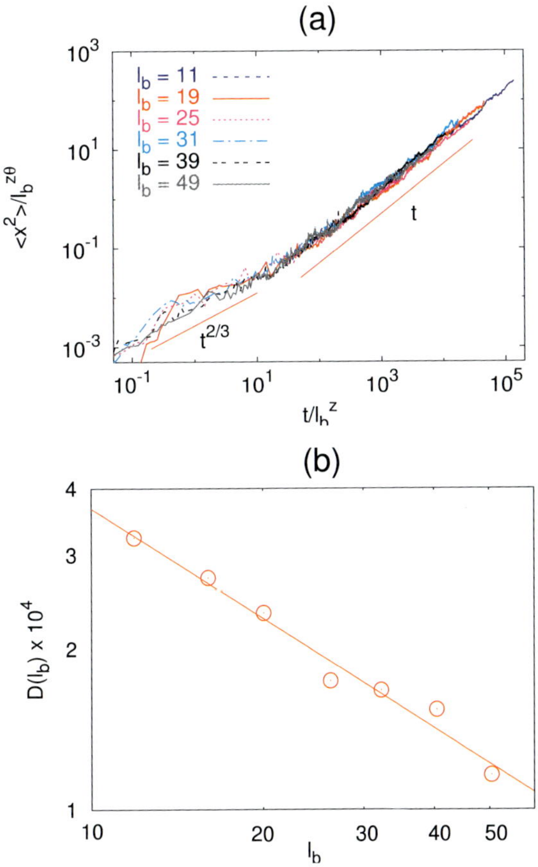

Fig. 6. (a): Mean-square deviation of blobs of size $l_b = 11, 15, 19, 23, 27, 31, 35, 39$ for a system size $L = 200$, $h = 7$ and the temperature $T = 0.3$. **(b)**: Shows the power low decay of the diffusion constant $\mathcal{D}(l_b)$ for different blob sizes. The exponent is found to be $\gamma = 0.68$.

temperature $T = 0.3$ and finally averaged over 1000 ensembles. It is evident from the figure that the onset of a true random-walk is delayed for bigger blobs. A scaling form for both small and large time scale regime is,

$$\langle x^2(l_b, t)\rangle \propto \tau^{\theta}(l_b)\ f(t/\tau(l_b)) , \tag{5}$$

with

$$f(x) \sim \begin{cases} x^{\theta} & \text{for } x \ll 1 \\ x & \text{for } x \gg 1 . \end{cases} \tag{6}$$

Assuming, $\tau(l_b) \sim {l_b}^z$, from Eq. (5), one can write for large time $t \gg \tau$,

$$\langle x^2(l_b, t)\rangle \propto {l_b}^{z(\theta-1)}\, t\,, \tag{7}$$

which readily gives,

$$\mathcal{D}(l_b) \sim l_b^{-z(1-\theta)} = {l_b}^{-\gamma}\,. \tag{8}$$

The scaling form in Eq. (5) gives quite reasonable data collapse for $\theta = 2/3$ and $z = 2$, as shown in Fig. 5a. Substituting these value in Eq. (8) we obtain the value of $\gamma = 2/3$. Diffusion constant $\mathcal{D}$ evaluated from the asymptotic behavior [*cf.* Fig. 5a] is plotted in Fig. 5b. From the slope of $\mathcal{D}$ *vs.* l_b curve in log–log scale we estimate the $\gamma = 0.68$ which agrees quite well with the previous value 2/3, obtained from the scaling. Note that the relation $\mathcal{D} \propto 1/l_b^{2/3}$ deviates from the naive expectation $\mathcal{D} \propto k_B T\eta/l_b$ [22, 23], where k_B, T, η are the Boltzmann's constant, temperature and viscosity coefficient of the medium, respectively.

2.5 Discussion

Using extensive Monte Carlo simulations we have shown how hysteresis arises in nano-pores for different pore structures. The characteristics of the sorption branches are influenced by the shape and size of the pore geometry. Since the hysteresis in nano-pores occur due to diffusion, the temperature has to be chosen very carefully to avoid a slow diffusion rate at low temperatures or a vanishing hysteresis-loop at high temperatures. The absence of a two-step desorption branch in 2d ink-bottle pores may be due to the effect of a small temperature as found in the experiment [14]. The choice of the pore diameters for one-end open ink-bottle geometry is also an important factor. Because of surface tension effect, a large curvature ratio $\frac{1}{h_1}/\frac{1}{h_2}$, can cause a huge pressure difference between the narrow and wide parts of the pore. Increase in curvature corresponds to a decrease in vapor pressure; described by the Kelvin equation. As an effect if the narrow part of the pore is filled early, can block the particle to diffuse into the wider part, unless a very high density is reached in the reservoir.

The stretched exponential decay of the pore density $\rho_f(t)$ also agrees qualitatively with the experiments [14]. One simple reason of such a behavior of $\rho_f(t)$ could be due to the formation of meniscus at the open end of the pore which eventually restricts the rate of evaporation depending upon the radius of its curvature.

For the phase separation of binary liquids in the pore environment, model B [29] corresponding to dynamics with conserved order parameter, is not well suited, as it does not account for the transport of the order parameter by hydrodynamic flow. Modifications to model B by adding an "advection" term describes the system quite well [30].

The late stage coarsening in the pore system that we studied is effectively driven by two mechanisms: (1) Transfer of holes from the surface of one blob to a neighboring blob and (2) the transfer of particles from one side of the blob to the other along the pore wall. Owing to the first mechanism a blob shrinks and the neighboring blobs grow, whereas due to the second one a blob moves to and fro as a whole and coalesce with another. Our numerical study for a single blob accounts for the contribution arises from (2) only and gives rise to the diffusion constant $\mathcal{D} \sim l_b^{-2/3}$ of a blob of size l_b. The first mechanism modifies the single-blob diffusion constant in a non-trivial way, which is difficult to estimate. Finally, the superposition of random motion of blobs due to mechanism (2) and the hole transfer mechanism in (1) leads to a late-stage growth law $R(t) \sim t^{1/3}$, independent of their individual contributions.

The domain growth exponent $1/\psi$, systematically investigated in [28], is significantly decreased in presence of random wall-particle potential W_{wp} which in practice is due to the effect of impurity atoms at the pore walls. The temperature dependence of the exponent may come from a logarithmic scaling [27] of barrier energy of the domains. On the other hand geometrical defects appear to slow down the growth process only in the early time regime. But in the late stage, as the blob size becomes larger than the wave length of the geometrical disorder, which in this case is the periodicity of the wall-defect, the domain growth displays the same behavior as in the case of smooth walls.

3. Frustrated antiferromagnetic six-state clock model on the triangular lattice

Two-dimensional systems cannot show long-range order by breaking a continuous symmetry at any finite temperature T. Nevertheless Kosterlitz and Thouless (KT) demonstrated that the two-dimensional (2d) XY-ferromagnet with nearest neighbor interaction and, equivalently, the planar rotator exhibit a phase transition from an ordered phase with an algebraic decay of the spin correlations to a disordered phase *via* the unbinding of vortex–antivortex pairs [31].

The KT-concept has been applied successfully to the onset of superfluidity [32, 33] and of superconductivity [34] in thin films and two-dimensional melting [35]. An application to layered magnetic systems with large ratios of the intra- and the inter-layer exchange appears to be more direct, but here the KT transition is masked by the transition to 3D behavior [36]. In Ref. [64] X-ray diffraction data on CF_3Br monolayers physisorbed on graphite where presented together with evidence that the orientational *pseudospin* ordering is of the KT-type.

Physisorbed monolayers are indeed good approximations to 2d systems, as evidenced by the fact that phase transitions occurring in such layers can be often described by the versions of elementary models of statistical mechanics such as the Ising or the 3-state Potts model [37–39].

The halomethane CF_3Br is a prolate C_{3v} molecule with a fixed dipole moment of about 0.5 D pointing along the threefold axis parallel to the C–Br bond. Clearly the pseudospin correlations are bound to a plane, thus the system is 2d with respect to the relevant degrees of freedom at the phase transition. The pseudospin is presumably not a strictly isotropic planar rotator but experiences a crystal field which breaks the continuous azimuthal symmetry. It has been shown however that – at least for the planar XY ferromagnet – the KT-behavior is stable with respect to small crystal fields of 6-fold symmetry [40]. The site symmetry of the monolayers is indeed 6-fold, but the ordered structure approached is not ferro- but antiferroelectric. Antiferro-type ordering on a triangular lattice is affected by frustration with the consequence that the helicity, a discrete 2-fold symmetry, enters into the problem. The order-disorder transition is then described by a confluence of the Ising and KT universality classes [41, 42] with an Ising type anomaly of the specific heat, but the spin correlations of the disordered phase still follow the KT-form of $\xi(T)$ as we will demonstrate now.

From the above considerations follows that the pseudo-spins formed by the CF_3Br dipoles are arranged on a triangular lattice, interact antiferromagnetically with nearest neighbors[1] and have 6 preferred orientations described by an azimuthal angle θ[2]. Hence the model that captures the *universal* features of the orientational ordering transition (*i.e.*, those that are independent of the microscopic details of the interactions between the molecules) should be the anti-ferromagnetic six-state clock model on a 2d $N = L_x \times L_y$ triangular lattice (see Fig. 7). The six-state clock spin is a planar spin pointing toward discrete six directions; $\mathbf{S} = (\cos\theta, \sin\theta)$ with $\theta = \frac{2\pi n}{6}$ $(n = 0, 1, \ldots, 5)$. The Hamiltonian of this model reads

$$\mathcal{H} = 2J \sum_{\langle i,j \rangle} \cos(\theta_i - \theta_j)\,, \tag{9}$$

where the sum is over nearest neighbor site pairs $\langle i, j\rangle$ and $J > 0$ is the antiferromagnetic coupling strength in which here all energies and temperatures (divided by k_B) are measured). The overall factor 2 is introduced for a computational convenience. It is a limiting case of an antiferromagnetic XY model with infinite anisotropy field $\cos p\theta$ $(p = 6)$.

The antiferromagnetic coupling on a triangular lattice induces a frustration. As a result the system has 12-fold degenerate ground states with three-sublattice structure. Global spin rotations by multiples of $2\pi/6$ and also Ising

[1] The character of the relevant orientation dependent interactions of CF_3Br is not known, but a comparison of monolayers of several polar methane derivatives [69] suggests that for CF_3Br the short range anisotropic part of the intermolecular van der Waals force and hard core repulsion are more important than the medium range dipole–dipole interaction.

[2] The polar orientation of the normal component of the dipole moments is not resolved by the X-ray diffraction data.

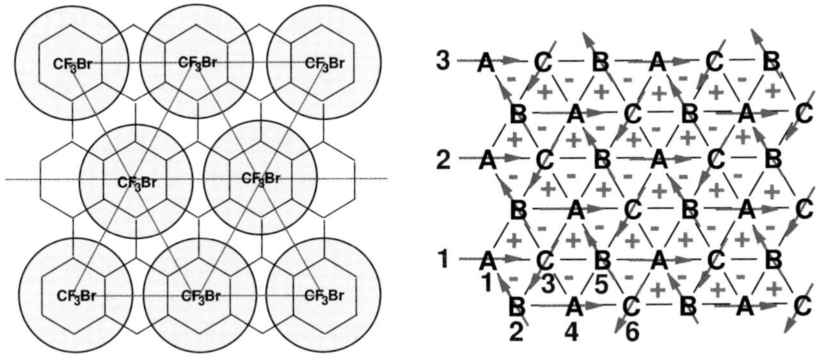

Fig. 7. *Left*: The spatial arrangement of the CF_3Br molecules on the underlying hexagonal graphite surface in the 2×2 structure that is present below 105 K. The resulting trigonal lattice is also shown. *Right*: A triangular lattice of size 12×3 with the 6-state clock spins indicated as arrows representing the 6 preferred directions of the CF_3B dipolar moments. The arrow configuration shown is one of the 12 ground states by which the lattice sites are separated into three A, B, and C sub-lattices. The helicity for each elementary triangle corresponding to this ground state is also indicated by $+$ and $-$.

like transformation that inverts all helicities formed by the spins on elementary triangles (see Fig. 7) transforms one ground state into another. Therefore the model possesses a C_6 (6-fold clock) $\times$ Z_2 (Ising) symmetry and we expect that there are two phase transitions associated with the spontaneous breaking of the C_6 and Z_2 symmetry. Despite a lot of efforts, there is no consensus on the nature of the phase transitions and it is not clear whether the two transitions happen at two different temperatures or at a single one. For instance, the fully frustrated XY (FFXY) model, describing an array of Josephson junctions under an external magnetic field [41, 43], has a continuous $U(1)$ symmetry *and* a discrete Ising or Z_2 symmetry that can be broken at low temperatures through a Kosterlitz–Thouless (KT) type transition and an Ising-like transition, respectively.

Renormalization group (RG) studies on the FFXY model drew the conclusion that the transitions occur at the same temperature [44]. Monte Carlo simulation studies on generalized FFXY models supported the single transition picture for the FFXY model [45]. Interestingly, it has been reported that the Z_2 symmetry breaking transition may not belong to the Ising universality class. Monte Carlo studies of the FFXY models on a triangular and a square lattice yielded the correlation length exponent $\nu = 0.83(4)$ (triangular lattice) and $\nu = 0.85(3)$ (square lattice), which is inconsistent with the Ising value $\nu = 1$ [46]. The specific heat appeared to follow a power-law scaling rather than the logarithmic scaling as one would expect for the Ising universality class. A non-Ising scaling behavior is also observed in the studies of the FFXY model *via* a Monte Carlo simulation [47] and Monte Carlo transfer matrix calculations [48, 49]. The same non-Ising critical behavior is also observed in the coupled XY-Ising model [46, 50, 51].

On the other hand, numerical evidence also in favor of two transitions at two different temperatures has been collected. Monte Carlo simulation studies of the frustrated Coulomb gas system, which is supposed to be in the same universality class as the FFXY model, showed that the KT type transition temperature T_{KT} and the Ising-like transition temperature T_{I} are different with $T_{\mathrm{KT}} < T_{\mathrm{I}}$ [52, 53]. Two transitions were also found in the FFXY model on a square lattice [54] and on a triangular lattice [55–57] using Monte Carlo simulations. A careful analysis of the RG flow of the FFXY model also led to a conclusion of the double transition scenario [58]. It is the general belief that the transitions at T_{KT} and at T_{I} belong to the KT universality class [59] and to the Ising universality class, respectively, if the transitions occur at different temperatures. However, the critical exponents associated with the Z_2 symmetry breaking, which are found by Monte Carlo simulations, turn out to be different from those of the Ising universality [53, 54, 57]. They are rather close to those obtained in the XY-Ising model [46]. Although there is an argument that the observed non-Ising exponents are due to a screening effect hindering the asymptotic scaling behavior [60], the controversy on the nature of the phase transition remains unsettled [61–63].

Here we present a thorough numerical study of a related model, the fully frustrated anti-ferromagnetic six-state clock model on a triangular lattice. We also report on an orientational ordering transition in a two-dimensional experimental system [64] that, as we argue, is in the same universality class as the model we study numerically. The experimental system actually motivates (besides numerical simplicity) our restriction to the six states of the spins rather than the continuum XY spins (and thus to a six-fold clock (C_6) symmetry rather than the $U(1)$ symmetry). In the unfrustrated (ferromagnetic) case it has been shown, however, that the KT-behavior is stable with respect to a crystal field of six-fold symmetry [40, 65, 66]. Hence we expect that our model displays the same critical behavior as the FFXY model.

3.1 Symmetries and order parameter

We investigate the phase transitions of the anti-ferromagnetic six-state clock model, defined by Eq. (9), on a two-dimensional $N = L_x \times L_y$ triangular lattice (see Fig. 7). The antiferromagnetic interaction on a triangular lattice induces a frustration. As a result the spins on each triangles should make an angle of $\pm 120^\circ$ with one another in the ground state. There are 12-fold degenerate ground states with three-sublattice structure. According to their chirality, we can categorize the ground states into $\mathcal{C}_n$ and $\mathcal{A}_n$ ($n = 0, 1, \ldots, 5$). The spin configurations in each ground state are given by

$$\theta_{i\in A} = \frac{2\pi n}{6}, \quad \theta_{i\in B} = \frac{2\pi(n\pm 2)}{6}, \quad \theta_{i\in C} = \frac{2\pi(n\pm 4)}{6}, \tag{10}$$

for $\mathcal{C}_n$ (upper sign) and $\mathcal{A}_n$ (lower sign). In other words, in the $\mathcal{C}(\mathcal{A})$ type ground state, the spin angles increase by $\frac{2\pi}{3}$ along the up-triangles in the (anti-) clockwise direction, and *vice versa* for down-triangles.

There are two different symmetries between them; those in the same class of $\mathcal{C}$ or $\mathcal{A}$ are related by the rotation $\theta_i \to \theta_i + \frac{2\pi m}{6}$ $(m = 0, \ldots, 5)$ and those in the others by the reflection $\theta_i \to -\theta_i$. Therefore the model possesses the C_6 (six-fold clock) symmetry and the Z_2 (Ising) symmetry. Note that the FFXY model on a triangular lattice has a $U(1)$ and a Z_2 symmetry [41]. Since the clock spin can have only six states, while the XY spin is a continuous one, the continuous $U(1)$ symmetry of the FFXY model is reduced to the discrete C_6 symmetry. The nature of the Z_2 symmetry in each model is identical.

From the symmetry consideration, we expect that there are two types of phase transitions associated with the spontaneous breaking of the C_6 symmetry and the Z_2 symmetry. The chirality at each elementary triangle is defined as

$$h_{\triangle_i,\nabla_i} = \frac{2}{3\sqrt{3}}\left[\sin(\theta_j - \theta_i) + \sin(\theta_k - \theta_j) + \sin(\theta_i - \theta_k)\right],$$

where $\triangle_i$ and ∇_i are up and down-triangles as depicted in Fig. 7. Then the ground state $\mathcal{C}_n$ ($\mathcal{A}_n$) has a checker-board pattern of the chirality with $h_\triangle = +1$ (-1) and $h_\nabla = -1$ $(+1)$. The staggered chirality

$$h = \frac{1}{2N}\sum_i \left(h_{\triangle_i} - h_{\nabla_i}\right), \tag{11}$$

plays a role of the order parameter for the Z_2 symmetry breaking transition. The order parameter for the C_6 symmetry breaking transition is the sublattice magnetization

$$m_A = \frac{1}{N}\sum_{i\in A}\exp(i\theta_i), \tag{12}$$

where the sum runs only over the sites in the A sublattice. m_B and m_C are defined analogously.

3.2 Transition temperatures

We performed Monte Carlo (MC) simulations on finite $N = L \times (L/2)$ lattices with a sublattice updating scheme; one of the three sublattices is selected randomly and then all spins in the chosen sublattice are flipped according to the Metropolis rule. One Monte Carlo step corresponds to three sublattice updates. Various observables are measured during the MC runs, such as the energy, the chirality, and the sublattice magnetization, from which we can measure the averaged quantities and their fluctuations. In some cases, histograms are

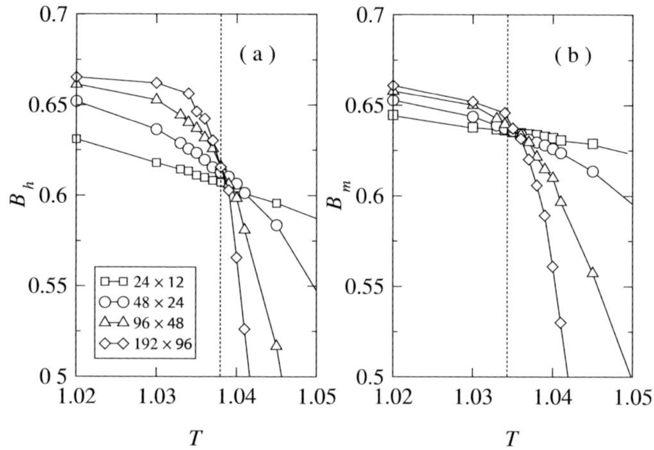

Fig. 8. Binder parameters for the staggered chirality in (a) and the magnetization in (b). The estimated transition temperatures are marked by broken lines.

constructed from particularly long MC runs to obtain the observables as continuous functions of the temperature.

The transition temperature is determined from the Binder parameter

$$B_h = 1 - \frac{\langle h^4 \rangle}{3\langle h^2 \rangle^2} \quad \text{and} \quad B_m = 1 - \frac{\langle m_A^4 \rangle}{3\langle m_A^2 \rangle^2} \tag{13}$$

for the chirality and the magnetization, respectively. Here the angle bracket denotes a thermal average, which can be done by a time average over the MC runs. The Binder parameter is scale-independent at the critical points and approaches $\frac{2}{3}$ (0) in the ordered (disordered) phase as L becomes larger. Hence the order–disorder transition temperature T_{I} related to the Z_2 symmetry breaking is obtained from the crossing point in the plot of B_h *versus* T at different system sizes. From Fig. 8a we estimate that

$$T_{\mathrm{I}} = 1.038 \pm 0.0005 \,. \tag{14}$$

B_m also displays the crossing behavior (Fig. 8b), from which we estimate that

$$T_{\mathrm{KT}} = 1.035 \pm 0.0005 \,, \tag{15}$$

for the C_6 symmetry breaking transition.

The Binder parameters show different size dependence at low temperatures. For the staggered chirality B_h converges to 2/3 rapidly as L increases. However, B_m appears to converge to values less than 2/3 at $T < T_{\mathrm{KT}}$. It indicates that there is a quasi-long-range order in spins at $T < T_{\mathrm{KT}}$. We confirm it from

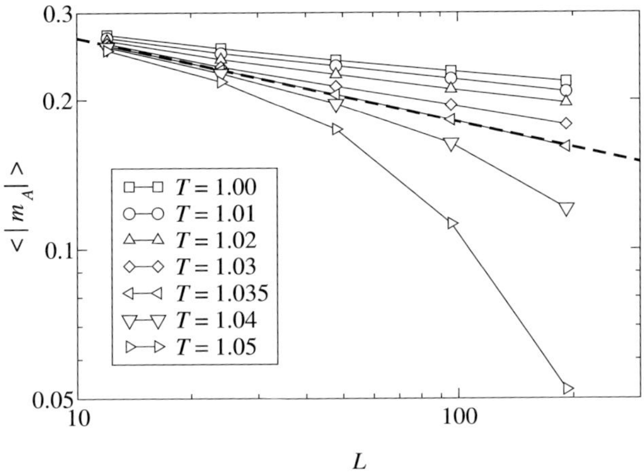

Fig. 9. Sublattice magnetization $\langle |m_A| \rangle$ near the critical temperature T_{KT}. At temperatures $T \leq T_{\mathrm{KT}}$ it shows a power law behavior. The broken line has a slope -0.17.

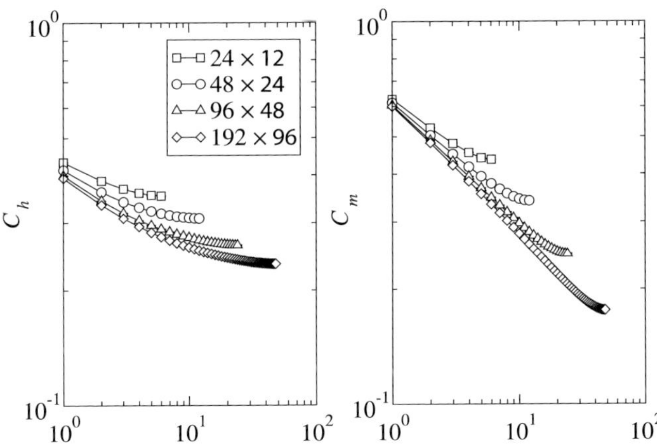

Fig. 10. Correlation functions of the staggered chirality, C_h, and the magnetization, C_m, at temperature $T = 1.037$.

finite-size-scaling (FSS) behaviors of the magnetization. The sublattice magnetization shows the power law scaling behavior, $\langle |m_A| \rangle \sim L^{-x}$, at $T \leq T_{\mathrm{KT}}$ with temperature dependent exponent x (see Fig. 9).

The two transition temperatures lie very close to each other. Nevertheless the accuracy of the data is sufficiently high that we can exclude the possibility that $T_{\mathrm{I}} = T_{\mathrm{KT}}$. Figure 10 demonstrates this *via* the correlation functions. We measure the correlations between the chirality of up-triangles and the magnetization of spins in the A sublattice displaced by a distance r in the vertical direction:

$$C_h(r) = \left\langle \frac{3}{N} \sum_{i \in A} h_{\triangle_i} h_{\triangle_{i+r}} \right\rangle , \tag{16}$$

$$C_m(r) = \left\langle \frac{3}{N} \sum_{i \in A} \cos(\theta_i - \theta_{i+r}) \right\rangle . \tag{17}$$

Here, $i+r$ denotes a site displaced vertically by a distance r from i, and $\triangle_i$ denotes an up-triangle whose left corner is i (see Fig. 7). In Fig. 10, we plot both correlation functions in the log–log scale at an intermediate temperature $T = 1.037$. Clearly one can see an upward curvature in the plot of $C_h(r)$, which implies that the chirality order has already set in. On the other hand, there is a downward curvature in the plot of $C_m(r)$ indicating that the magnetic order has not set in yet. Therefore we conclude that $T_{\mathrm{KT}} < T_{\mathrm{I}}$.

3.3 Universality class

With $T_{\mathrm{KT}} \neq T_{\mathrm{I}}$, one expects that both symmetry breakings take place independently. Then, the Z_2 symmetry breaking transition should be in the Ising universality class. The C_6 symmetry is equivalent to that of the XY model perturbed by the $(p = 6)$-fold anisotropy field. The unperturbed XY model displays a KT transition which separates a disordered high-temperature phase and a quasi-long-range ordered low-temperature phase. A renormalization group (RG) study shows that an anisotropy field with $p > 4$ does not change the KT nature of the transition [40]. The same is true for the extreme case of the ferromagnetic six-state clock model [66]. So the C_6 symmetry breaking phase transition at $T = T_{\mathrm{KT}}$ is expected to belong to the KT universality class.

The interplay of the Ising type and the KT type ordering has long been studied in the context of the fully frustrated XY models on a square or a triangular lattice. Frequently, it has been claimed [44–49] that the Ising-like transition and the KT-type transition occur at the same temperature. On the other hand, more recent studies have reported the transitions to occur at two different temperatures [52–58, 60, 61, 63]. Surprisingly, some of recent high-precision Monte Carlo simulation studies suggested that the Z_2 symmetry breaking transition does not belong to the Ising universality class in spite of the double transition [54, 57, 63]. They estimated that $\nu \simeq 0.8$ for the correlation length exponent and $\alpha/\nu \simeq 0.46$ with the specific exponent α, which are incompatible with the Ising values of $\nu = 1$ and $\alpha = 0$. The critical exponents are consistent with the values observed along the single transition line of the coupled XY-Ising model [46, 50, 51]. Olsson [60] argued that the measured non-Ising critical exponents are artifacts of unusual finite size effects originated from non-critical spin-wave fluctuations. However, apparently there is no general agreement yet on this subject [61].

In our model, we observe the phase transitions occurring at two *different* temperatures. However, the controversy existing in the FFXY model tells us that it does not guarantee necessarily that each transition will belong to the KT universality class and the Ising universality class, respectively. Therefore we perform a thorough finite-size-scaling (FSS) analysis to understand the nature of the phase transition. We believe that the antiferromagnetic six-state clock model can be used to resolve the existing controversy for the FFXY model. It has the proper symmetry properties as discussed previously. And, from a practical point of view, larger system sizes are available since it is a discrete model, with which one can reduce the finite size effects.

3.4 Z_2 symmetry breaking

We investigate the nature of the Z_2 symmetry breaking transition by exploring the FSS property of the specific heat, defined by

$$c = N\left[\langle e^2\rangle - \langle e\rangle^2\right], \tag{18}$$

with e the energy per site. For a finite system ($N = L \times L/2$) the specific heat has a peak $c_* \sim L^{\alpha/\nu}$ near the critical temperature, diverging with system size $L \to \infty$. Since the specific heat does not diverge at a KT type transition, the divergence is due to the Z_2 symmetry breaking phase transition. The Ising universality class has $\alpha = 0$, and the specific heat shows a logarithmic divergence, $c_* \sim \ln L$.

From a detailed analysis [67] of the specific heat and the chirality correlation length $\xi_h = \sqrt{\sum_{\mathrm{r}} r^2 C_h(r)/\sum_{\mathrm{r}} C_h(r)}$, with $C_h(r)$ defined in Eq. (16), we conclude that the Z_2 symmetry breaking transition belongs to the Ising universality class. Our results for the *effective* specific heat exponent $\alpha/\nu \simeq 0.5$ and the correlation length exponent $\nu \simeq 0.8$ appearing at small length scale are comparable with the results of previous MC works claiming the non-Ising nature in the frustrated XY models [46, 47, 50, 51, 54, 57, 62, 63]. Our results are fully consistent with Olsson's argument claiming that the non-Ising exponents reported by others are explained by a failure of finite-size-scaling at small length scale due to the screening length associated with the nearby KT transition [60]. The strong finite size effect is overcome in this work since we could study large systems. We expect that one could observe the same crossover to the asymptotic Ising type scaling behavior in larger scale simulations in the frustrated XY systems.

3.5 C_6 symmetry breaking

The KT transition is characterized by the essential singularity of the correlation length and the susceptibility approaching the critical temperature. We measure

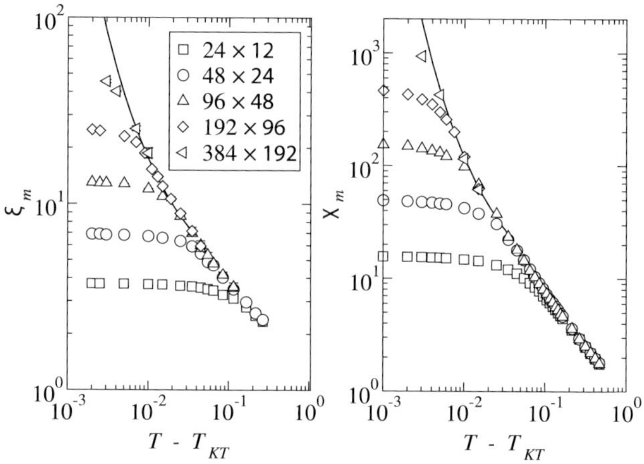

Fig. 11. Correlation length and the susceptibility of the magnetization. The solid lines are the fits to the KT form of $\xi_m = 2.54e^{0.193(T-T_{\mathrm{KT}})^{-1/2}}$ and $\chi_m = 4.20e^{0.333(T-T_{\mathrm{KT}})^{-1/2}}$.

the correlation length ξ_m as

$$\xi_m = \sqrt{\frac{\sum_r r^2 C_m(r)}{\sum_r C_m(r)}}, \tag{19}$$

from the magnetic correlation function C_m in Eq. (17). One can also estimate the correlation length by fitting $C_m(r)$ to a form with e^{-r/ξ_m}. We tried with various function forms but found that Eq. (19) yields the most stable estimate against statistical errors. The magnetic susceptibility is measured from the fluctuation of the magnetization, $\chi_m = N\langle m_A^2\rangle$. Figure 11 shows the plot of ξ_m and χ_m *versus* $T - T_{\mathrm{KT}}$ in log–log scale. Upward curvature in both plots indicates the stronger divergence than the algebraic divergence. We fit those quantities with the KT scaling form, $\xi_m = a\, e^{b(T-T_{\mathrm{KT}})^{-y}}$ and $\chi_m = a' e^{b'(T-T_{\mathrm{KT}})^{-y}}$ with $y = \frac{1}{2}$ fixed, whose results are drawn with solid lines in Fig. 11. It shows that the KT scaling behaviors are emerging only after $L \geq 96$. If y is also taken as a fitting parameter, one obtains $y < 0.5$ whose value depends on the fitting range; it increases as approaching the critical temperature.

In the XY phase ($T < T_{\mathrm{KT}}$) the spins have the quasi-long range order; the magnetization scales algebraically as $\langle |m_A| \rangle \sim L^{-x}$ with temperature dependent exponent x (see Fig. 9). The RG theory predicts that $x = 1/8$ at the KT transition point [40]. Our numerical data show that $x \simeq 0.17$ at $T = T_{\mathrm{KT}}$. We think that the discrepancy originates from a sensitive dependence of x on T. In summary, in spite of a quantitative disagreement, the C_6 symmetry breaking phase transition is in qualitative agreement with the KT universality class; we observe the essential singularity in the correlation length and the susceptibility and quasi-long range order in the low temperature phase.

3.6 Experimental realization: CF_3Br on graphite

In this section we describe a possible experimental realization of the theoretical model we investigated above. In a recent work we reported results on X-ray powder diffraction study on a monolayer of halomethane CF_3Br adsorbed on exfoliated graphite [64]. CF_3Br is a prolate molecule and has a dipole moment of about 0.5 D. The coverage ρ, temperature T phase diagram is rather complex [68, 69]. In [64] we concentrated on a coverage which is representative of the extended monolayer regime in which the monolayer lattice is commensurate with the graphite lattice. This yields a 2×2 triangular lattice arrangement of the CF_3Br molecules below a temperature of 105 K [70], which is the melting temperature of the commensurate layer. The inter-molecular distance is $a = 4.92$ Å. Note that the lateral size of the graphite crystallites is only around 180 Å, which confines any spatial correlation length to this value.

An isolated CF_3Br would prefer to lie flat on the substrate, but the 2×2 mesh is too tight to accommodate the molecules in this orientation. Therefore the individual CF_3Br molecules stand on the substrate, presumably with the F_3 tripod down, with a maximum tilt angles of the molecular axis up to 30° with respect to the substrate normal due to steric repulsion. A tilt leads to a non-zero in-plane component of the dipole moment. We regard this component as planar pseudospin $\mathbf{S}_i = (\cos\theta_i, \sin\theta_i)$ with the azimuthal angle θ_i of the molecule. In this sense the 2×2 state is disordered with a zero time average of every $\mathbf{S}_i$, and is stabilized at higher temperatures by a libration and/or a precession of the molecular axis about the substrate normal.

As the temperature is decreased additional features develop in the diffraction pattern which finally, below 40 K, can be identified [64] as Bragg peaks (with a finite width of around $(180\,\text{Å})^{-1}$ due to the lateral size of the crystallites) indicating an orientational order in the dipole moments identical to the one depicted in Fig. 7. The temperature dependence of the correlation length ξ, determined from the intrinsic width of this peak, can be fitted with the KT-expression

$$\xi = A\exp(B(T/T_{\mathrm{KT}} - 1)^{-1/2}) \qquad (20)$$

to the data for $T > 40$ K (see Fig. 3 of Ref. [64]). The fit parameters are $A = 9 \pm 2$ Å, $B = 1.5 \pm 0.4$, $T_{\mathrm{KT}} = 30 \pm 3$ K. Note that the value of A is reasonably close to the lattice parameter of the 2d mesh. Thus ξ is expected to diverge at a KT-critical temperature T_{KT} of about 30 K, but the growth of the correlated regions is interrupted when ξ reaches the size of the graphite crystallites. This happens at about 40 K.

We think that the model of Eq. (9) is a good description of the orientational ordering process described in this physical system: Clearly the pseudospin correlations are bound to a plane, thus the system is 2d with respect to the relevant degrees of freedom at the phase transition. Moreover, as mentioned before, below 105 K the CF_3Br molecules are arranged in a triangular lat-

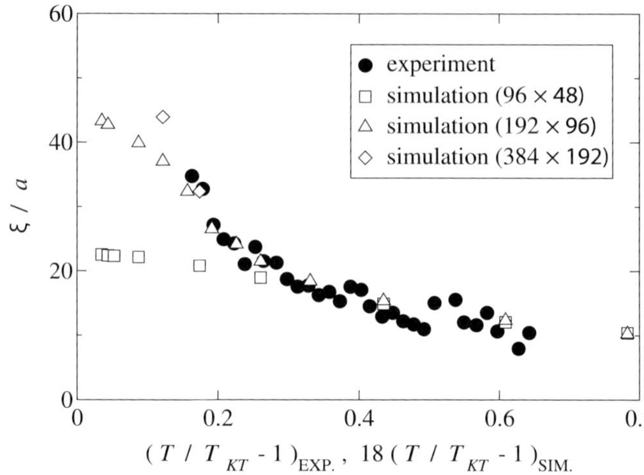

Fig. 12. Comparison of the magnetic correlation length (scaled by the lattice constant) from the experimental data (from Fig. 3 of Ref. [64]) and the Monte Carlo simulations (from Fig. 11). The reduced temperature $(T/T_{\mathrm{KT}} - 1)$ for the simulation data are rescaled by a factor of 18 in order to achieve an acceptable data collapse for linear lattice sizes L between 96 and 192.

tice. The pseudospin representing the CF_3Br dipole moment is presumably not a strictly isotropic planar rotator but experiences a crystal field from the graphite substrate which breaks the continuous azimuthal symmetry into the six fold symmetry of the monolayer. This is reflected by the six-state clock variables of model (9). The ordered structure of the CF_3Br dipoles is antiferroelectric, which is taken into account by the antiferromagnetic couplings between the pseudospins in Eq. (9). Finally, the character of the relevant orientation dependent interactions of CF_3Br is not known, but a comparison of monolayers of several polar methane derivatives [69] shows that fully halogenated molecules including CF_3Br with small dipole moments around 0.5 D have structures different from partially halogenated molecules such as CH_3Cl with strong dipole moments around 1.7 D. This suggests that for CF_3Br the short range anisotropic part of the intermolecular van der Waals force and hard-core repulsion are more important than the medium range dipole–dipole interaction. Thus the interactions can be assumed to be short ranged. Thus one expects that model (9) and the physical system discussed here are in the same universality class.

In Fig. 12 a comparison of the magnetic correlation length from the experiment [64] and the simulation (from Fig. 11) is shown. Since the factor B in the KT-form (20) of the correlation length is a non-universal number the reduced temperature $(T/T_{\mathrm{KT}} - 1)$ has to be rescaled by an appropriate factor in order to achieve an acceptable data collapse. The rescaling factor turns out to be quite large, namely 18, which is not unusual for microscopically different systems in the KT universality class (see *e.g.*, [71]). Note that the finite linear size of the crystallites plays a similar role as the finite lattice sizes in the simulations and sets the saturation value for the correlation length (divided by the lattice constant for the triangular lattice of the CF_3Br molecules, which is $a = 4.92$ Å). The nice collapse of the experiment and simulation data supports our claim that the two physical systems are in the same universality class and that the orientational ordering of CF_3Br molecules on graphite is in the KT universality class.

3.7 Summary

To summarize we have studied the phase transitions in the anti-ferromagnetic six-state clock model on a triangular lattice, which is fully frustrated. As a result the ground states have a C_6 (six-state clock) symmetry and a Z_2 (Ising) symmetry. Through extensive Monte Carlo simulations we found that the model undergoes a Kosterlitz–Thouless transition at T_{KT} and an Ising transition at T_{I}. The two transitions correspond to the C_6 and the Z_2 symmetry breaking transition, respectively. High-precision Monte Carlo data indicate that the two transitions take place at different temperatures, $T_{\mathrm{KT}} < T_{\mathrm{I}}$ (Eqs. (14) and (15)). This has been checked explicitly by analyzing the behavior of the spin and the chirality correlation function at temperatures between T_{KT} and T_{I} (Fig. 10). Furthermore, we have shown that the Z_2 symmetry breaking transition belongs to the Ising universality class. For small system sizes, the scaling property of the specific heat and the correlation length deviates apparently from the Ising universality class. However, simulation results for larger system sizes indicate that the model belongs asymptotically to the Ising universality class. As for the transition at T_{KT}, we have found that the magnetization correlation length and the susceptibility diverges at $T = T_{\mathrm{KT}}$ according to the KT scaling form (Fig. 11). We have also found that the spins have a quasi long range order below T_{KT} (Fig. 9). Combining these, we conclude that the transition at T_{KT} belongs to the KT universality class.

Our model is a variant of the fully frustrated *XY* models where the KT type ordering and the Ising type ordering interplay interestingly. Our numerical results support a scenario that there are two separate phase transitions with $T_{\mathrm{KT}} \neq T_{\mathrm{I}}$; one at T_{KT} in the KT universality class and the other at T_{I} in the Ising universality class. Our results are consistent with those in Ref. [60] very well.

We have proposed that our theoretical model describes the orientational ordering transition of CF_3Br molecules on graphite since the model has the same symmetry as the experimental system. We argue that the orientational ordering transition belongs to the KT universality class [64] by comparing the magnetic correlation length obtained from the experiment [64] with the correlation length obtained numerically in the six-state clock model. With a suitable rescaling of parameters, we have shown that the correlation lengths in both systems have the same scaling behavior (Fig. 12). It gives more evidence that the orientational ordering transition is indeed the KT transition.

4. Growing length scales during aging in 2d disordered systems

The non-equilibrium dynamics of disordered, in particular of glassy systems has become a very rich field in recent years and despite many efforts the understanding of non-equilibrium dynamics of disordered and glassy systems in finite dimensions remains a challenging problem. In particular in glasses and spin glasses the aging process displays a very rich phenomenology demanding new theoretical concepts [72]. But already less complex – and apparently less glassy – systems, like disordered but non-frustrated systems [73] or even pure systems [74] reveal interesting and unexpected aging phenomena. One of the most intriguing questions in this context is whether the out-of-equilibrium dynamics is essentially fully determined by a coarsening process (a question that even arises in the more complex spin glass situation [75]), describable by a growing length scale that characterizes essentially all out-of-equilibrium processes. In this paper we will consider three paradigmatic models for two-dimensional systems with quenched disorder with a focus on existence and analysis of a growing length scale during aging at low temperatures: the random bond Ising ferromagnet, the Edwards Anderson model for a spin glass, and the solid-on-solid model on a disordered substrate which is equivalent to the sine-Gordon model with random phase shifts.

4.1 The random bond Ising ferromagnet

As the first example for two-dimensional disordered system we consider the random bond Ising ferromagnet. It is defined by the Hamiltonian

$$H = -\sum_{(ij)} J_{ij} S_i S_j, \qquad S_i = \pm 1\,, \tag{21}$$

where the couplings J_{ij} are non-negative quenched random variables of variance ε and the sum is over all nearest neighbor pairs (ij) on a square lattice of size $L \times L$ with periodic boundary conditions. This paradigmatic model for a disordered magnetic system (with bond- or temperature randomness) with an Ising symmetry has a second order phase transition from a paramagnetic to a ferromagnetic phase at a critical temperature $T_c(\epsilon)$ that decreases with increasing disorder strength ϵ. For temperatures T below T_c the magnetization $\overline{\langle m_i \rangle_T}$, where $\langle \cdots \rangle_T$ means the thermal average and $\overline{\cdots}$ the average over the disorder, takes on a non-vanishing value. Temperature is measured in the same unit as energy (divided by k_B).

Non-equilibrium dynamics at temperatures below T_c arises for instance *via* an instantaneous quench of the systems from the paramagnetic phase to

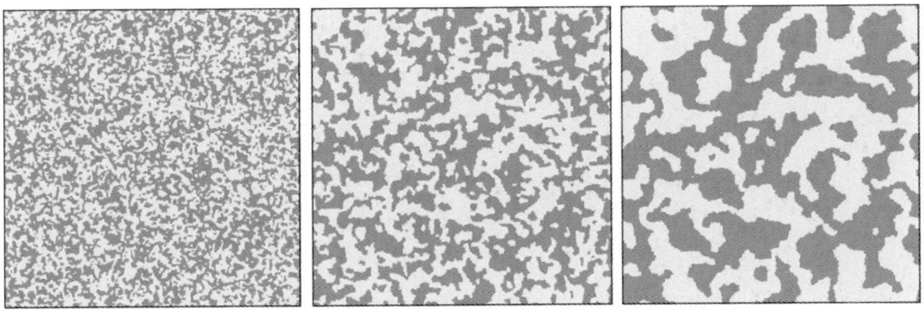

Fig. 13. Domain growth in the RBIM with Glauber kinetics. We show evolution pictures at $t = 10^2$, 10^4 and 10^6 MCS for a 512×512 lattice, after a quench from $T = \infty$ to $T = 0.5$ and J_{ij} uniformly distributed between 0 and 2 ($J_{ij} \in [0, 2]$). The up spins are marked in black, and the down spins are marked in grey.

a temperature below T_c. A stochastic process defined by single spin–flip transition rates defined for instance by the Metropolis rules $w(S_i \to -S_i) = 1/(1 + \exp(-\beta(H(S_i) - H(-S_i))))$ models a non-conserved order parameter dynamics and can be studied by computer simulations. For a quench below T_c the dynamics is a coarsening process during which ferromagnetic domains of a typical lateral extension $R(t)$ form, where t is the time elapsed after the quench (see Fig. 13). A standard way to extract this time dependent length scale is *via* the spatial two-point correlation function $C(r, t) = \overline{\langle m_i(t) m_{i+r}(t)\rangle_T}$, which is expected to scale like $C(r, t) = \tilde{c}(r/R(t))$.

An important study of the non-conserved RBIM is due to Huse and Henley (HH) [76]. HH argued that coarsening domains are trapped by energy barriers $E_B(R) \simeq E_0 R^{\psi}$, with exponent $\psi = \chi/(2-\zeta)$, where χ and ζ are the pinning and roughening exponents. For $d = 2$, these exponents are known to be $\chi = 1/3$ and $\zeta = 2/3$ [77], yielding $\psi = 1/4$. As a consequence of the HH scenario one expects the following scaling scenario for the length scale $R(t)$:

$$R(t)/R_0 = h(t/t_0)\,, \quad \text{with} \quad h(x) \sim \begin{cases} x^{1/2} & \text{for } x \ll 1 \\ (\ln x)^4 & \text{for } x \gg 1 \end{cases}, \tag{22}$$

where $R_0 \sim T^4$ and $t_0 \sim T^8$. Instead we find (*via* an extensive Monte Carlo study, see [78]) that $R(t)$ grows algebraically with a temperature and disorder strength dependent exponent $1/z(T, \epsilon)$:

$$R(t) \sim t^{1/z(T,\epsilon)} \quad \text{for} \quad t \gg t_0 \quad \text{with} \quad z(T, \epsilon) = 2 + \epsilon/T\,, \tag{23}$$

where the time t_0 does not depend on T and ϵ (see Fig. 14). This algebraic growth law with a temperature and disorder strength dependent growth exponent θ indicates a logarithmic barrier scaling form $E_B(R) \sim \epsilon \ln(1 + R)$ in contrast to the algebraic form $E_B \propto R^{\psi}$ assumed in the HH picture.

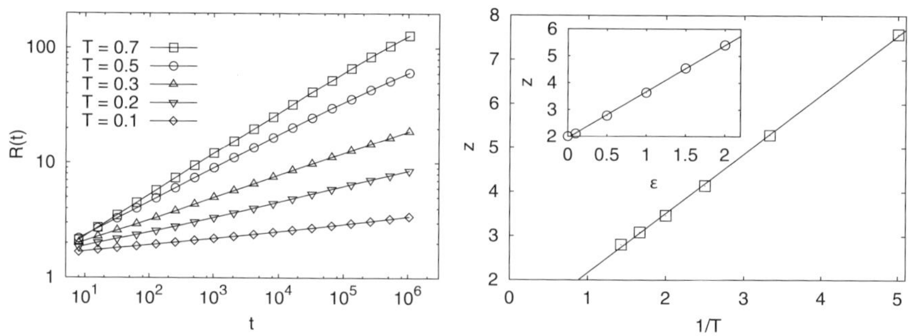

Fig. 14. *Left:* Log–log plot of the correlation length R *vs.* time t in the 2d random bond Ising model for different temperatures and $J_{ij} \in [0, 2]$. *Right*: Estimates of the exponent z *vs.* $1/T$ for the data shown left (note that for $T > 1 \approx T_c$, *i.e.* for $1/T < 1$ the system is in the paramagnetic phase). Inset shows z *vs.* the disorder strength ϵ for fixed temperature $T = 0.5$ (the distribution $P(J)$ is chosen to be uniform over $[1 - \epsilon/2, 1 + \epsilon/2]$. The straight lines (in the main figure and in the inset) represent the analytical prediction $z = 2+\epsilon/T$.

4.2 2d EA spin glass

Here we consider the two-dimensional Ising spin glass with nearest-neighbor interactions distributed according to a Gaussian with zero mean and variance one

$$H = -\sum_{\langle ij \rangle} J_{ij} S_i S_j \,, \quad \text{with} \quad P(J_{ij}) = \frac{1}{\sqrt{2\pi}} \exp\left(-\frac{J_{ij}^2}{2}\right) . \tag{24}$$

This model is in a paramagnetic state for all temperature $T > 0$ but displays a very slow dynamics at low temperatures which can be observed for instance in the non-equilibrium dynamics occurring after a quenched from high temperatures. It turns out [79] that this aging process can be characterized by a coarsening process up to a maximum domain size given by the equilibrium correlation length $\xi_{\rm eq}$.

It is possible to calculate exactly the ground state (GS) of this system using for instance a minimal weight perfect matching algorithm [80]. Denoting the GS for a particular disorder realization with $\{S_i^0\}$ we define the local overlap with it as $q_i^{\rm gs}(t) = S_i(t) S_i^0$. For a ferromagnetic system (*i.e.* $J_{ij} = J > 0$) the GS obviously $S_i^0 = 1$ and therefore q_i corresponds to the (time dependent) local magnetization. In Fig. 15 snapshots of the time evolution of the local GS overlap are depicted, showing an increasing average domain size. In contrast to the time evolution of a random bond ferromagnet shown in Fig. 13 even for very large waiting times very small domains exist. These are either very stable clusters because strong bonds have to be broken to flip the spins or new domains within the bigger ones appear since less strongly bound spins initialize the formation of a new domain.

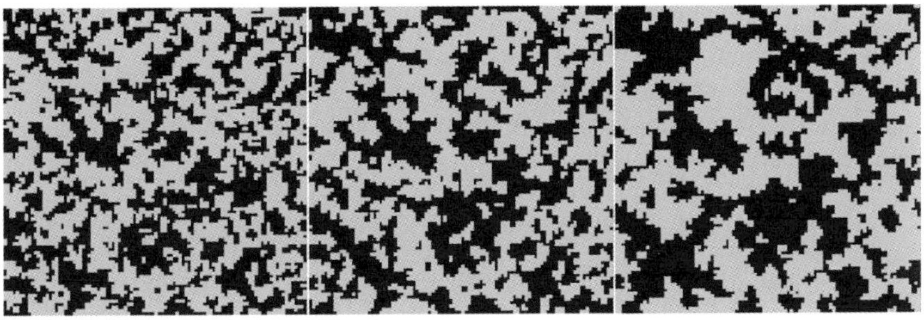

Fig. 15. Domain growth in the 2d EA-model with Gaussian couplings for $T = 0.3$. The system size is $L = 100$, *i.e.* much smaller than in Fig. 13. The snapshots show the domains relative to the ground state after $t = 10^2$, 10^4 and 10^6 Monte Carlo sweeps.

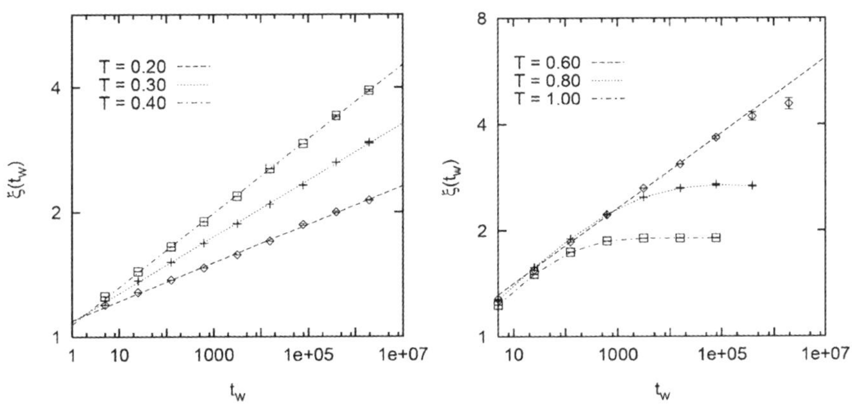

Fig. 16. Correlation length in the 2d EA SG model as a function of time t_w for different temperatures. *Left*: Fits to an algebraic growth law $\xi(t_w) \sim t^{1/z(T)}$ with $1/z(T = 0.2) = 0.046$, $1/z(T = 0.3) = 0.068$ and $1/z(T = 0.4) = 0.090$. *Right*: For sufficiently large temperatures the time dependent correlation length $\xi(t)$ saturates at the equilibrium correlation length $\xi_{eq}(T)$ within the accessible time window.

The spatial correlation function

$$G(r, t) = \overline{\langle q_i^{gs}(t) q_{i+r}^{gs}(t) \rangle} \tag{25}$$

allows for a quantitative analysis of the domain size evolution. It turns out that it scales like $G(r, t) = g(r/\xi(t))$ and we can obtain an estimate for the correlation length (or typical domain size) *via* an integral of $G(r, t)$ over r. The result is shown in Fig. 16. Note that for increasing temperatures a) the domain growth speeds up, b) the equilibrium correlation length gets smaller. As a consequence of both tendencies one can observe the saturation of the time dependent correlation length at the finite equilibrium correlation length for higher temperatures on the right panel.

We observe that the data for $\xi(t)$ (in the non-equilibrium regime $\xi(t) \ll \xi_{eq}(T)$) can very well be fitted by an algebraic growth law with a temperature dependent exponent $z(T)$:

$$\xi(t) \propto t^{1/z(T)} \quad \text{with} \quad z(T) \approx (0.23T)^{-1} , \tag{26}$$

which displays again the $1/T$ behavior that we have encountered already in the last section for the random bond ferromagnet, indicating also here the presence of logarithmic barriers.

4.3 2d SOS model on a disordered substrate

Here, we investigate the non equilibrium relaxational dynamics of a solid-on-solid (SOS) model on a disordered substrate, defined on a two dimensional square lattice and described by the following elastic Hamiltonian in terms of height variables h_i

$$H_{\text{SOS}} = \sum_{\langle ij \rangle} (h_i - h_j)^2 , \quad h_i \equiv n_i + d_i , \tag{27}$$

where n_i are unbounded discrete variables, *i.e.* $n_i \in \{0, \pm 1, \pm 2, \dots\}$ and $d_i \in [0, 1[$ are uniformly distributed quenched random offsets, uncorrelated from site to site. In the absence of disorder, *i.e.* $d_i = 0$, the model exhibits a roughening transition in the same universality class as the Kosterlitz–Thouless transition [81], at a temperature T_r separating a flat phase at low T from a logarithmically (thermally) rough one above T_r. The presence of disorder is known to modify significantly the nature of the transition [82]. The so-called *super-roughening* transition occurs at a temperature $T_g = T_r/2 = 2/\pi$. Above T_g, where the disorder is irrelevant on large length scales, the surface is logarithmically rough again, although below T_g the system exhibits a glassy phase where the pinning disorder induces a stronger roughness of the interface.

The spatial (2-point) connected correlation function is defined as

$$C(r, t) = \frac{1}{L^2} \sum_i \overline{\langle h_i(t) h_{i+r}(t) \rangle - \langle h_i(t) \rangle \langle h_{i+r}(t) \rangle} , \tag{28}$$

which scales as

$$C(r, t) = \mathcal{F}\,(r/\mathcal{L}(t)) \quad \text{with} \quad \mathcal{L}(t) \sim t^{1/z} . \tag{29}$$

Therefore one can estimate $\mathcal{L}(t)$ by integrating $C(r, t)$ over r. In Fig. 17 we show the value of $\mathcal{L}(t)$ computed in this way for different temperatures. One obtains a rather good fit by a power law $\mathcal{L}(t) \sim t^{1/z(T)}$, thus obtaining a value

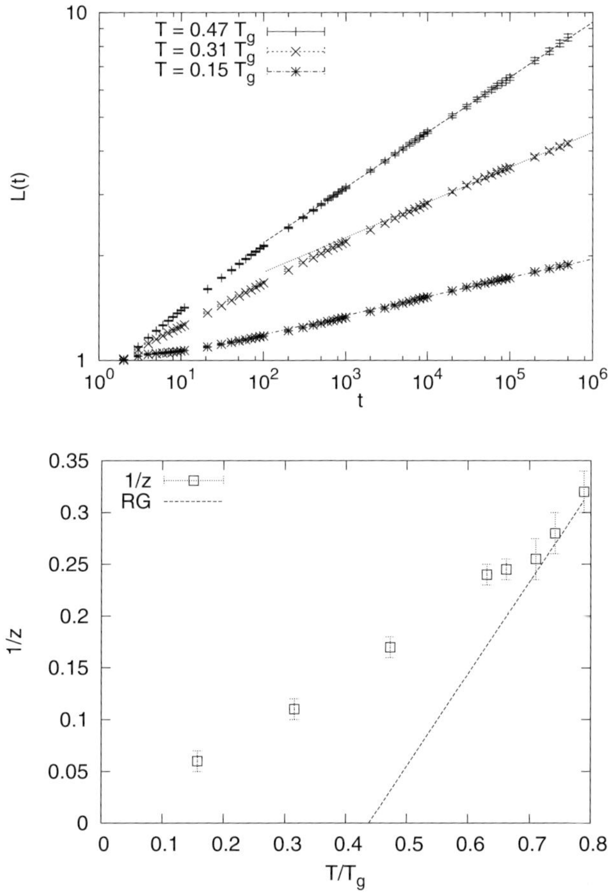

Fig. 17. *Left*: Growing length scale $\mathcal{L}(t)$ for different temperatures. The solid lines are guides to the eyes. *Right*: $1/z(T)$ as a function of T/T_g. The dashed line which shows the result of the one loop RG is drawn without any fitting parameter.

of the T dependent dynamical exponent. One notices also that $\mathcal{L}(t)$ approaches an algebraic growth after a pre-asymptotic regime which increases with decreasing temperature. Figure 17 shows our estimate for $1/z(T)$ as a function of T (for details see [84]. As expected, the dynamical exponent is a decreasing function of the temperature. One expects that $z = 2$ for $T > T_g$ and that it becomes T-dependent below T_g with $z = 2 + 2e^{\gamma_E}\tau + \mathcal{O}(\tau^2)$ as predicted by a one loop RG calculation [83]. At high temperature $T > T_g$ and in the vicinity of T_g^-, it is numerically rather difficult to extract a reliable estimate for the dynamical exponent due to finite size effects. Therefore we restrict ourselves here to lower temperatures $T < 0.8\,T_g$. For temperature $T \gtrsim 0.7\,T_g$, the value of z is still in reasonable agreement with the RG prediction. Around the value

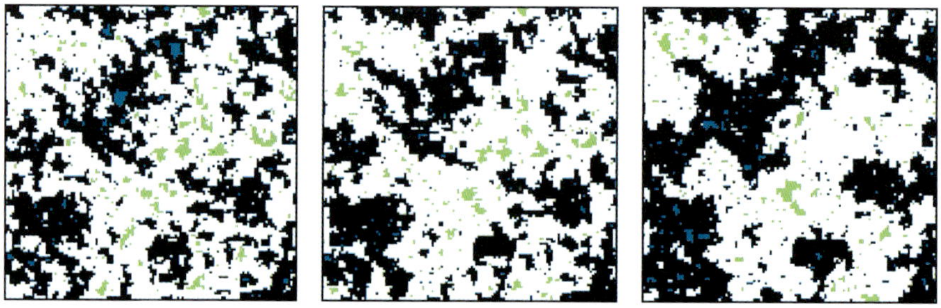

Fig. 18. Snapshot of the height field of the random SOS model relative to the ground state $m_i(t) = n_i(t) - n_i^0$ for $T = 0.47\,T_g$. The system site is $L = 128$. Different colors correspond to different values of $m_i(t)$: $m_i(t) = -2$ in green, $m_i(t) = -1$ in white, $m_i(t) = 0$ in black and $m_i(t) = +1$ in blue. Note that large domains in white and black persist and change only slowly in time.

$T^* \simeq 0.63\,T_g$, where $z \simeq 4$, the curve $1/z(T)$ shows an inflection point, below which $1/z$ decreases linearly with T. In this regime, $z(T)$ is well fitted by

$$z(T) \approx 4T^*/T \quad \text{for} \quad T \leq T^* \approx 0.63T_g\,. \tag{30}$$

The behavior $z \propto 1/T$ is compatible with an activated dynamics over logarithmic barriers, *i.e.* an Arrhenius type behavior $t_{\text{typ}} \sim e^{B_{L\text{typ}}/T}$ with $B_{L_{\text{typ}}} \sim \log L_{\text{typ}}$.

One may try to relate the length scale $\mathcal{L}(t)$ to the size of spatially correlated structures like domains or droplets. We first explored the idea that at low temperature, the nonequilibrium dynamics could be understood as a coarsening process reflected in a spatially growing correlation with the ground state (GS). Interestingly, computing the GS of the SOS model on a disordered substrate Eq. (27) is a minimum cost flow problem for which a polynomial algorithm exists and can therefore be computed exactly [85]. After determining one GS n_i^0 (note that the GS, which is computed with free boundary conditions, is infinitely degenerate since a global shift of all heights by an arbitrary integer is again a GS), we define for each time t the height difference $m_i(t) = n_i(t) - n_i(0)$ and identify the connected clusters (domains) of sites with identical $m_i(t)$ using a depth-first search algorithm. Notice that for the comparison with the ground state, the Monte Carlo simulations are performed here using free boundary conditions.

In Fig. 18 we show snapshots of these domains for $T < T_g$. Starting from a random initial configuration one can for $T < T_g$ very quickly ($t \lesssim 100$) identify large domains that evolve only very slowly at later times. On the other hand for $T > T_g$ the configurations decorrelated very quickly in time. To make this analysis more quantitative, we determined the cluster size distribution $P_{\text{th}}(S, t)$ for one realization of the disorder (and for different realizations of the thermal noise).

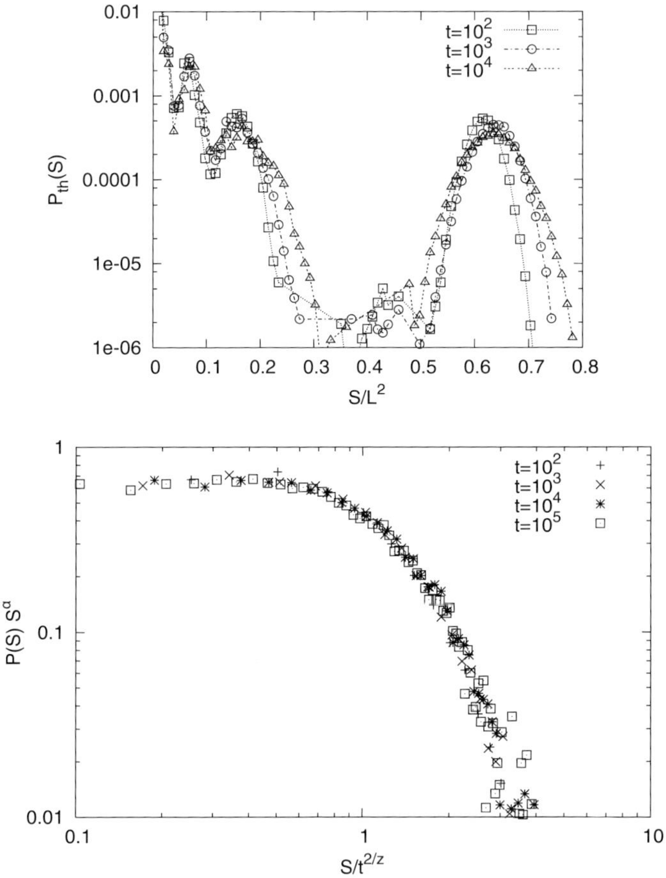

Fig. 19. *Left*: Size distribution $P_{\text{th}}(S, t)$ (see definition in the text) for different times t. Here $T = 0.47T_g$. *Right*: $S^\alpha P^{\text{GS}}_{\text{droplet}}(S, t)$ with $\alpha = 1.9 \pm 0.1$ as a function of $S/t^{2/z}$ with $2/z = 0.26 \pm 0.03$. Here the initial condition is the ground state and $T = 0.3\, T_g$.

As shown on Fig. 19, $P_{\text{th}}(S, t)$ starts to develop a peak at a rather large value $S^*(t)$ on the earlier stage of the dynamics (this peaks also develops if we start with a random initial configuration). It turns out that $S^*(t)$ is the size of the largest connected flat cluster of the ground state configuration $n_i^0 = C^{\text{st}}$. On the time scales presented here, as time t is growing, this peak remains stable $S^*(t) \simeq C^{\text{st}}$, implying that the system is *not* coarsening. At later times, as suggested by simulations on smaller systems, this peak progressively disappears and the distribution becomes very flat. We also checked that the mean size of these connected clusters is not directly related to $\mathcal{L}(t)$. One has however to keep in mind that we are computing the *connected* correlation functions, *i.e.* we measure the thermal fluctuations of the height profile around its mean (typical) value $\langle h_i(t) \rangle$. Therefore, we believe that these connected correlations are instead related to the broadening of this "stable" peak (Fig. 19), *i.e.* the fluctuations around this typical state at time t.

To characterize more precisely the fluctuations around this cluster, we identify "droplets" by initializing the system in the ground state itself $n_i(t=0) = n_i^0$. At low temperature, and on the time scales explored here, one expects that the ground state represents a good approximation of a typical configuration, *i.e.* $\langle n_i(t) \rangle \simeq n_i^0$. We compute the distribution $P^{\mathrm{GS}}_{\mathrm{droplet}}(S,t)$ of the sizes of the connected clusters with a common value of $m_i(t) \neq 0$. It turns out, as shown in Fig. 19, that $P^{\mathrm{GS}}_{\mathrm{droplet}}(S,t)$ obeys the scaling form

$$P^{\mathrm{GS}}_{\mathrm{droplet}}(S,t) = \frac{1}{S^{\alpha}} \mathcal{F}^{\mathrm{GS}}_{\mathrm{droplet}} \left(\frac{S}{\mathcal{L}^2(t)} \right), \qquad \alpha = 1.9 \pm 0.1\,, \tag{31}$$

where α is independent of T within the accuracy of our data and $\mathcal{L}(t) \sim t^{1/z}$. The value of z in (31) is in good agreement with the one extracted from the 2-point correlation function $C(r,t) = F(r/\mathcal{L}(t))$ Eq. (29).

4.4 Conclusion

We studied the time dependent correlation length $R(t)$ in three models of disordered systems in two dimensions that are characterized by distinct features: 1) The random bond Ising ferromagnet as an example for a random system that has long range order at low temperatures and is expected to perform a simple (but slow) coarsening process after a temperature quench into the ordered phase. 2) The Edwards–Anderson spin glass model as an example for a frustrated system without a finite temperature critical point and an ordered phase but with an extremely slow dynamics and a large correlation length at very low temperatures. 3) The disordered SOS model as a model with a critical point and a low temperature phase without long range order but infinite correlation length. Surprisingly, in spite of the pronounced differences between these systems we find that all three show an algebraic dependence of the correlation length on the age t of the system $R(t) \sim t^{1/z(T)}$ and that the exponent z (which would be identical to the dynamical exponent if the system is critical) depends linearly on the inverse temperature:

$$z(T) \propto 1/T\,. \tag{32}$$

IF the dynamics at low temperatures in all 3 systems is thermally activated this behavior hints at a logarithmic scaling behavior of the energy barriers as a function of their size. In the disordered SOS model one would actually expect such a scaling [84]. For the EA spin glass (in 2d) the situation is complicated by the fact that the ground state is not expected to be stable with respect to thermal fluctuations, *i.e.* in principle excitations of increasing size would cost less and less energy – therefore a logarithmic barrier scaling comes a bit as a surprise. Finally for the random bond Ising model a simple scaling picture [76] based on the scaling behavior of the domain walls in this model would predict an algebraic energy scaling – resulting in a formally infinite value for z, which is not confirmed by our results [78]. Hence we have to conclude with the observation that the common behavior Eq. (32) of the growth exponent z in 2d disordered models indicates a more complicated and yet hidden mechanism that is active in the non-equilibrium dynamics of these systems at low temperatures at least during the first 10 decades of the aging process.

We also would like to emphasize the fact that the physical interpretation of the growing length scale in the three systems under consideration in this paper is quite different: In the random bond ferromagnet it is simply the typical transverse domain size, where domains are easily identified as connected clusters of common magnetization sign. In the EA spin glass model the length scale is also determined by a domain size – where domains where defined as connected clusters of spins with common orientation with respect to one of the two ground states. These domains grow steadily up to a maximum size set by the equilibrium correlation length. In the disordered SOS model, however, the growing length scale is *not* connected to growing domains – actually the system settles quite fast after the temperature quench into a configuration that has a pretty large overlap with one of the ground states. Instead of growing further these initially very large domains thermal fluctuations of increasing size destroy these domains – and it is the spatial extent of these fluctuations that is characterized by the growing length scale studied here.

References

1. F. Rouqerol, J. Rouqerol, and K. Sing, *Adsorption by Powders and Porous Solids*. Academic Press, New York (1999) Chapt. 7.
2. K. S. W. Sing, D. H. Everett, R. A. W. Haul, L. Moscou, R. A. Pierotti, J. Rouqerol, and T. Siemieniewska, Pure Appl. Chem. **57** (1985) 603.
3. L. D. Gelb, K. E. Gubbins, R. Radhakrishnan, and M. Sliwinska-Bartkowiak, Rep. Prog. Phys. **62** (1999) 1573.
4. D. H. Everett, in *The Solid Gas Interface*. E. A. Flood (Ed.), Marcel Dekker, New York (1967) Vol. 2, p. 1055.
5. W. F. Saam and M. W. Cole, Phys. Rev. B **11** (1975) 1086.
6. F. Celestini, Phys. Lett. A **228** (1997) 84.
7. A. Papadopoulou, F. van Swol, and U. M. B. Marconi, J. Chem. Phys. **97** (1992) 6942.
8. R. Evans, U. Marini Bettolo Marconi, and P. Tarazona, J. Chem. Phys. **84** (1986) 2376.
9. R. A. Guyer and K. R. McCall, Phys. Rev. B **54** (1996) 18.
10. G. Mason, Proc. R. Soc. London Ser. A **415** (1988) 453.
11. L. Sarkisov and P. A. Monson, Langmuir **17** (2001) 7600 and references therein.
12. J. H. Page, J. Liu, B. Abeles, H. W. Deckman, and D. A. Weitz, Phys. Rev. Lett. **71** (1993) 1216.
13. P. Huber and K. Knorr, Phys. Rev. B **60** (1999) 12 657.

14. D. Wallacher, N. Künzner, D. Kovalev, N. Knorr, and K. Knorr, Phys. Rev. Lett. **92** (2004) 195 704.
15. D. Stauffer and R. B. Pandey, J. Phys. A Math. Gen. **25** (1992) L1079.
16. M. C. Goh, W. I. Goldburg, and C. M. Knobler, Phys. Rev. Lett. **58** (1987) 1008.
17. S. B. Dierker and P. Wiltzius, Phys. Rev. Lett. **58** (1987) 1865.
18. M. Y. Lin, S. K. Sinha, J. M. Drake, X.-L. Wu, P. Thiyagrajan, and H. B. Stanley, Phys. Rev. Lett. **72** (1994) 2207.
19. F. Brochard and P. G. de Gennes, J. Phys. Lett. (Paris) **44** (1983) 785.
20. P. G. de Gennes, J. Phys. Chem. **88** (1984) 6469.
21. R. J. Birgeneau, R. A. Cowley, G. Shirane, and H. Yosthizawa, J. Stat. Phys. **34** (1984) 817.
22. L. Monette, A. J. Liu, and G. S. Grest, Phys. Rev. A **46** (1992) 7664; A. J. Liu, D. J. Durian, E. Herbolzheimer, and S. A. Safran, Phys. Rev. Lett. **65** (1990) 1897.
23. E. D. Siggia, Phys. Rev. A **20** (1979) 595.
24. I. M. Lifshitz and V. V. Slyozov, J. Phys. Chem. Solids **19** (1961) 35.
25. J. C. Lee, Phys. Rev. B **46** (1992) 8648.
26. M. Kardar, G. Parisi, and Y. C. Zhang, Phys. Rev. Lett. **56** (1986) 889.
27. R. Paul, S. Puri, and H. Rieger, Europhys. Lett. **68** (2004) 881.
28. R. Paul and H. Rieger, J. Chem. Phys. **123** (2005) 024 708.
29. P. C. Hohenberg and B. I. Halperin, Rev. Mod. Phys. **49** (1977) 435.
30. A. J. Bray, Adv. Phys. **51** (2002) 481.
31. J. M. Kosterlitz and D. J. Thouless, J. Phys. C **6** (1973) 1181; J. M. Kosterlitz, J. Phys. C **7** (1974) 1046.
32. D. J. Bishop and J. D. Reppy, Phys. Rev. Lett. **40** (1978) 1727.
33. I. Rudnick, Phys. Rev. Lett. **40** (1978) 1454.
34. M. R. Beasley, J. E. Mooij, and T. P. Orlando, Phys. Rev. Lett. **42** (1979) 1165.
35. B. I. Halperin and D. R. Nelson, Phys. Rev. Lett. **41** 121 (1978); D. R. Nelson and B. I. Halperin, Phys. Rev. B **19** (1979) 2457.
36. L. P. Regnault and J. Rossat-Mignod, in *Magnetic Properties of Layered Transition Metal Compounds*. L. J. deJongh (Ed.), Kluwer Academics Publ., Dordrecht (1990) p. 271.
37. M. Bretz, Phys. Rev. Lett. **38** (1977) 501.
38. Y. P. Feng and M. H. W. Chan, Phys. Rev. Lett. **71** (1993) 3822.
39. S. Fassbender, C. Steimer, D. Arndt, and K. Knorr, Phys. Rev. Lett. **75** (1995) 2526; Surf. Sci. **388** (1997) 201.
40. J. V. José, L. P. Kadanoff, S. Kirkpatrick, and D. R. Nelson, Phys. Rev. B **16** (1977) 1217.
41. D. H. Lee, J. D. Joannopoulos, J. W. Negele, and D. P. Landau, Phys. Rev. Lett. **52** (1984) 433; Phys. Rev. B **33** (1986) 450.
42. S. Lee and K.-C. Lee, Phys. Rev. B **49** (1994) 15 184; P. Olsson, Phys. Rev. Lett. **75** (1995) 2758; H.J. Xu and B. W. Southern, J. Phys. A **29** (1996) L133.
43. S. Teitel and C. Jayaprakash, Phys. Rev. B **27** (1983) 598.
44. M. Y. Choi and D. Stroud, Phys. Rev. B **32** (1985) 5773; M. Yosefin and E. Domany, Phys. Rev. B **32** (1985) 1778.
45. B. Berge, T. Diep, A. Ghazali, and P. Lallemand, Phys. Rev. B **34** (1986) 3177; H. Eikmans, J. E. van Himbergen, H. J. F. Knops, and J. M. Thijssen, Phys. Rev. B **39** (1989) 11 759.
46. J. Lee, J. M. Kosterlitz, and E. Granato, Phys. Rev. B **43** (1991) 11 531.
47. G. Ramirez-Santiago and J. V. José, Phys. Rev. Lett. **68** (1992) 1224; Phys. Rev. B **43** (1994) 11 531.
48. E. Granato and M. P. Nightingale, Phys. Rev. B **48** (1993) 7438.
49. Y. M. M. Knops, B. Nienhuis, H. J. F. Knops, and H. W. J. Blöte, Phys. Rev. B **52** (1994) 7402.

50. E. Granato, J. M. Kosterlitz, J. Lee, and M. P. Nightingale, Phys. Rev. Lett. **66** (1991) 1090.
51. M. P. Nightingale, E. Granato, and J. M. Kosterlitz, Phys. Rev. B **52** (1995) 7402.
52. G. S. Grest, Phys. Rev. B **39** (1989).
53. J.-R. Lee, Phys. Rev. B **49** (1994) 3317.
54. S. Lee and K.-C. Lee, Phys. Rev. B **49** (1994) 15 184.
55. S. Miyashita and H. Shiba, J. Phys. Soc. Jap. **53** (1984) 1145.
56. H.-J. Xu and B. W. Southern, J. Phys. A **29** (1996) L133.
57. S. Lee and K.-C. Lee, Phys. Rev. B **57** (1998) 8472.
58. G. S. Jeon, S. Y. Park, and M. Y. Choi, Phys. Rev. B **55** (1997) 14 088.
59. J. M. Kosterlitz and D. J. Thouless, J. Phys. C **6** (1973) 1181; J. M. Kosterlitz, *ibid.* **7** (1974) 1046.
60. P. Olsson, Phys. Rev. Lett. **75** (1995) 2758.
61. J. V. José and G. Ramirez-Santiago, Phys. Rev. Lett. **77** (1995) 4849; P. Olsson, Phys. Rev. Lett. **77** (1995) 4850.
62. E. H. Boubcheur and H. T. Diep, Phys. Rev. B **58** (1998) 5163.
63. D. Loison and P. Simon, Phys. Rev. B **61** (2000) 6114.
64. S. Faßbender, M. Enderle, K. Knorr, J. D. Noh, and H. Rieger, Phys. Rev. B **65** (2002) 165 411.
65. M. Itakura, J. Phys. Soc. Jpn. **70** (2001) 600.
66. M. S. S. Challa and D. P. Landau, Phys. Rev. B **33** (1986) 437.
67. J.-D. Noh, H. Rieger, M. Enderle and K. Knorr, Phys. Rev. E **65** (2002) 026 111.
68. E. Maus, Ph.D. thesis, Universität Mainz (1991).
69. K. Knorr, Phys. Rep. **214** (1992) 113.
70. K. Knorr, S. Faßbender, A. Warken, and D. Arndt, J. Low Temp. Phys. **111** (1998) 339.
71. S. W. Pierson, S. M. Ammirata, J. C. Hunnicutt, and L. A. Gorham, Phys. Rev. B **60** (1999) 1309.
72. J. L. Barrat, M. V. Feigelman, J. Kurchan, and J. Dalibard (Eds.), *Slow Relaxation and Nonequilibrium Dynamics in Condensed Matter*, Springer Berlin, Heidelberg, New York (2003).
73. G. Parisi, F. Ricci-Tersenghi, and J. J. Ruiz-Lorenzo, Eur. Phys. J. B **11** (1999) 317.
74. L. F. Cugliandolo, J. Kurchan, and G. Parisi, J. Phys. I (France) **4** (1994) 1641.
75. N. Kawashima and H. Rieger, *Recent Progress in Spin Glasses*. In *Frustrated Spin Systems*. H. T. Diep (Ed.), World Scientific, Singapure (2004); cond-mat/0312432.
76. D. A. Huse and C. L. Henley, Phys. Rev. Lett. **54** (1985) 2708.
77. D. Forster, D. R. Nelson, and M. J. Stephen, Phys. Rev. A **16** (1977) 732; D. A. Huse, C. L. Henley, and D. S. Fisher, Phys. Rev. Lett. **55** (1985) 2924.
78. R. Paul, S. Puri, and H. Rieger, Europhys. Lett. **68** (2004) 881; Phys. Rev. E **71** (2005) 061 109.
79. H. Rieger, B. Steckemetz, and M. Schreckenberg, Europhys. Lett. **27** (1994) 485; J. Kisker, L. Santen, M. Schreckenberg, and H. Rieger, Phys. Rev. B **53** (1996) 6418.
80. A. Hartmann and H. Rieger, *Optimization Algorithm in Physics*. Wiley VCH, Berlin (2002).
81. P. Nozières and F. Gallet, J. Phys. (Paris) **48** (1987) 353.
82. J. Toner and D. P. DiVincenzo, Phys. Rev. B **41** (1990) 632; Y.-C. Tsai and Y. Shapir, Phys. Rev. Lett. **69** (1992) 1773; Phys. Rev. B **50** (1994) 3546; T. Hwa and D. S. Fisher, Phys. Rev. Lett. **72** (1994) 2466.
83. Y. Y. Goldschmidt and B. Shaub, Nucl. Phys. B **251** (1985) 77.
84. G. Schehr and H. Rieger, Phys. Rev. B **71** (2005) 184 202.
85. C. Zeng, A. A. Middleton, and Y. Shapir, Phys. Rev. Lett. **77** (1996) 3204; H. Rieger and U. Blasum, Phys. Rev. B **55** (1997) R7394.

Quantitative Evaluation of Elastic Properties of Nano-Crystalline Nickel Using Atomic Force Acoustic Microscopy

By M. Kopycinska-Müller[1,*], A. Caron[1,**], S. Hirsekorn[1], U. Rabe[1], H. Natter[2], R. Hempelmann[2], R. Birringer[3], and W. Arnold[1,***]

[1] Fraunhofer Institute for Non-Destructive Testing (IZFP), Campus E 3.1, University, D-66123 Saarbrücken, Germany
[2] Physical Chemistry, Saarland University, Campus B 2.2, D-66123 Saarbrücken, Germany
[3] Technical Physics, Saarland University, Campus D 2.2, D-66123 Saarbrücken, Germany

Nanocrystalline Nickel / Atomic Force Acoustic Microscopy / Contact Mechanics / Indentation Modulus / Nanoindentation

Atomic force acoustic microscopy (AFAM) is a near-field technique, where the vibration behavior of a micro-fabricated elastic cantilever beam in contact with a sample surface is sensitive to its local elastic properties. The resolution of this technique is given by the contact radius a_c of the atomic force microscope sensor-tip on the sample surface. Taking into account only the Hertzian forces, a_c depends on the static load applied by the cantilever, on the elastic constants of the tip and the sample and on the geometry of the contacting bodies. The shape of the sensor tip used in atomic force acoustic microscopy is between a sphere and a flat punch. Hence a_c extends from just below 10 nm to a few tens of nanometers. In this review, we give an overview of the AFAM technique, present data on the indentation moduli of nanocrystalline nickel, and discuss some of the error sources in quantitative AFAM. The AFAM indentation moduli measured are comparable to the values obtained by nanoindentation and lower than the indentation moduli calculated from ultrasonic velocity measurements. There seems to be a decrease of the indentation modulus with decreasing grain size for grain sizes below 30 nm. The data are discussed taking into account X-ray diffraction and electron back-scattering data revealing some texture and macro-strain due to internal stresses in the samples investigated.

---

* *Present address*: Department of Electrical Engineering and Information Technology, Electronics Packaging Laboratory (IAVT), Technical University Dresden, D-01602 Dresden, Germany

** *Present address*: Institute of Micro- and Nanomaterials, University of Ulm, D-89081 Ulm, Germany

*** Corresponding author. E-mail: arnold@izfp.fraunhofer.de

Z. Phys. Chem. **222** (2008) 471–498

1. Introduction

With the development of micro-electromechanical systems and materials with a nano-scale structure the assessment of elastic and mechanical properties with high spatial resolution has become crucial. To this end several techniques have been developed, based on dynamic atomic force microscopy (AFM), such as tapping mode [1], force modulation mode [2], atomic force acoustic microscopy (AFAM) [3], and ultrasonic atomic force microscopy (UAFM) [4]. AFAM and UAFM are contact-resonance spectroscopy techniques exploiting the coupling of the vibrating cantilever with its tip to the elastic restoring forces of the sample surface. The spatial resolution is given by the tip–sample contact radius a_c. The measured quantities are the contact-resonance frequencies of the cantilever and the widths of the resonance curves. Using a suitable mechanical model to describe the vibration of the AFM cantilever beam in contact with the sample surface, it is possible to obtain quantitative values of the contact stiffness k^* and the contact damping γ^* from the shift of the contact-resonance frequencies and the increase in widths of the resonance curves, respectively, relative to the free cantilever beam-resonances [5]. In order to gain quantitative information on the sample elasticity, one has to make use of a contact-mechanics model which relates k^* to the local elasticity.

The Hertzian contact model of a spherical indenter or of a flat tip acting as a punch is often used [6, 7]. This contact model can be applied for AFAM provided the tip radius is known. The shapes of the sensing tips have been investigated by scanning electron microscopy (SEM) [8], and test structures consisting of sharp needles or roofs exist allowing one to scan tip shapes [9]. The images obtained in this way revealed tip shapes between the limiting cases of a flat punch and a sphere. Moreover, the tip radius and its shape change during a measurement due to wear, in particular on hard surfaces [10]. To account for the actual tip shape, various optimizations of the contact-mechanics model have been proposed [4, 11]. These optimized models are based on the dependence of the contact stiffness and thus the resonance frequency on the applied static load.

In this work we discuss evaluation models for the AFAM contact resonances accounting for the actual tip shape and size. We apply these models to obtain elastic data by AFAM measurements on nano-crystalline nickel (nc-Ni) with various grain sizes using a dual reference method. The results are then compared to elastic moduli data for single-crystalline nickel, nano-indentation measurements, and finally ultrasonic data in the nc-materials.

2. Atomic force acoustic microscopy

Atomic force acoustic microscopy combines ultrasonics with atomic force microscopy. As shown in Fig. 1, an ultrasonic transducer injects longitudinal waves from the bottom surface into the sample causing ultrasonic vibrations

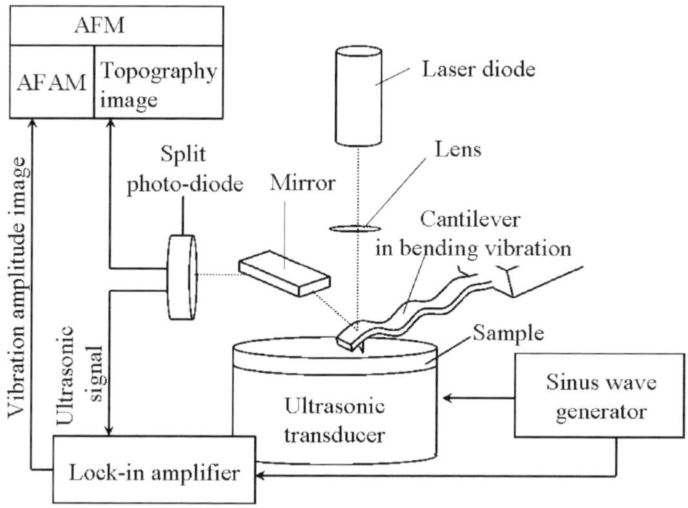

Fig. 1. Experimental set-up of atomic force acoustic microscopy. An ultrasonic transducer excited by a sinusoidal voltage sets the sample surface into vertical vibrations and *via* the tip–sample forces the cantilever is forced into bending vibrations. The cantilever vibrations are detected by a beam-deflection detector consisting of a laser diode, lens, mirror, and a split photo-diode. The low-frequency components of the cantilever deflection are used for the topography image. The high-frequency components are processed by a lock-in amplifier and displayed as an acoustic image.

of its opposite (top) surface. These vibrations couple *via* the tip into the AFM cantilever and excite it to bending vibrations. The high-frequency cantilever vibrations can be detected with the same split photo-detector as used for topography measurements, as long as the bandwidth of the optical detector amplifier is broad enough.

The free bending resonances of the cantilever depend on its geometry, elastic constants and density. The corresponding resonance curves are characterized by Q-factors of typically 200–1000 in ambient air. Bringing the cantilever into contact, a shift of the resonance frequencies to higher values and Q-factors of typically 20–100 are observed, due to the stiffening and damping effects of the contact. The still high Q-values of the contact resonances lead to resonance amplification and therefore also allow one to detect the cantilever resonances in contact with a high signal-to-noise ratio.

AFAM can be used as an imaging instrument where the amplitude of vibration is mapped over the sample surface. In this case the cantilever is excited at a fixed frequency near its resonance. Depending on the local contact stiffness, the resonance frequency shifts and leads to changes in the vibration amplitude at the working frequency (Fig. 2). Likewise, one can use the contact-resonance frequency value as an imaging quantity [12]. As an advantage over the ampli-

Fig. 2. *Left*: AFM topography image of polycrystalline nickel, height range: 20 nm. Due to mechanical and electro-polishing, grains become visible because the abrasion and the current density are stronger at grain boundaries; *Right*: AFAM image recorded at a frequency $f = 699$ kHz near the first contact bending-mode of the cantilever. Here, the contrast between the grains is due to elastic anisotropy. The image size is $20 \times 20\,\mu\text{m}^2$. The image was recorded using a diamond-coated silicon-cantilever (type dt-nclr Nanoworld, Erlangen, Germany).

tude imaging, the recording of contact-resonance curves allows the quantitative evaluation of the local elastic properties.

2.1 Properties of an AFM cantilever-beam with its sensor-tip

In order to describe the vibrations of the cantilever, it is convenient to treat it as a micro-scale rectangular beam. Figure 3 shows scanning electron micrographs of a cantilever typically used in AFAM measurements. This type of commercial sensor is microfabricated by etching from single-crystalline silicon. The side-view of the cantilever is shown in (a). As one can see, the tip is not placed exactly at the end of the beam. The distance from the base of the cantilever to the position of the tip is designated as L_1. The SEM image in (b) shows that the thickness T of this particular beam was about 7 μm. The SEM image in (c) presents the top-view of the cantilever. The cross section of the cantilever beam is trapezoidal for most of its length, as visible in (c) and (d). The mean value of the bottom and top base of this trapezoid gives the mean width W, here about 40 μm. The cantilever becomes triangular at the very end, where the tip is located. The length L of the cantilevers ranges from 220 to 250 μm, their static spring constants k_C are about 40 N/m or higher.

Very often cantilevers contain a surface coating that can change their behavior while vibrating due to stresses between the cantilever and the coating layer. Typical layers are reflective coatings, conductive coatings, and hard magnetic coatings. The thickness of the coating layers is typically a few tens of nm. The tip of the AFM cantilever is usually made from silicon. Depending on the application, the tip radius can range from 2 to 15 nm. Knowledge of the elastic

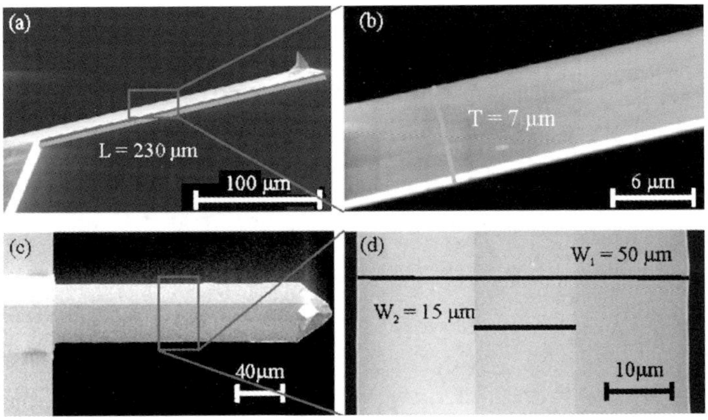

Fig. 3. SEM images of a cantilever typically used in AFAM experiments; (a) Side-view of the cantilever and (b) higher magnification image that allows precise measurement of the thickness; (c) Top-view image of the cantilever and (d) higher magnification image showing a change of the cantilever width in the cantilever thickness direction. The tip is not placed at the end of the beam, and its thickness varies between the position of the tip and the free end of the cantilever. In addition, the suspension of the cantilever is not symmetric, which plays a role in the theoretical description of beam vibrations. The images were obtained by Dr. R. Geiss at NIST, Boulder, Co, USA (from [13]).

properties of the tip is indispensable to calculate the contact area or penetration depth. For a silicon tip the elastic properties are known. The presence of a silicon-oxide layer on the tip surface which is only a few atomic layers thick does not play a role for AFAM measurements. However, the silicon tips wear. A layer intended to reduce tip wear is a diamond layer. This kind of coating is approx. 100 nm thick significantly increasing the tip radius. Such coatings are relatively rough, but due to the roughness at the end of the tip, the lateral resolution can be improved inadvertently. The coating layers can change the effective elastic properties of the silicon tip and its shape, so that additional care must be taken to characterize the properties of the coated tips.

2.2 Free and contact resonance frequencies of an AFM cantilever

The mechanical model of the vibrating cantilever is presented in Fig. 4. Vibrations of beams are treated in textbooks of acoustics and engineering mechanics [14, 15] and with regard to application to AFAM in [3, 5, 13, 16]. For a beam of uniform cross section, the equation of motion for flexural vibrations is the fourth-order differential equation

$$EI\frac{\partial^4 y}{\partial x^4} + \rho A\frac{\partial^2 y}{\partial t^2} = 0 \tag{1}$$

if damping is neglected. Here, E is the Young's modulus, ρ is the density, $A = W \times T$ is the cross-section area, and I is the area moment of inertia of

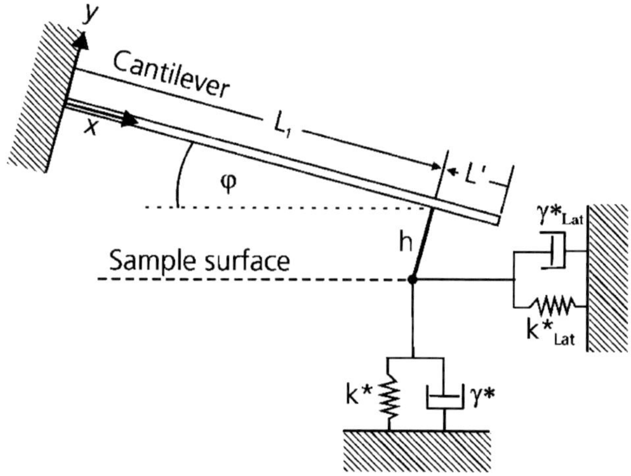

Fig. 4. Mechanical model of the AFM cantilever as a rectangular elastic beam. The contact itself is described as two parallel linear springs and two dashpots for both vertical and lateral contact forces. In the figure $L = L_1 + L'$ is the cantilever length, L' is the distance between the tip position and the free end of the cantilever, $h(\approx 15\,\mu\mathrm{m})$ is the length of the tip, k^* and k^*_{lat}, γ and γ^*_{lat} are the contact stiffnesses and the damping constant for vertical and lateral forces, respectively. Due to technical reasons the cantilever is inclined by an angle ϕ.

the beam, respectively. Furthermore, $y(x)$ is the deflection of the beam from the equilibrium position at the length element x, $\partial y/\partial x$ is the slope, $EI\partial^2 y/\partial x^2$ is the torsional moment, and $EI\partial^3 y/\partial x^3$ is the shear force at x. The boundary conditions for the clamped-free beam must be as follows: At the clamped end ($x = 0$), the deflection and the slope must be zero, *i.e.* $y = 0$ and $\partial y/\partial x = 0$. At the free end of the beam ($x = L$) no moment or shear force can be present which entails $\partial y^2/\partial x^2 = 0$ and $\partial y^3/\partial x^3 = 0$. Taking into account these boundary conditions, one obtains the characteristic equation

$$\cos k_n L \cosh k_n L + 1 = 0\,, \tag{2}$$

for which the solutions $k_n L$, $n = 1, 2, 3 \ldots$. give the wave numbers k_n of an infinite set of flexural modes. The dispersion relations are used to calculate the resonance frequency f_n of the nth mode [3].

$$f_n = \left(\frac{k_n L}{c_{\mathrm{c}}}\right)^2 \quad \text{where} \tag{3}$$

$$c_{\mathrm{c}} = L\sqrt{2\pi}\sqrt[4]{\frac{\rho A}{EI}}\,. \tag{4}$$

Here, c_{c} is a constant dependent on the geometry of the cantilever and its mechanical properties.

The boundary conditions change when the tip is in contact with the sample surface. Then the tip senses long-range attractive forces as well as repulsive tip–sample forces. The static load P applied to the tip by a cantilever deflection d_c is $P = k_c \times d_c$, where the spring constant k_c can be calculated assuming a cantilever with rectangular cross-section:

$$k_c = \frac{EWT^3}{4L_1^3} \,. \tag{5}$$

Though the geometry of the commercial cantilevers (Fig. 3) deviates from the simple model according to Fig. 4, Eq. (5) leads to good approximations for k_c. The discrepancies between theoretical and measured values of k_c were reported in [17]. The exact determination of spring constants of AFM cantilevers is a very active research field [18].

The superposition of vertical and lateral tip–sample interaction forces is represented by a set of springs and dashpots as shown in Fig. 4. For example, the interaction forces normal to the sample surface are represented as a linear spring with a characteristic spring constant k^*, which is the negative derivative of the tip–sample force in the equilibrium position:

$$k^* = -\left.\frac{\partial F(z)}{\partial z}\right|_{z=z_e} , \tag{6}$$

where z is the tip–sample distance, $F(z)$ is the tip–sample interaction force, and z_e is the equilibrium position. The lateral contact stiffness k_{Lat}^* is defined analogously. The high Q-values of the resonances even in contact mode allow one to assume that the damping has only little influence on the resonance frequencies of the system.

Therefore the dashpots γ^* and γ_{Lat}^* are omitted in the further calculations. For the spring-coupled cantilever one obtains a new characteristic equation instead of Eq. (2) [5]:

$$\begin{aligned} &A_0 c_P \left(\tfrac{h}{L_1}\right)^2 \left(3\tfrac{k^*}{k_C}\right)^2 + 3A_1 \tfrac{k^*}{k_C}(\cos^2\varphi + c_P \sin^2\varphi) + 3A_2 \tfrac{h}{L_1}\tfrac{k^*}{k_C} \\ &\times \sin\varphi\cos\varphi(c_P - 1) + 3A_3 \left(\tfrac{h}{L_1}\right)^2 \tfrac{k^*}{k_C}(\sin^2\varphi + c_P\cos^2\varphi) + A_4 = 0 \,. \end{aligned} \tag{7}$$

In this equation the parameters A_0–A_4 are defined as follows:

$$A_0 = (1 - \cos k_n L_1 \cosh k_n L_1)(1 + \cos k_n L' \cosh k_n L') \,, \tag{8a}$$

$$\begin{aligned} A_1 = {} & k_n L_1[-(1 - \cos k_n L_1 \cosh k_n L_1) \\ & \times (\sin k_n L' \cosh k_n L' - \sinh k_n L' \cos k_n L') \\ & + (1 + \cos k_n L' \cosh k_n L')(\sin k_n L_1 \cosh k_n L_1 - \sinh k_n L_1 \cos k_n L_1)] , \end{aligned} \tag{8b}$$

$$A_2 = 2(k_n L_1)^2[\sin k_n L_1 \sinh k_n L_1 (1 + \cos k_n L' \cosh k_n L') + \sin k_n L' \sinh k_n L'(1 - \cos k_n L_1 \cosh k_n L_1)] , \tag{8c}$$

$$A_3 = (k_n L_1)^3[(\sin k_n L_1 \cosh k_n L_1 + \sinh k_n L_1 \cos k_n L_1) \times (1 + \cos k_n L' \cosh k_n L') - (\sin k_n L' \cosh k_n L' - \sinh k_n L' \cos k_n L') \times (1 - \cos k_n L_1 \cosh k_n L_1)] , \tag{8d}$$

$$A_4 = 2(k_n L_1)^4(1 + \cos k_n L \cosh k_n L) . \tag{8e}$$

For technical reasons the beam is slightly inclined by an angle φ (Fig. 4). For the calculations reported in this study, it was assumed that $f = 11°$ and $c_p = 0.85$. The parameter $c_p = k^*_{\text{lat}}/k^*$ describes the ratio of the vertical and lateral contact stiffnesses. From our own experience and from results of other authors [19], $c_p = 0.85$ is indeed a reasonable value. Let us assume momentarily that the tip is located exactly at the end of the cantilever ($L' = 0$ and $L_1 = L$), and that the beam is parallel to the surface ($\varphi = 0$). Furthermore let us neglect the lateral forces ($k^*_{\text{lat}} = 0$). In this simplified case, one obtains the following characteristic equation for the spring-coupled cantilever [3]:

$$\frac{3k^*}{k_c}(\sinh k_n L \cos k_n L - \sin k_n L \cosh k_n L) = (k_n L)^3 \times (1 + \cos k_n L \cosh k_n L) . \tag{9}$$

Equation (9) reduces to Eq. (2) for $k^* = 0$, which is equivalent to an infinitely soft spring. In the opposite case, namely for $k^* \to \infty$, the cantilever end at $x = L$ is pinned to the surface and the cantilever becomes insensitive to the elastic properties of the surface. The resonance frequencies of the spring-coupled system shift to higher values, called contact-resonance frequencies, due to the tip–sample interactions. The value of k^* can be calculated if the resonance frequencies of the clamped-spring coupled system are known. In the simplified case the solution is straightforward:

$$k^* = \frac{k_c}{3} \frac{(k_n L)^3(1 + \cos k_n L \cosh k_n L)}{\sinh k_n L \cos k_n L - \sin k_n L \cosh k_n L} . \tag{10}$$

The normalized wave number $k_n L$ of each contact resonance is related to its experimentally measured frequency f_{cont} by the dispersion relation Eq. (3). However, the dimensions of the beam required in Eq. (4) are in most cases not known with sufficient accuracy. As the dispersion relation holds for all flexural modes of one beam independently of its boundary conditions, the exact knowledge of the individual constants can be circumvented by measuring the resonant frequencies of the clamped-free beam f_{free}:

$$k_n L = (k_n L)_{\text{free}} \sqrt{\frac{f_{\text{cont}}}{f_{\text{free}}}} . \tag{11}$$

The values of $(k_n L)_{\text{free}}$ are known, and for example, the value of $(k_1 L)_{\text{free}} = 1.8751$ holds for the first free resonance [3]. The position L_1 of the tip is a very important parameter in the AFAM technique which may not be neglected in practical application. If lateral forces and inclination of the cantilever are further taken into account, Eq. (7), which is a quadratic equation in k^*, has to be solved for k^*. One obtains for k^* [5]:

$$k^* = \frac{k_c}{3}\left(z \pm \sqrt{z^2 - \frac{A_4}{A_0 c_p}\left(\frac{L_1}{h}\right)}\right), \tag{12}$$

where z is given by:

$$z = \frac{1}{2A_0 c_P \left(\frac{h}{L_1}\right)^2}\left(A_1(\cos^2\varphi + c_P \sin^2\varphi) + A_2\frac{h}{L_1}(c_P - 1)\right.$$
$$\left.\times \sin\varphi\cos\varphi + A_3\left(\frac{h}{L_1}\right)^2(\sin^2\varphi + c_P\cos^2\varphi)\right). \tag{13}$$

If the frequencies of two modes are known, one can write two equations containing the two unknown parameters L_1 and k^*. By plotting k^* as a function of L_1 for the two modes, one obtains two curves, the cross-point of which will yield the value of $k^*$of the system. The tip position parameter L_1 can vary within a certain range. Therefore different procedures were developed, which allowed for the determination of the optimal value of L_1/L. First a range of values for L_1/L is defined, using micrographs of the cantilevers. Then the values of k^* are calculated for the two measured modes as a function of L_1/L. As it was mentioned above, in the case of only one pair of the contact-resonance frequencies, the optimal value of L_1/L will be defined by the cross-point of the two functions. In the case of many pairs of the contact-resonance frequencies, the differences in the values of k^* determined for the two modes are averaged and plotted as a function of L_1/L. The minimum of this function will yield the optimal value of the tip-position parameter.

2.3 The AFAM set-up used

The experimental set-up used to excite and to detect AFAM contact-resonance frequencies is presented in Fig. 1. As mentioned above, the sample of interest is placed on an ultrasonic transducer that emits longitudinal waves. Typical transducers employed are broadband transducers with a center frequency of 1, 2.5, and 5 MHz, for example, from Olympus/Panametrics, Inc. The signal from a waveform generator is applied to the transducer, creating longitudinal waves that cause out-of-plane surface vibrations. When the AFM cantilever tip is in contact with the sample surface, the surface movement is transferred into vibrations of the cantilever beam. The ensuing signal of the split photodiode is sent to a lock-in amplifier and compared with a reference signal from the waveform generator. The amplitude and/or phase of the resultant signals are sent to a digital oscilloscope (LeCroy 9350) and *via* a computer interface back to the computer, where they are stored as contact-resonance spectra. The AFAM experiments carried out in the course of the research work reported here were performed with a modified Veeco/DI Dimension 3000 (S. Barbara, Ca, USA) and a MT-NDT "Solver" (Zelenograd, Russia).

2.4 AFAM spectroscopy mode

Computer programs were written in Labview in order to measure contact-resonance spectra in either a "step-by-step" or a "sweep" mode. In the step-by-step mode, the waveform generator incrementally changes its output frequency from a start to an end value. In the sweep mode, the frequency is swept in 0.5 s over a defined frequency range. This approach allows a larger number of spectra to be collected, so that the statistics of the measurement can be taken into account. All of the spectra can be very similar, but in each case conditions may be slightly different [13, 20]. Mechanical drift of the sample on the AFM stage or the drift of the AFM scanner-head may change the tip–sample contact conditions, the applied static load may vary within certain limits, or additional forces like adhesion may change during the experiment. Using a set of contact-resonance spectra, one can judge the stability of the system and the reproducibility of the experiment. The contact-resonance frequencies are measured as a function of the static load. If the end of the tip has a geometry different from a flat punch, then an increase in the static load will lead to an increase in the tip–sample contact area and hence the tip–sample contact stiffness k^*. This causes the contact resonance frequency to shift to higher values. For a tip with a flat punch shape, the tip–sample contact area is constant and does not depend on the static load. Therefore, the resonance frequencies should not shift with static load. Figure 5 shows contact-resonance spectra, measured on a sapphire reference sample for different static loads. The cantilever used in these experiments had a silicon tip and a spring constant $k_c = 44$ N/m. The free resonance frequencies were 164 kHz and 1017 kHz for the first and for the second mode, respectively. The applied load P was varied from 41 nN to 1663 nN. It can be seen that the value of the contact-resonance frequency for both modes increases systematically with P.

The spectra presented in Fig. 5 have symmetrical shapes, especially for the first mode. Such spectra can be measured when the amplitude of the surface vibration is small (< 0.02 nm). In this case, the tip–sample interaction forces behave linearly (the contact or repulsive region of the force-distance curve). When the amplitude of the surface vibrations increases, however, the tip–sample forces shift from the contact region to the highly nonlinear non-

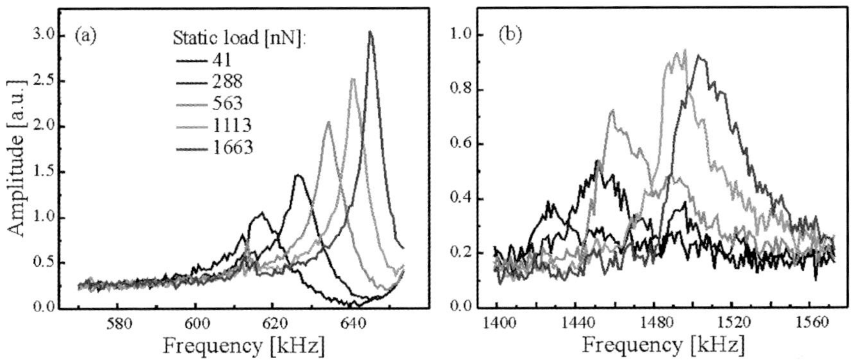

Fig. 5. Contact-resonance spectra measured on the sapphire sample for the (a) first and (b) second bending mode. The cantilever used in these measurements had a spring constant of $k_c = 44\,\mathrm{N/m}$ and free resonance frequencies of 164 kHz and 1017 kHz for the first and second mode, respectively. The resonance frequencies increase as the static load increases. This set of spectra was measured with the MT-NDT "Solver" instrument (from [13]).

contact region. In this case, the tip is sensitive to repulsive forces and adhesion forces that influence the shape of the contact resonance spectrum [21, 22].

2.5 AFAM imaging mode

In AFAM amplitude imaging, ultrasound at a fixed frequency close to a contact resonance frequency is applied to the sample while the cantilever is scanning across its surface. At frequencies close to the measured resonance frequency, the amplitude of the cantilever vibrations changes when there is a change in stiffness, see Fig. 2, right. This mode of imaging has been used in many applications. Amplitude imaging provides only qualitative information about sample properties. In order to obtain a quantitative image of elastic properties, the value of the contact-resonance frequency must be measured at each point in the image and used as an imaging quantity. Instead of using a single ultrasonic excitation frequency, one must sweep over a defined frequency range [12, 23, 24]. The amplitude of the cantilever vibration is then analyzed and the resonance frequency is detected with the help of the peak detector. Either the resonance frequency or the whole spectra are digitally recorded in every image point. If the tip properties such as the modulus and radius are known, then a contact stiffness k^* image (Fig. 6a) can be converted into an indentation-modulus image (Fig. 6b) as discussed below.

3. Contact mechanics and calibration procedure

The tip–sample contact stiffness k^* calculated from the free and the contact-resonance frequencies of a cantilever depends on the diameter of the surface

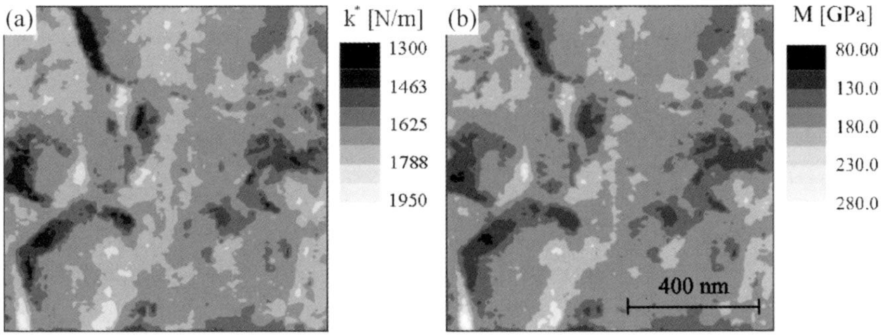

Fig. 6. (a) Image of local tip–sample contact stiffness obtained for a polycrystalline Ni sample with a grain size of 167 nm (sample Ni_1, see Table 2). The values of k^* varied from 1300 N/m to 1950 N/m, depending on the local elastic properties of the sample and the sample topography; (b) Calibrated image of the indentation modulus M calculated from the values of k^* in (a). The values of M varied from 80 GPa to 280 GPa. The main part of the image is coded in grey scale that corresponds to the values of 180 GPa and 205 GPa. The extreme values of M were found in the regions where the inaccuracies in k^* were the largest [13].

area involved in the interaction and the elastic moduli of the tip and the sample. It has been shown that the quantity measured in the AFAM experiments is the indentation modulus M measured also by nanoindentation [12]. The relation of the contact stiffness to the elastic properties has been the subject of intensive research, due to the development of depth-sensing indentation techniques. Based on the work of Sneddon [25], Oliver and Pharr formulated the contact stiffness k^* for the case of the elastic contact between a symmetric indenter of revolution and a flat surface as [7]:

$$k^* = \frac{2}{\sqrt{\pi}} E^* \sqrt{A_c}\,, \tag{14}$$

where $A_C = \pi a_c^2$ is the contact area and E^* is the reduced modulus of elasticity. For isotropic solids, the relation

$$\frac{1}{E^*} = \frac{1-\nu_s^2}{E_s} + \frac{1-\nu_{tip}^2}{E_{tip}} \tag{15}$$

holds. Here, $E_{s,tip}$ are the Young's moduli of the sample and of the tip, respectively, and $\nu_{s,tip}$ are the corresponding Poisson ratios. In the case of anisotropic solids an indentation modulus M is introduced, which can be calculated from the elastic single-crystal constants, if the single-crystal orientation of the surface area considered is known [25]. If there exists a three- or fourfold rotational symmetry axis perpendicular to the boundary, the contact area is circular. The required symmetry holds for silicon sensor tips, which are oriented in (001) crystallographic direction. In this case Eq. (15) can be replaced by:

$$\frac{1}{E^*} = \frac{1}{M_S} + \frac{1}{M_{tip}}, \tag{16}$$

where M_S and M_{tip} are the indentation moduli of the sample and the tip, respectively [26, 27]. For isotropic bodies the indentation modulus is calculated as $M = E/(1-\nu^2)$, and Eq. (16) becomes equivalent to Eq. (15). Equation (14) is not suitable for directly evaluating AFAM measurements. The main reason for this is the lack of knowledge of the tip shape of the AFM sensor and thus of the contact area. Therefore, in order to determine the contact radius a_C, the tip geometry must be known. The tip shape can be determined using SEM and/or by imaging with the tip under investigation the topography of calibration samples with reference structures much smaller than the tip radius R_1. Furthermore, the tip shape can be characterized in the AFAM mode using reference samples with known values of the indentation modulus. For spherical tip geometry, assuming Hertzian contact conditions, R_1 and the contact stiffness k^* are given by

$$R_1 = \frac{k^*3}{6PE^*2} \tag{17}$$

or conversely

$$k^* = \sqrt[3]{6PR_1E^{*2}}. \tag{18}$$

In the case of a flat punch, the radius of the punch R_p equals the contact radius ($a_c = R_p$). In this case the contact stiffness k^*, which is no longer load-dependent, is given by:

$$k^* = 2R_pE^*. \tag{19}$$

In the case of a silicon tip the value of its indentation modulus can be calculated using the known elastic constants. For a coated tip, however, it is necessary to determine the effective indentation modulus M_t of the tip. Then, two reference samples must be used in order to determine M_t [13, 20]. Equations (16) and (18) are used to obtain the following relation:

$$M_t = \frac{M_1M_2\left(1-\sqrt{k_1^{*3}/k_2^{*3}}\right)}{\sqrt{k_1^{*3}/k_2^{*3}}\,M_2 - M_1}, \tag{20}$$

Table 1. Reference samples used for tip characterization and the corresponding values of elastic constants, Poisson ratios, and calculated indentation moduli.

Sample	Structure	Elastic constants [GPa]	Indentation modulus [GPa]
Fused silica I	amorphous	$E = 74.8;\ \nu = 0.17$	$M_{\text{iso}} = 77.1$
Fused silica II	amorphous	$E = 73.6;\ \nu = 0.17$	$M_{\text{iso}} = 75.8$
Nickel	cubic	$c_{11} = 252.8;\ c_{12} = 152;\ c_{44} = 123.8$	$M_{\langle 111\rangle} = 256.2$
Zerodur™	amorphous	$E = 91;\ \nu = 0.24$	$M_{\text{iso}} = 96.1$
Silicon	cubic	$c_{11} = 165.7;\ c_{12} = 63.9;\ c_{44} = 79.6$	$M_{\langle 100\rangle} = 165.0$
Sapphire	hexagonal	$c_{11} = 496;\ c_{12} = 164;\ c_{13} = 115;$ $c_{33} = 498;\ c_{44} = 148$	$M_{\langle 101\rangle} = 423.2$
Strontium titanate	cubic	$c_{11} = 318.1;\ c_{12} = 102.5;$ $c_{44} = 123.6$	$M_{\langle 100\rangle} = 298.8$

where M_1, M_2, k_1^*, and k_2^* are the indentation moduli and the measured contact stiffnesses for the reference samples 1 and 2, respectively. After measurements are made on the reference samples, the unknown sample is measured. We assume that the geometry and mechanical properties of the tip remain the same for both samples. The elastic properties of the sample can be extracted by comparing measurements:

$$E_{\text{S}}^* = E_{\text{r}}^* \left(k_{\text{r}}^* / k_{\text{s}}^*\right)^m . \tag{21}$$

Here, r and s refer to the reference sample and the unknown sample, respectively, and m describes the tip geometry. For a flat punch $m = 1$ and for a spherical tip $m = 3/2$ which are the two tip geometries considered so far.

4. Reference samples

Several reference samples were used in order to characterize the cantilever tip and to determine elastic properties of the unknown samples. The reference samples included fused silica, Zerodur™ glass, (100)-oriented silicon wafers, (101)-sapphire single-crystal, (100)-single crystal strontium titanate, and (111)-single crystal Ni. From literature values of the elastic constants for each material [28, 29], the indentation moduli were calculated. The values of the elastic constants and the corresponding indentation moduli are presented in Table 1. The literature values used for the elastic constant of Ni single crystal were in good agreement with the ultrasonic measurements we carried out in the [111] direction of our Ni single crystal. We obtained for the longitudinal and the transverse velocity $v_{\text{L}} = 6.47$ mm/µs and $v_{\text{T}} = 2.97$ mm/µs, respectively. These values are compatible with $c_{11} = 252.8$ GPa, $c_{12} = 152$ GPa, and

Table 2. Average grain-size diameters obtained from XRD-spectra and EBSD images and range of the grain sizes given by the cross-section analysis of the tapping mode images of the electrodeposited nickel samples. The sample Ni_2 possessed a bimodal grain-size distribution with many small and very large grains. Therefore the mean value was less than the deviation.

Sample	XRD	EBSD	Tapping mode (range)
Ni_1	167.1	155 nm ± 74 nm	30 nm–600 nm
Ni_2	96.4	478 nm ± 610 nm	120 nm–1300 nm
Ni_3	64.8	–	20–30 nm
Ni_4	48.2	156 nm ± 99 nm	40 nm–1100 nm
Ni_5	30.4	274 nm ± 142 nm	50 nm–550 nm

$c_{44} = 123.8$ GPa from which one obtains $M_{\langle 111\rangle} = 256.2$ GPa based on the procedure described in [27].

5. Nickel samples

Nickel (Ni), a metal with cubic structure, is an attractive material for industrial applications, due to its high oxide and corrosion resistance. The nanocrystalline (nc) phase of Ni enhances these properties. In a first series of experiments, six samples of nc-Ni fabricated by pulsed electrodeposition were investigated [30, 31]. Each of the samples was intended to possess a different average grain diameter. The volume-weighted grain diameter was evaluated from XRD spectra using a modified Warren–Averbach method [32, 33]. Two surface evaluation techniques, namely the electron backscattered diffraction (EBSD) and tapping-mode, were used to determine the average grain size at the samples surface. The results are shown in Table 2.

The Ni samples were polished mechanically, first with a diamond paste and then with an alumina solution in order to remove remaining polishing grooves. The surface obtained in this way had a height difference of less than 10 nm as determined from AFM topography images. All samples investigated were between 0.6 and 1.5 mm thick and could be evaluated with ultrasonic measurements to determine their elastic moduli. Furthermore, EBSD was used to characterize the microstructure of the nc-Ni samples investigated in this study [13] which indicated some texture in some of the samples. The XRD spectra measured for the nickel samples suggested a preferred orientation for all of the samples, either in orientation (111) or (100). From the EBSD data, a larger grain size was obtained than from the AFM tapping-mode images and the XRD spectra. The density measurements proved the nc samples to be porosity free. Porosity leads to a reduction of their Young's modulus [34].

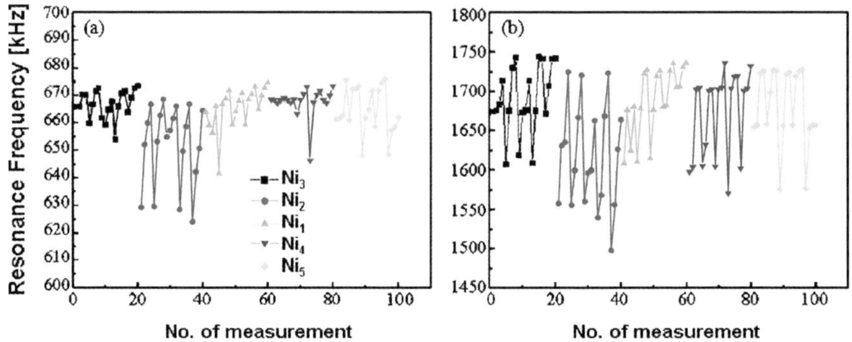

Fig. 7. (a) First and (b) second contact-resonance frequencies measured for the nc-Ni samples. The data were taken at different static loads. The static load was increased consecutively in four steps and then decreased again. Five load cycles were performed on each sample. The obtained values are very similar for both the first and second modes with one exception.

6. Experimental results for electrodeposited nanocrystalline nickel

6.1 Measurement of the local indentation modulus by AFAM and nanoindentation

We measured the contact resonance frequencies for the nc-Ni samples at five different, randomly chosen locations with four different static loads [13]. Figure 7 presents the measured frequencies for the first and second modes. The contact-resonance frequencies were very similar for all the nickel samples. The first resonance frequency varied from 650 kHz to 675 kHz. The range of the second contact-resonance frequency was much broader: 1650 kHz–1750 kHz. Figure 8 presents the average tip–sample contact stiffness, which was calculated using the first and the second contact resonance frequency. The values of k^* obtained for four nc samples were in a very similar range. This indicates that from the point of view of AFAM measurements these samples have similar elastic properties. As can be seen in Fig. 8, the values of the error Δk^* are relatively low, in most cases smaller than 10%. The accuracy of the measurements was estimated using data on the calibration samples [13].

Values of k^* were calculated from each pair of contact-resonance frequencies in 30 separate measurements. In order to check the agreement between the calculated value of k^* and the value predicted from an analytical model for a vibrating beam, values of k^* were calculated separately for the first and the second modes, assuming the same position of the tip ($L_1/L = 0.987$). The difference Δk^* between the values of k^* calculated for the first and second modes is taken as the mean error. The values of M for the nc-Ni samples were calculated from the data of k^* with help of the reference measurements obtained on the strontium titanate. The averaged values of M with the corresponding

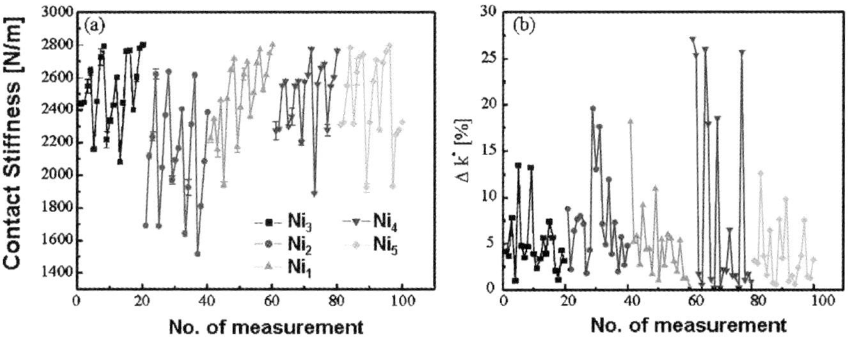

Fig. 8. (a) Contact stiffness k^* calculated from the contact-resonance frequencies for the samples Ni_1–Ni_5. (b) The corresponding Δk^* were for most of the measurements lower than 10% indicating a good agreement between the measured contact-resonance frequencies and the theory of a vibrating beam [13].

Table 3. Average values of the indentation modulus M obtained by AFAM and nanoindentation techniques.

Sample	D_{volume} [nm]	M_{AFAM} [GPa]	M_{NI} [GPa]
Ni_1	167.1	232.8 ± 53.0	278.7 ± 13.1
Ni_2	96.4	150.4 ± 44.1	217.5 ± 17.2
Ni_3	64.8	303.3 ± 63.0	198.9 ± 12.0
Ni_4	48.2	218.1 ± 46.4	189.7 ± 15.5
Ni_5	30.4	198.7 ± 55.5	152.9 ± 6.8

standard deviation are presented in Table 3, which contains also the averaged values of M obtained with nanoindentation. As can be seen from table and from Fig. 9, the values of M obtained for samples Ni_1, Ni_3, Ni_4, and Ni_5 with AFAM are similar to those provided by nanoindentation within the limits of uncertainty of both techniques. For the AFAM measurements presented in this section, the mean values of the standard deviation are relatively large ranging from 20 to 30%. Concerning Ni_2, it is likely that its surface was rougher than those of the other samples on the scale of the tip–sample contact area which leads to softening [13].

6.2 Comparison of the results with macroscopic ultrasonic data

We compared the experimental results above with the indentation moduli calculated from single crystal constants and with ultrasonic data obtained for the nc-Ni samples. Nickel is a material with a high elastic anisotropy factor $A_z \approx 2.4$. The values of the Young's moduli E vary from 120 GPa in the $\langle 100 \rangle$ direction to 304 GPa in $\langle 111 \rangle$ direction [28]. Table 4 displays the indentation

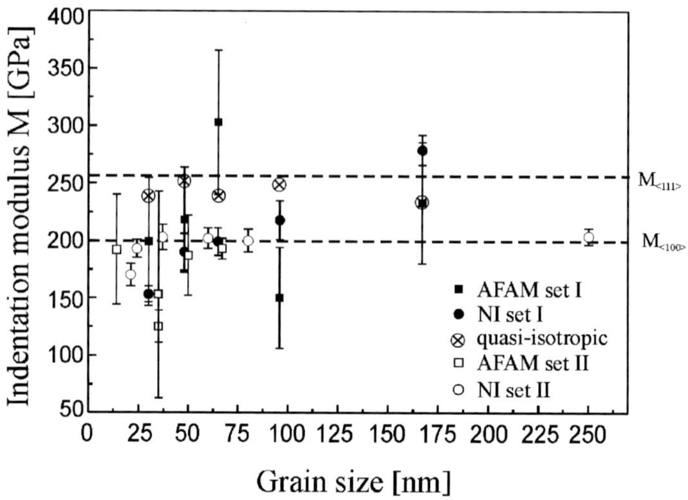

Fig. 9. Indentation modulus M_{AFAM} obtained with the AFAM technique for both sample series. The data obtained with nanoindentation (NI set I) are plotted as well. It is interesting to note that the nanoindentation data taken by another research group fit well into our data (open circles, from [36, 37]). The dashed and dashed-dotted lines are the indentation moduli for the single-crystal Ni calibration material in two directions.

Table 4. Elastic constants and calculated indentation moduli in different single-crystal orientations for Ni. There are several sets of single-crystal elastic constants of Ni at room temperature (No. 1–6). The anisotropy factor varies from 2.1 to 2.6.

	Elastic constants [GPa]			A_z	Indentation modulus M [GPa]		
	c_{11}	c_{12}	c_{44}		(111)	(110)	(100)
1	244	158	102	2.4	223	218	202
2	252	151	104	2.1	234	230	216
3	251.2	157.1	121.3	2.6	249.9	243.9	224.2
4	243.6	149.4	119.6	2.5	245.9	239.9	220.6
5	252.8	152	123.8	2.5	256.2	250.2	230.6
6	246.5	147.3	124.7	2.5	254.3	248.2	228.0
7	250	160	118.5	2.6	244.6	238.6	219.2

moduli calculated from elastic constants of single-crystal Ni. Depending on the crystal orientation, the indentation moduli vary from 202 GPa to 256 GPa. The effective Young's modulus of coarse-grained polycrystalline nickel, based on the Hill approximation for averaging the different grain orientations [28, 35], is 213.1 ± 21.2 GPa. Using the pulse-echo technique, the time-of-flight for longitudinal and shear waves in the nc-Ni samples was measured at 5 MHz and at 10 MHz in the direction perpendicular to the largest sample surface

Table 5. Young's and shear moduli E and G, and Poisson's ratio υ, and indentation modulus M_{qi} calculated from bulk ultra-sonic velocities measured at 5 and 10 MHz, assuming quasi-isotropic properties of the polycrystalline samples.

Sample No.	Grain size [μm]	E [GPa]	G [GPa]	υ	M_{qi} [GPa]
Ni_1	167	209.7	79.38	0.3211	233.8
Ni_2	96	229.0	89.52	0.2790	248.3
Ni_3	65	208.65	76.99	0.3549	238.7
Ni_4	48	238.2	96.16	0.2396	252.3
Ni_5	30	215.2	82.06	0.3115	238.3

with three different polarizations. The obtained values were independent of the shear-wave polarization indicating that the nc-Ni samples were transversally isotropic. The elastic constants including the quasi-isotropic indentation moduli $M_{qi} = E/(1-\upsilon^2)$ were calculated from the sound velocities and are presented in Table 5. A dependence of Young's modulus E on the grain size cannot be deduced. As can be seen, all values of E obtained from the bulk ultrasonic measurements correspond within the error margins to the effective Young's modulus of coarse-grained nickel. The quasi-isotropic indentation moduli M_{qi} fit well between the bounds given by $M_{(111)}$ and $M_{(100)}$ as can be seen in Table 4 and in Fig. 9, and the indentation moduli M_{AFAM} are in good agreement with the values M_{qi} obtained by bulk ultrasound with the exception of sample Ni_2.

According to computer simulations [38, 39], the reduction of the elastic moduli due to the influence of the grain boundaries and triple junctions is expected to be small for grain sizes larger than 20 nm and may not be noticeable in AFAM measurements. The volume deformed by the sensing tip is about $3\pi a_c^3$ [6] with $a_c \approx 10$ nm (see Sect. 8), and hence at most one grain with its boundaries will be deformed. However, in nanoindentation a larger volume is deformed and hence one could interprete the grain-size dependence for grain sizes less than 30 nm (Fig. 9) as an effect due to the grain boundaries, despite the measurement inaccuracies. Finally, we think that the samples topography is responsible for the largest discrepancies between the values of M_{AFAM} and M_{qi}.

7. Contact stiffness as a function of load

For the first set of AFAM measurements discussed so far, contact stiffness was measured at four different static loads of the AFM cantilever. Additional insight into tip–sample forces can be obtained by recording the contact resonance frequencies and hence the contact stiffness continuously as a function of static load. Such curves depend on the elasticity of tip and sample, on the sample topography, and on the tip shape. Above we showed that an AFM tip typically

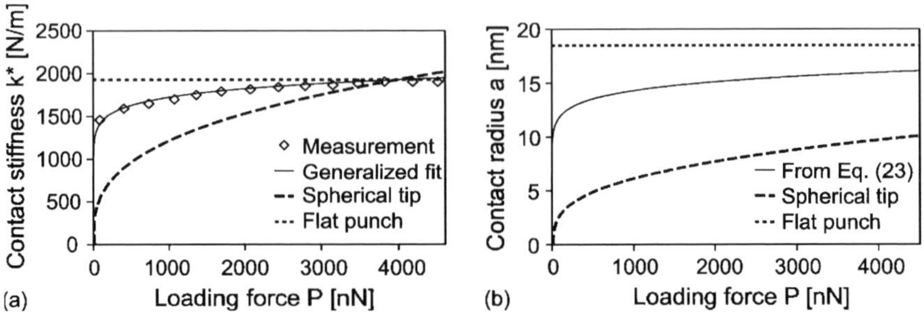

Fig. 10. (a) Contact-stiffness *versus* loading force P for a spherical tip and a flat punch and the corresponding fits to the experimental data measured on the calibration material fused silica; (b) Contact-radius *versus* loading force for a spherical tip and a flat punch on fused silica according to Eq. (24) with $n = 0.92$, $R = R' = R_2 = 18.5$ nm, and $E^* = 60.5$ GPa.

has a geometry between a sphere and a flat punch whose contact stiffness k^* is given by Eqs. (18) and (19), respectively. In order to account for deviations of the two limiting tip geometries, we propose the following contact stiffness k^*:

$$k^* = 2^{(n+2)/3} R_2^{2n-1} P^{1-n} E^{*n}, \tag{22}$$

where n is a geometrical parameter having the value 1 for a flat punch and $2/3$ for a spherical tip. These two limiting cases retrieve Eqs. (18) and (19), if the geometrical quantity R_2 is the radius of a sphere or a flat punch, respectively. It can be seen in Fig. 10, solid line, that the measured contact-stiffness on the calibration material fused silica *versus* loading force can be well described by Eq. (22) by setting $R_2 = 18.5$ nm, $n = 0.92$, and using the indentation moduli $M_{\text{Quartz}} = 75.8$ GPa and $M_{\text{tip}} = 300$ GPa, which results in $E^* = 60.5$ GPa. In order to obtain the contact radius a_c, we use k^* from Eq. (22) and divide it by $2E^*$ according to Eq. (14). This yields

$$a_c = \frac{k^*}{2E^*} = 2^{(n-1)/3} R_2^{2n-1} P^{1-n} E^{*n-1}. \tag{23}$$

Setting $n = 1$ or $2/3$ in Eq. (23) retrieves also the contact radius for the flat punch and the sphere, respectively. The dashed and the dotted lines in Fig. 10 are obtained by setting the contact radius a_c in the Hertzian contact or the radius of the flat punch R_p such that at $P = 4\,\mu$N the contact stiffness are the same, here $k^* \approx 1930$ N/m, and $a_c \approx 16$ nm. The experimental curve shows a clear deviation from the two limiting cases. This typical behavior of AFAM data can be explained by a deviation of the tip shape from the spherical and the flat punch geometries. We found that in quite a number of measurements, also in different materials than the ones examined here, Eq. (22) fits the contact stiffness *versus* load well. Despite this, the validity of Eq. (22) must be further examined, in particular the physical meaning of R_2 and n.

Table 6. Grain size as a function of the annealing time at 200 °C. The grain size was calculated from the width and the angular position of the X-ray diffraction profiles, using Scherrer's equation [32]. The samples not designated were not measured by AFAM.

Sample No.	Annealing time at 200 °C [min]	Grain size [nm]
Ni_{10}	0	14 ± 3.2
Ni_9	15	34.5 ± 22
Ni_8	30	35.1 ± 19
Ni_7	45	50 ± 7
–	60	54.6 ± 6
–	75	61.3 ± 5
Ni_6	90	66.8 ± 4
–	105	64.4 ± 4
–	120	77.5 ± 3
–	150	78.1 ± 3

8. Experimental results for phosphor-stabilized and annealed electrodeposited nanocrystalline nickel

In a second series of experiments another attempt was made in order to find out whether the indentation modulus or the Young's modulus of nc-Ni is grain-size dependent [40]. The investigated material was again pulse electro-deposited nanocrystalline nickel with 3 wt. % phosphor [30]. The samples were obtained from a single deposit and subsequently annealed at 200 °C in order to obtain a variety of grain sizes.

8.1 Characterization of the microstructure by AFM tapping mode, EBSD, and XRD

The samples were characterized by X-ray diffraction using Cu $K_{\alpha 1}$ radiation with respect to grain size, lattice parameter, strain, and texture. The grain size was calculated from the width of the (111) Bragg-peak profiles and their angular position, using the Scherrer equation [32]. Table 6 reports the change in grain size after annealing at 200 °C for times from 15 to 150 min. From these data we observed an incubation time of 30 min before the microstructure started to coarsen. After 150 min the grain size reached 78 nm. The considered Bragg profiles did not exhibit a symmetric shoulder typical for abnormal growth [41]. From the angular position of the (100)-, (111)-, and (311)-reflexes the macro-strain was calculated according to the following equation [42]:

$$\varepsilon = \frac{\Delta d}{d_0} = -\cot\theta\,\Delta\theta\,, \tag{24}$$

where d_0 is the unperturbed lattice parameter, Δd is the change in lattice constant caused by a macro-strain and $\Delta\theta$ is the change of the Bragg angle θ.

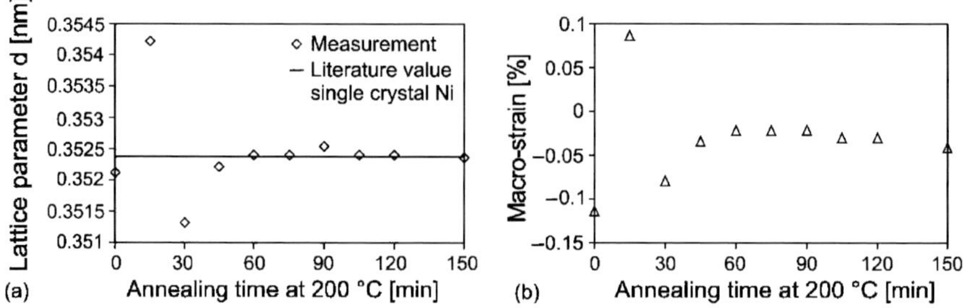

Fig. 11. (a) Lattice parameter as a function of the annealing time at 200 °C. The lattice constant was calculated from the position of the (311) reflex; (b) macro-strain calculated from the position of the (311)-reflex as a function of annealing time.

During the incubation time the lattice relaxed as shown in Fig. 11a and b. After 30 min the strain almost vanished. The lattice parameter changed in the first 30 min of annealing and then remained the same after the strain vanished. Furthermore, pole figures were made for the (111) crystallographic plane in the as-received sample and after annealing for 90 min and 150 min. Like the samples from the first series, the as-received sample already exhibits some texture in the [111] direction which is attributed to the sample-preparation procedure. Further investigation using EBSD revealed, however, that after 90 min annealing at 200 °C all grain orientations were present. The EBSD orientation map also revealed a much coarser microstructure than the one measured by XRD in the first set of samples. However, EBSD only images grains at the surface whereas XRD probes the interior. The microstructure depicted on the orientation map is similar to an abnormally grown microstructure. Still, as mentioned above, no symmetric shoulders were observed on the XRD reflexes. Parallel AFM images were carried out on our nc-Ni 3 wt. % phosphor samples after annealing at 200 °C for different time periods by using the tapping-mode phase-contrast (Fig. 12). Considering all results from the investigation of the microstructure obtained in this work we assume, that before growing by grain boundary-migration, the nano-sized grains first rearrange and agglomerate during a certain incubation time. The mechanisms involved in these microstructural changes during annealing are not well-understood and deserve further investigation. As seen in the image made after 60 min annealing, we observe that these agglomerates consist of single nano-sized grains.

8.2 Measurement of the local indentation modulus by AFAM using stiffness as a function of load

In this second series of experiments the calibration procedure consists in a successive continuous recording of contact resonance curves as a function of static load on the calibration samples and the material under test. For these measurements we used stiff silicon single-crystal cantilever-beams, $k_c = 39$–41 N/m

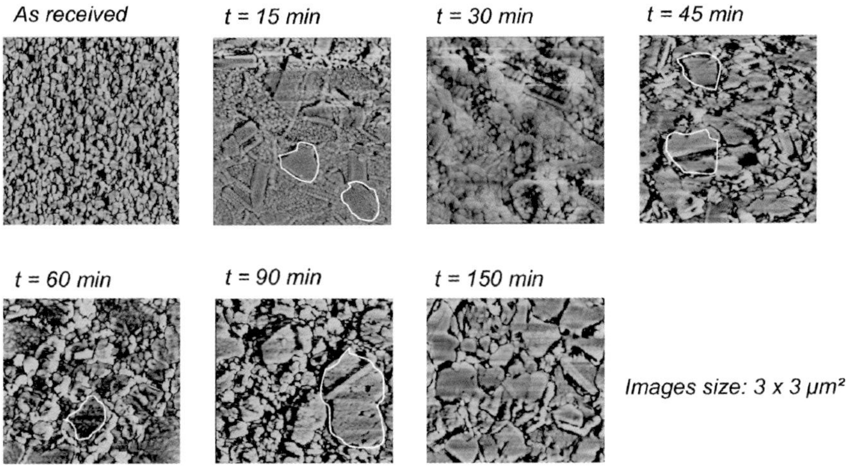

Fig. 12. Phase-contrast AFM tapping mode image of the microstructure of nc-Ni after annealing at 200 °C. With increasing annealing time we observe a rearrangement of the microstructure. Initially homogeneously distributed, the nano-sized grains form coarser agglomerates (as framed in white).

with diamond-coated tips (type dt-nclr, Nanosensor), and a new AFM-sensor was used for each nc-Ni sample. As calibration samples, a single Ni crystal oriented in the [111] direction with an indentation modulus of $M_{\langle 111\rangle} = 255$ GPa and a fused silica sample with an indentation modulus of $M_{\mathrm{iso}} = 75.8$ GPa (see Sect. 4) were used. The ratio L_1/L of tip position and cantilever length was determined for each cantilever from the first and the second contact-resonance bending mode recorded on fused silica. The indentation modulus M_{tip} of the cantilever tips were evaluated using the two calibration samples. To this end the $k^*(P)$-plots of the reference samples were fitted using Eq. (22) assuming a common value of the tip indentation modulus M_{tip} and individual parameters R_2 and n for fused silica and (111)-oriented single-crystal nickel, respectively. Assuming a starting value $M_{\mathrm{tip}} = 250$ GPa the parameters were optimized till slight variations in the value for M_{tip} did not lead anymore to a significant change of R_2 and n (see Figs. 13a and b). For the measurements depicted in Figs. 13a and b the n was found to vary from 0.88 to 0.93 after measurements on fused silica and (111)-oriented Ni single crystal which corresponds to a change of 15%. The radius R_2 was found to vary from 11.2 nm to 18.8 nm. The indentation modulus of the diamond-coated silicon tip was found to vary from $M_{\mathrm{tip}} = 210$ GPa to $M_{\mathrm{tip}} = 297$ GPa with a mean value of $M_{\mathrm{tip}} = 236.1$ GPa. This variation is quite reasonable considering the data published for diamond-like carbon-layers [10, 43, 44].

Based on the model outlined in Eqs. (2)–(13) and described in detail including damping in [5], contact resonance spectra as measured by a beam-

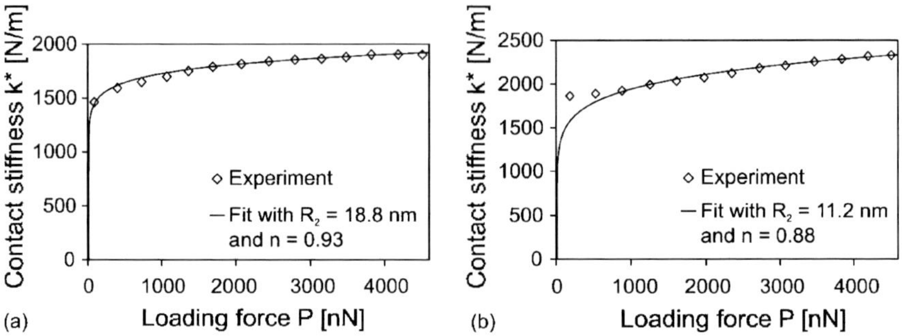

Fig. 13. Measured contact stiffness as a function of static load and fit using Eq. (23) for fused silica (a) and (111) oriented single-crystal nickel (b). From this measurement the value of the tip indentation modulus was calculated to be $M_{\text{tip}} = 243$ GPa. As can be seen from both figures, Eq. (23) leads to a very good fit of the measured data for the calibration samples.

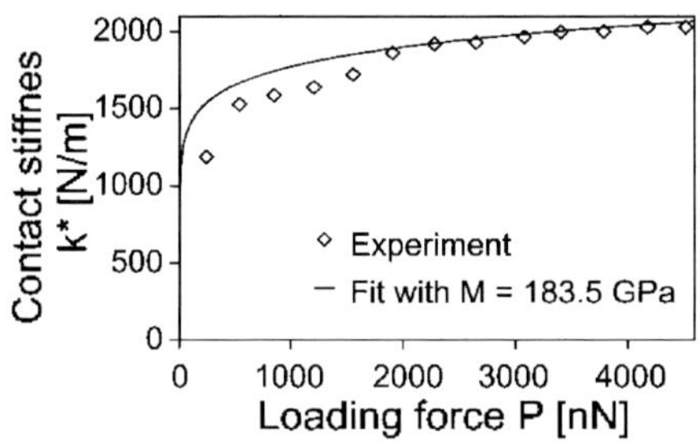

Fig. 14. Fit of the contact stiffness *versus* loading force P in nc-Ni with a grain size of 14 nm based on Eq. (23). The fit parameter was the indentation modulus of $M =$ 183.5 GPa.

deflection detector were simulated. The ratio L_1/L of tip position and cantilever length previously determined for each cantilever on fused silica was used. On the various nc-Ni samples k^* and γ^* were obtained by fitting the contact-resonance curves of the first bending mode with the theoretical response function of the cantilever beam. The damping constant contributes to the stiffness of the contact [45] but as already mentioned above, the Q-value of the contact resonances measured was about 50, allowing one to neglect the influence of the contact damping on the contact stiffness k^* [40].

Using the same procedure as for the calibration samples, the reduced elasticity moduli of the nc-Ni samples were obtained by fitting the experimental results for $k^*(P)$ with Eq. (22) (Fig. 14), where R_2 and n were selected within the values ranges determined for the reference measurements. Table 7 shows the indentation moduli of five of the nc-Ni samples of the second series (designated Ni_6–Ni_{10}) of different grain sizes obtained with AFAM. In Table 7

Table 7. Indentation modulus of nc-Ni with different grain sizes. A new AFM-sensor was used for each nc-Ni sample measurement with different tip indentation moduli M_{tip}.

Sample	Grain-size nc-Ni [nm]	Δk^* [%]	M_{tip} [GPa]	$M_{nc\text{-}Ni}$ [GPa]
Ni_6	14 ± 3.2	5.1	255.6 ± 37	192 ± 48
Ni_7	34.5 ± 22	2.8	248 ± 3	124.6 ± 14
Ni_8	35.1 ± 19	5.1	248 ± 3	152.9 ± 90
Ni_9	50 ± 7	5.9	224.2 ± 4	186.9 ± 35
Ni_{10}	66.8 ± 4	6.1	216.6 ± 6	193 ± 9

the errors given for M_{tip} and $M_{nc\text{-}Ni}$ were calculated from the standard deviation obtained for the three different calibration measurements yielding errors varying from 4.4% to 58%. Finally the error average associated with the fit to Eq. (22) was calculated from the difference between the measurement and fitting curve yielding Δk^*. This yielded errors in M varying from 2.8% to 6.1%. These errors are mainly given by the deviation of the fit from the measured data at loads $P \approx 1$–$2\ \mu$N (Fig. 14). This deviation was also accompanied by a maximum in the contact damping, and its origin will be the subject in a forthcoming paper. If we plot the data for Ni_6 to Ni_{10} into Fig. 9, one can see that the data for $M_{nc\text{-}Ni}$ of the second series are lower than the values from the first series.

In the second series of measurements on nc-Ni, the as-received sample had the smallest grain size, 14 nm. Using Eq. (23) with $R_2 = 18.8$ nm, $n = 0.93$, and $M = 192$ GPa, leads to a contact radius $a_c = 8$ nm at a loading force $P = 4.5\ \mu$N. The stress-field of the tip extends about $3a_c$ into the depth certainly stressing the nearest grain boundary. The measured values for M are for all nc-Ni samples below the indentation moduli of single-crystal nickel: $M_{100} = 219$ GPa, $M_{110} = 249$ GPa, and $M_{111} = 255$ GPa (Table 4), yet they are comparable to the first series of measurements (Fig. 9), except in two cases where appreciably lower values were measured.

A reduced elastic modulus for nc-Ni has been reported in [36, 46–49]. In [46] the elasticity measurements were performed by using AFAM on nc-Ni thin films with grain sizes ranging from 8 nm to 50 nm. As the contact radius was comparable to the grain size, the probed volume included several grains and grain boundaries. The reduced elasticity modulus of nc-Ni compared to (111)Ni single crystal was attributed to the effect of the grain boundaries. For the nanoindentation data reported in [36], the situation was the same, several grain boundaries were deformed, and hence the reduced moduli for the smallest grain size measured (21 and 24 nm) are attributed to the same cause. The most detailed measurements in this respect have been performed by Zhou *et al.* [48]. A clear dependence of the Young' modulus as a function of grains size was found for grain sizes below 20 nm which was attributed to the effect of softening of the grain boundaries and triple junction. A rule of mixture was

39. S. J. Zhao, K. Albe, and H. Hahn, Scripta Mater. **55** (2006) 473.
40. A. Caron, PhD thesis, Science and Technical Faculty III, Saarland University (2007) in preparation.
41. J. Weissmuller, W. Krauss, T. Haubold, R. Birringer, and H. Gleiter, Nanostruct. Mater. **1** (1992) 439.
42. R. Glocker, *Materialprüfung mit Röntgenstrahlung*. Springer, Berlin (1985).
43. B. Bhushan, Diam. Relat. Mater. **8** (1999) 1985; X. Li and B. Bhushan, J. Mater. Res. **14** (1999) 2328.
44. D. Schneider, T. Schwarz, H. J. Scheibe, and M. Panzner, Thin Solid Films **295** (1997) 107.
45. J. A. Turner, S. Hirsekorn, U. Rabe, and W. Arnold, J. Appl. Phys. **82** (1997) 966.
46. G. Kopycinska-Müller, R. H. Geiss, J. Müller, and D. C. Hurley, Nanotechnology **16** (2005) 703.
47. K. Fujiwara, H. Tanimoto, and M. Mizubayashi, Mater. Sci. Eng. A **442** (2006) 336.
48. Y. Zhou, U. Erb, K. T. Aust, and G. Palumbo, Z. Metallk. **94** (2003) 10.
49. T. D. Shen, C. C. Koch, T. Y. Tsui, and G. M. Pharr, J. Mater. Res. **6** (1991) 1012.
50. J. Tian, H. Ogi, and M. Hirao, Appl. Phys. Lett. **87** (2005) 204 107.

Mechanical Properties of Nanomaterials Examined with a NI-AFM

By H. Vehoff[1,*], B. Yang[1], A. Barnoush[1], H. Natter[2], and R. Hempelmann[2]

[1] Saarland University, Department of Materials Science, Bldg. D2.2 P.O. Box 151150, D-66041 Saarbruecken, Germany
[2] Saarland University, Department of Physical Chemistry, Bldg. B2.2, D-66123 Saarbruecken, Germany

Nanomaterial / Nanoindentation / Dislocation / Strain Rate / Hydrogen Embrittlement

The mechanical properties and strain rate sensitivity of nanocrystalline nickel was studied as a function of grain size at different temperatures in tensile tests and with a nanoindenter in order to examine the different deformation mechanisms of nanocrystalline materials. The effect of lateral boundaries and hydrogen on the nucleation of dislocations was studied in detail. For the first time it was possible to observe the reduction of the dislocation nucleation stress on a nanoscale. In addition the experiments yielded, depending on temperature and strain rate, the strain rate sensitivity, the activation volume and the creep exponents as a function of stress and grain size. From the creep experiments the transition between grain boundary sliding and dislocation climb as a function of temperature was obtained. The strain rate jump tests gave extremely small activation volumes, nearly a factor of 100 smaller than in conventional nickel as a function of grain size. To help in understanding this behaviour the strain rate sensitivity of single grains was tested with a nanoindenter. The results clearly showed that the primary interaction of dislocations with grain boundaries is the reason for the observed strong rate effects and small activation volumes.

1. Introduction

The measurement of the mechanical properties on the scale of the microstructure and of nanomaterials is still an ambitious topic in modern research. In the last years mainly two techniques have been established, compression and bending tests on micro-specimens cut by FIB (focused ion beam) and loaded by a nanoindenter und instrumented nanoindentation combined with imaging

* Corresponding author. E-mail: Vehoff@matsci.uni-sb.de

Z. Phys. Chem. **222** (2008) 499–525

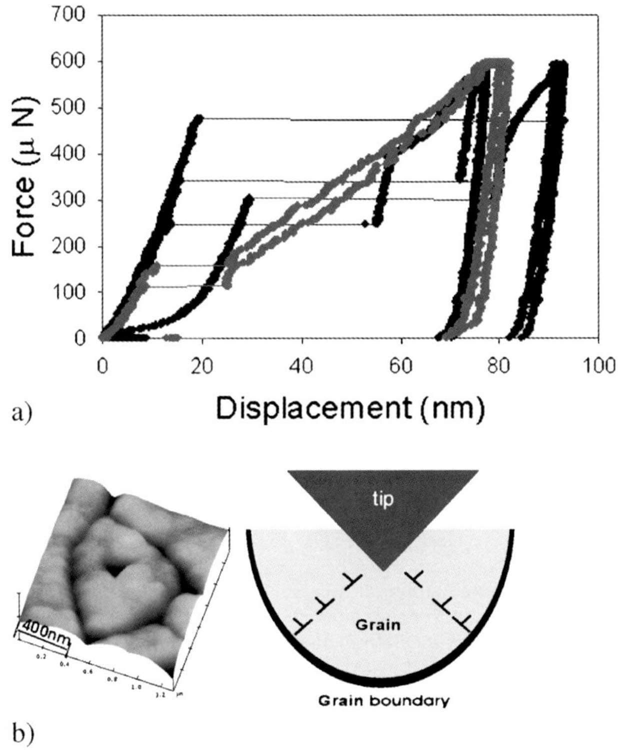

Fig. 1. a) Force–displacement curves on oxide films with different thicknesses in Ni, b) image and schematic drawing of an indent positioned in the centre of a grain.

within an AFM (atomic force microscope). In this short report recent results on local mechanical properties measured with a NI-AFM (nanoindenting AFM) obtained or based on the interaction within the SFB are reported as the effect of interfaces on the hardness of single grains, strain rate sensitivity of single grains and the effect of hydrogen on dislocation nucleation.

2. Testing of microstructures with an NI-AFM

Nanoindentation is a well-established technique to study the mechanical properties of thin films. For the study of materials with a complicated microstructure a NI-AFM, a nanoindenter combined with an AFM, is needed. When a nano-sized tip indents a surface several effects must be considered, which alter the measured hardness. These can be surface roughness, surface films, and the underlying microstructure like the dislocation cell structure, as well as interfaces perpendicular and below the surface [1]. All these effects result in a large scatter of hardness and alter the load-displacement curve. Figure 1 shows as an example the effect of oxide films in single crystal Ni on the load displacement curve. The grey curves correspond to the freshly electropolished surface, the black curves to the same surface after being exposed to air for one day. Hardness and pop-in behaviour of the fresh surface are almost the same but scatter significantly after longer exposure to air.

Therefore, we put much effort into the specimen preparation described in detail elsewhere [2, 3]. We even make nanoindentation in a protective atmosphere or under controlled electrochemical condition to have a complete control of the oxide layer thickness [4].

In addition, the evaluation of hardness data for nanoindentation is only developed so far for single crystals. However, underlying interfaces as well as lateral interfaces which have a distance or size in the order of the tip interaction radius can affect the results drastically, which on the other hand, can be used to study the influence of interfaces on hardness directly, which is the topic of this report. Without interfaces, in materials which deform by dislocations nucleation and motion, the hardness decreases with increasing indentation depth [5, 6] known as indentation size effect (ISE). This decrease is associated with geometrically necessary dislocations (GNDs). Nix and Gao (NG) [7] developed a simple model for the ISE of crystalline materials by considering the density of GNDs around a conical indenter. It is worth to note that for example Swadener *et al.* showed that the nanoindentation results should not fit this model [8]. On the other hand, Durst *et al.* [9] using a correction factor for the plastic zone radius improved the Gao–Nix model and showed a very good agreement for indentation depths as low as 20 nm. The analytical formulation of NG-model in combination with the modification given by Durst is used in this study to discuss the effect of grain boundary on hardness. In the NG-model, it is assumed that the indentation is accommodated by circular loops of GNDs with Burgers vector b normal to the surface. The total line length of dislocation loops necessary to form the shape of the conical indenter is:

$$\lambda = \frac{\pi h a}{b} = \frac{\pi \tan\theta}{b} a^2 = \frac{\pi}{b} \frac{h^2}{\tan\theta} \tag{1}$$

where h is the indentation depth, the contact radius, and θ the angle between the surface of the indenter and the plane of the sample surface. It is assumed that all the GNDs are contained in a hemisphere, namely the plastic zone, with a radius equal to the contact radius a. The volume of the plastic zone is defined by the contact radius

$$V = \frac{2}{3}\pi a^3 = \frac{2}{3}\pi \frac{h^3}{\tan^3\theta}\,. \tag{2}$$

Thus the density of the GNDs is given by:

$$\rho_G = \frac{\lambda}{V} = \frac{3}{2}\frac{\tan^2\theta}{bh}. \tag{3}$$

By taking Tabor's factor to be 3 and applying the von Mises flow rule and the Taylor relation, the relation between hardness and total dislocation density ρ was expressed as:

$$H = 3\sigma = 3\sqrt{3}\tau = 3\sqrt{3}\alpha_1 Gb\sqrt{\rho} = 3\sqrt{3}\alpha_1 Gb\sqrt{\rho_G + \rho_s} \tag{4}$$

where σ is the normal stress, τ the shear stress, α_1 a constant, G the shear modulus, and ρ_s the density of statistically stored dislocations which determines the hardness H_0 in the absence of any GNDs:

$$H_0 = 3\sqrt{3}\alpha_1 Gb\sqrt{\rho_s}. \tag{5}$$

From these equations, the measured hardness H related with the indentation depth h can be described as:

$$\frac{H}{H_0} = \sqrt{1 + \frac{\rho_G}{\rho_s}} = \sqrt{1 + \frac{h_0}{h}} \tag{6}$$

where h_0 is a length scale depending on ρ_s through H_0. Rewriting this equation as

$$H^2 = H_0^2\left(1 + \frac{h_0}{h}\right) \tag{7}$$

a linear relationship would be found in the plot of H^2 *vs.* h^{-1}. The deviation from this relationship is used below to discuss the effect of grain boundaries on hardness since for small grain sizes the effect of depth and grain size on hardness cannot be measured separately.

The NG-model overestimates the hardness since it underestimates the plastic zone, especially at small indentation depths [10, 11]. In the NG-model the plastic zone is assumed to be a hemisphere with a radius equal to the contact radius a. However, GNDs will spread beyond this hemisphere, due to the strong repulsive force between GNDs for very small indentation depths. Therefore the plastic zone volume should be corrected. The correction proposed by Durst *et al.* recently [10], where the radius of the plastic zone is assumed to be fa with f ranging from 0 to 3.5 resulted from finite element simulations. The factor f plays an important role to explain the ISE, but a measurement of f is not straight forward. In our case, however, we measure in a grain size range for which the plastic zone is confined by grain boundaries which act as strong barriers for dislocation transmission.

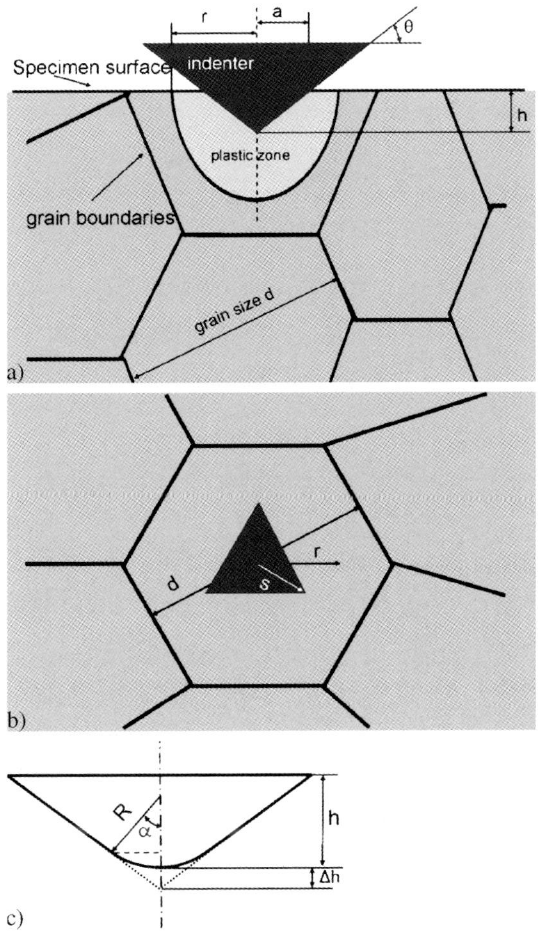

Fig. 2. Schematics of nanoindentation performed in the centre of a grain. The plastic zone size is assumed to be a hemisphere with a radius r. The grain sized is denoted as d, the contact radius is denoted as a, the indentation size is denoted as s, the indentation depth is denoted as h, θ is the angle between the surface of the indenter and the plane of the sample surface, c) enlarged view of the tip.

A schematic diagram and a picture of a nano-indent in the centre of a single grain is shown in Fig. 1b. The grain size is denoted as d, the contact radius as a and the radius of indentation size as s (Fig. 2. For simplicity we take

$$s = a\,. \tag{8}$$

The plastic zone size scales with the contact radius

$$r = fa \tag{9}$$

where r is the radius of the plastic zone size and $f \geq 1$, as treated by other authors [11].

The relation between the indentation depth h and the radius of indentation size s can be deduced from the area function of an ideal Berkovich indenter (without tip bluntness):

$$A_c = 24.56h^2 = \frac{3\sqrt{3}}{4}s^2 \tag{10}$$

$$s \approx 4.35h\,. \tag{11}$$

For a blunt tip with radius R, (Fig. 2), a height supplement Δh has to be considered and Eq. (11) then changes to:

$$s \approx 4.35(h + \Delta h)\,. \tag{12}$$

Taking the end of blunted pyramidal indenter as a pommel with the same area function of an ideal Berkovich indenter (Fig. 2), we have

$$24.56(R - R\cos\alpha)^2 = \pi(R\sin\alpha)^2\,. \tag{13}$$

This results in an apex angle $\alpha = 20°$ and

$$\Delta h = R/\cos\alpha - R = 0.064R\,. \tag{14}$$

Equation (12) is

$$s \approx 4.35(h + 0.064R)\,. \tag{15}$$

This gives the relationship between the plastic zone size $2r$ and the indentation depth h:

$$2r = 2fa = 2fs = 8.7f(h + 0.064R)\,. \tag{16}$$

In this work, the tip radius of the indenter is about 200 nm, thus Eq. (16) is

$$2r = f2s = 8.7f(h + 13) \tag{17}$$

2.1 Indentation size effect *vs.* grain boundary effect

When now the indenter is positioned in a centre of a grain, as demonstrated in Fig. 1b, h can be measured by varying the size relation between indent and grain. Figure 3 shows that the hardness of coarse grains with a size of about 80 μm decreases with increasing indentation depth. This is the ISE observed in single crystals. Considering the small indentation depth (maximum 100 nm),

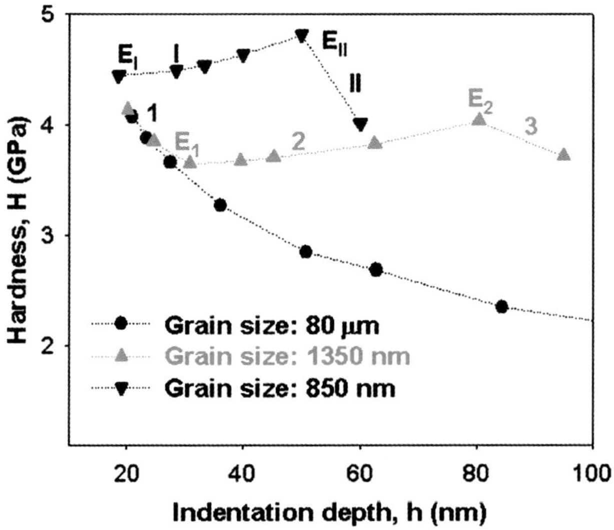

Fig. 3. Nanohardness (H) *vs.* indentation depth (h).

grains larger than 80 μm can be treated as single crystals, therefore the ISE observed here is not surprising. For a double check, H^2 *vs.* h^{-1} is plotted with the data from Fig. 3. A linear relationship between H^2 and h^{-1} is obtained as predicted by Eq. (7).

Figure 4 also shows that at low indentation depths (smaller than 31 nm) the data of grains with a size of about 1350 nm and 80 μm fell roughly on the same line. This means that for this smaller indentation depth range, the hardness depends mainly on the indentation size and not on the grain size. For a nanoindenter, a large grain behaves like a single crystal.

At a larger indentation depth or for lower grain sizes, however, the indenter sees the boundaries and the hardness starts to increase again with increasing indenter depth. This is due to the strong blocking of dislocations by grain boundaries.

In the NG-model, the effect of interfaces is not taken into account. The density of GNDs ρ_G depends on the total length λ of the GNDs and the plastic volume V in which the GNDs are contained. In addition, the volume of indent formed in the grain during indentation, V_{indent}, which also increases with indentation depth h, is ignored. Since λ is proportional to h^2 (Eq. (1)) and the plastic zone volume V is proportional to h^3 (Eq. (2)), hardness decreases with increasing h in the model.

Boundaries do not alter λ, however, Eq. (2) does not hold any longer. The plastic zone is now confined by the grain size, altering the relation between the plastic zone volume V and indentation depth h. The volume of the plastic zone V, according to Fig. 2, is now defined by the radius of the grain size and can be

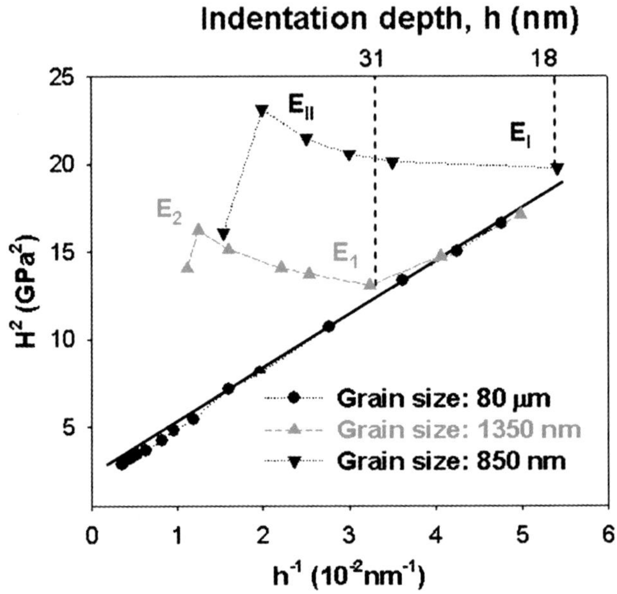

Fig. 4. Examination of the indentation size effect in single grains with plots of H^2 *vs.* h^{-1}.

described as:

$$V = \frac{\pi}{12} d^3 \tag{18}$$

where d is the grain size. In addition, it is also necessary to take V_{indent} into account, and the actual plastic zone volume V should be:

$$V = \frac{\pi}{12} d^3 - V_{\text{indent}} \tag{19}$$

$$V_{\text{indent}} = \gamma h^3 \tag{20}$$

where h is the indentation depth and γ a constant depending on the real shape of the indenter (bluntness).

Assuming that the indented single grain has a real volume V_{grain} and that the plastic zone is confined to the grain, the actual plastic zone volume V can than be written as:

$$V = \beta \frac{\pi}{12} d^3 - \gamma h^3 \tag{21}$$

where β is a factor increasing from 1 to a constant value smaller than $V_{\text{grain}}/(\frac{\pi}{12} d^3)$. Then the density of GNDs ρ_G can thereby be written as:

$$\rho_G = \frac{\lambda}{V} = \frac{\frac{\pi}{b} \frac{h^2}{\tan\theta}}{\beta \frac{\pi}{12} d^3 - \gamma h^3} . \tag{22}$$

In this case, Eq. (6) changes to:

$$\frac{H}{H_0} = \sqrt{1+\frac{\rho_G}{\rho_s}} = \sqrt{1+\frac{h^2}{h_0^* \beta d^3 - \gamma^* h^3}} \qquad (23)$$

where h_0^* still depends on ρ_S through H_o and γ^* is a constant.

Equation (23) now indicates that hardness increases with increasing h when the grain boundaries hinder the spreading of the plastic zone. This explains why hardness increases again with increasing indentation depth h after a critical indentation depth at which the plastic zone riches the grain boundary. Equation (23) also indicates that at a constant indentation depth h, the smaller grains will exhibit higher hardness.

2.2 Indentation size *vs.* plastic zone size

Assuming that at the critical indentation depth (corresponding to the extreme point E_1 in Fig. 3) the produced plastic zone reaches the surrounding grain boundaries, then according to Fig. 4 we have

$$r = fa = fs = d/2\,. \qquad (24)$$

Combining Eq. (17) results in $f = 3.52$ for grains with a size of 1350 nm at an indentation depth of 31 nm. This means that the radius of the plastic zone is about 3.5 times larger than the contact radius.

According to Eq. (17) and taking $f = 3.52$, at an indentation depth of 18 nm the plastic zone size reaches about 950 nm, which is larger than the indented grain size. This means that the dislocations below the indenter tip interact directly with the surrounding interfaces for this grain sizes and the plastic zone is confined. In this case, ρ_G increases with the indentation depth h (Eq. (22)), hence an increase in the nanohardness occurs.

2.3 The spreading of the plastic zone out of the indented single grains

When single grains are indented the hardness decreases again when the plastic zone starts to spread in adjacent grains. This results in an extreme point in hardness (E_2 and E_{II} in Fig. 3). A decrease in hardness with increasing h indicates a decrease in the dislocation density. This would happen, considering that at a sufficient h the indentation size would eventually exceed the indented grain size. In this case the factor β in Eq. (21) would become larger than $V_{\text{grain}}/(\frac{\pi}{12}d^3)$, depending how many surrounding grains are indented and the grain sizes of these indented grains. This increases the plastic zone volume V and thus decreases the dislocation density, leading to a decrease in hardness.

The extreme point corresponds to an indentation depth of about 48 nm for the grains with size of about 850 nm, and 80 nm for the grain with size of about 1350 nm. According to Eq. (17), the indentation size $2s = 530.7$ nm at $h = 48$ nm, which is smaller than the indented grain size 850 nm; at $h = 80$ nm the indentation size $2s = 809$ nm, also smaller than the indented grain size 1350 nm. According to Eq. (23), hardness will increase with increasing h if only the plastic zone is confined within the indented grain. Therefore, the decrease in hardness with increasing h observed here, *i.e.* the increase in the plastic volume V, must result from the spreading of the plastic zone out of the indented grains to surrounding grains. This is a direct evidence of the activation of new dislocation sources in the neighbouring grains caused by the pile-up of dislocations. Before this happens, a higher maximum hardness for the smaller grains (850 nm) than that of the larger grains (1350 nm) was found, indicating that the smaller the grain size, the higher the dislocation density in the grain. It also means that for a larger grain, a smaller external applied load is needed to activate the new dislocation sources in neighbouring grains. This agrees with the well-known dislocation pile-up model of the Hall–Petch effects. In coarse grained materials dislocations primarily interact which each other. In ultra fine grained materials, however, the dislocations nucleated below the indenter primarily interact with the grain boundary which results in new properties as will be discussed below.

3. Hardness as a function of grain size

A simple way to test deformation mechanisms is using hardness measurements. Figure 5a shows the hardness of nickel (Ni) measured by a typical Vickers indenter as a function of grain size. The results clearly demonstrate that the grain size dependence of hardness changes around 1 μm and around 20 nm [12]. TEM examinations in this grain size range show, that approximately at grain sizes below 1 μm, the typical dislocation cell structure cannot develop.

With the nanoindenter a different behaviour is observed. For large grain sizes only single grains are tested, a large scatter is observed in the range 2 (Fig. 5b) where the indenter size fell into the range of the grain size. When the grains become smaller the nanoindenter behaves like a microindenter, however, the hardness is still a function of the indent depth as discussed above. The experiments given below consider mainly range 1, hence the hardness of single grains.

4. Interaction between dislocations and interfaces

The microstructure obtained after a special *in-situ* heat treatment in a SEM [13] and typical force displacement curves are shown in Fig. 6a–d. The indentation was always made in the centre of the largest grains for a given grain size distribution obtained from a specific heat treatment in order to diminish the probability of hitting boundaries directly under the surface. The good specimen

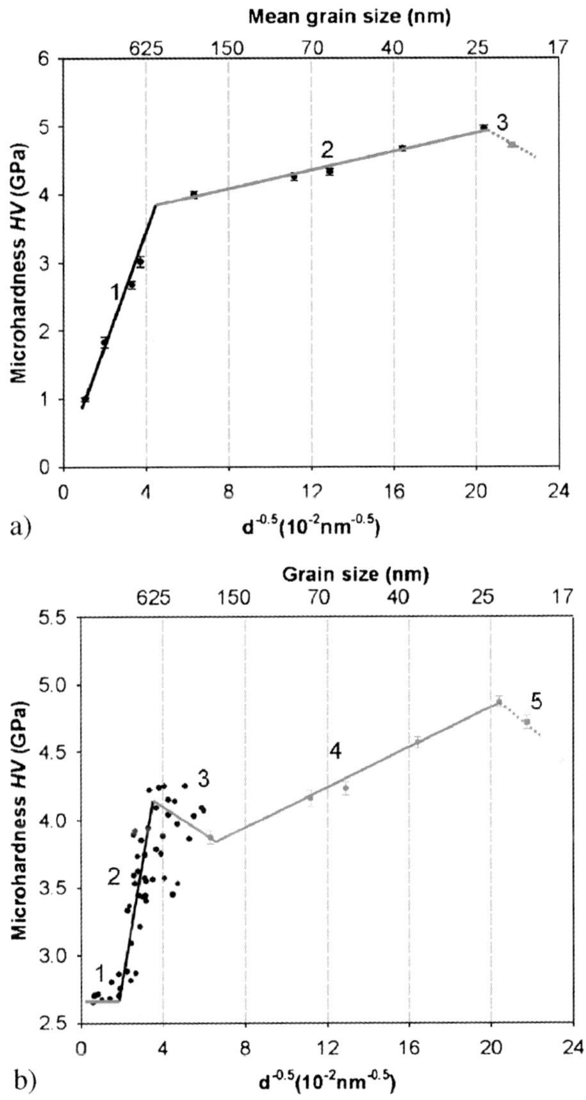

Fig. 5. Vickers-Hardness (maximum loading force: 4900 mN for 15 s) versus mean grain size. Different grain sizes are produced by heat treatments from as deposited Ni, b) the same plot for nanohardness.

preparation guaranteed constant pop-in loads in the grain size range examined in Fig. 6. The load displacement curves (Fig. 6e) had distinct properties, the pop-in width (labelled 1) as well as the indentation depth increased with grain size until a critical grain size is reached (Fig. 7).

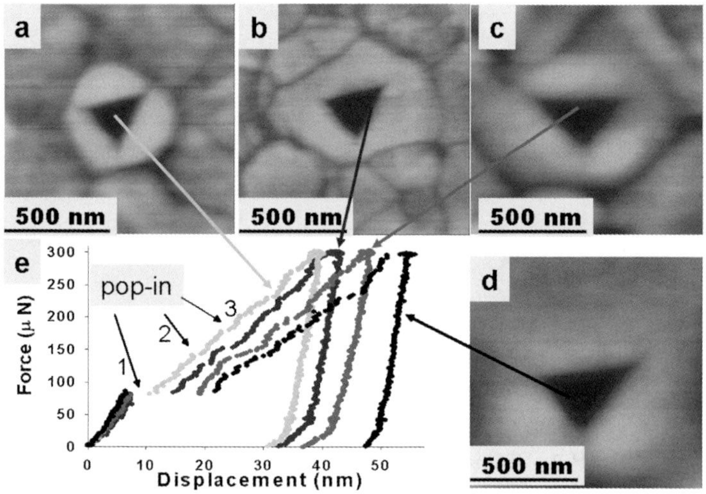

Fig. 6. (a)–(d) nanoindents in the centre of grains of UFG-Ni with different mean grain sizes; (e) corresponding load–displacement curves.

Before we discuss the results presented in Fig. 6, some remarks on the effect of grain boundaries on hardness are needed. Indenting into a single crystal results in decreasing hardness with increasing depth (indentation size effect, straight line in Fig. 4), and for well polished surfaces, pop-ins occur with similar width independent of the indent position (black curve in Fig. 6e) which indicates the pop-in width in single crystals. However when the dislocations start to interact with the grain boundaries (decreasing pop-in width) the effect of the grain boundaries clearly override the indentation size effect, even better seen in the Nix–Gao plot [7] given in Fig. 4. In Fig. 7a and b the effect of grain size on pop-in width and hardness is plotted. The data clearly show that the hardness increases as long as the dislocations could pile up against the boundary (Fig. 7b). When the indentation depth is further increased, dislocation sources in neighbouring grains are activated, which results in a decrease of hardness (Fig. 7a).

Figure 8 shows the effect of adjacent grains in more detail for a peak force of 250 μN. An increase in hardness with decreasing grain size can be seen, when the indent size is smaller than the indented grain size (Fig. 8a–c), while the plastic zone scales with the grain size. When the indent size is larger than the indented grain, the plastic zone spreads over to the surrounding grains and a decrease in hardness was found (Fig. 8d).

Note the pop-ins (discontinuity in the force–displacement curve) in Fig. 8. Besides the first pop-in, later pop-ins appear with increasing applied force. The displacements and loads for the different pop-ins are plotted in Fig. 9. The results can be summarized:

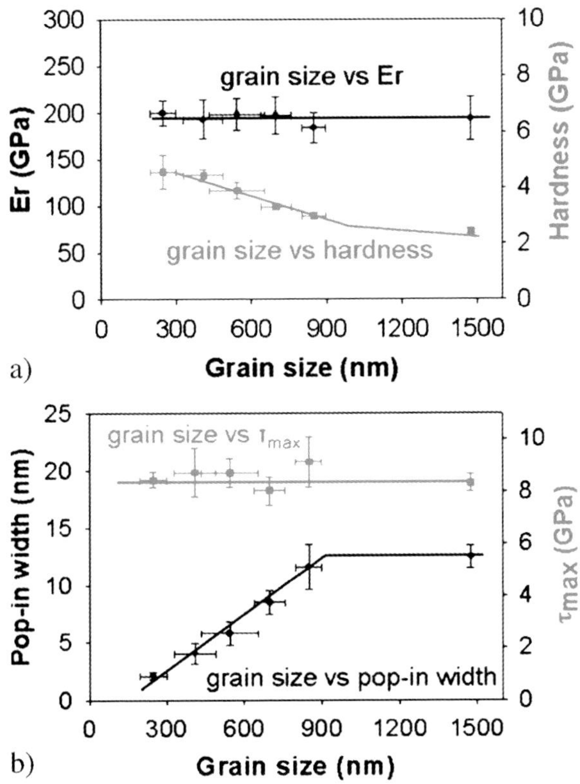

Fig. 7. (a) Hardness *vs.* indentation depth for nanoindents in the centre of grains corresponding to Fig. 6. (b) First pop-in width and maximum shear stress at pop-in *vs.* grain size from the curves in Fig. 6e.

- The first pop-in always appears at similar displacements (indent depth around 6 nm) at nearly the same applied force (around 70 μN) for all individual grains (Figs. 7b, 9)
- The later pop-ins appear also at similar displacements (about 13 nm for the second and 25 nm for the third), but only for the grain sizes smaller than 1000 nm. For grains larger than 1000 nm, no later pop-ins are detected in this displacement range. Here, it can be assumed, that the dislocation pile-ups formed under the tip under load control are not interacting directly with the grain boundary any more.
- With decreasing grain size an increasing force and a decreasing displacement for later pop-ins were found.
- The first pop-in corresponds to the initial yield point of nickel (the same pop-in effect was also found in single crystalline nickel). The pop-ins which occur at higher loads are assumed to be an evidence for the activation

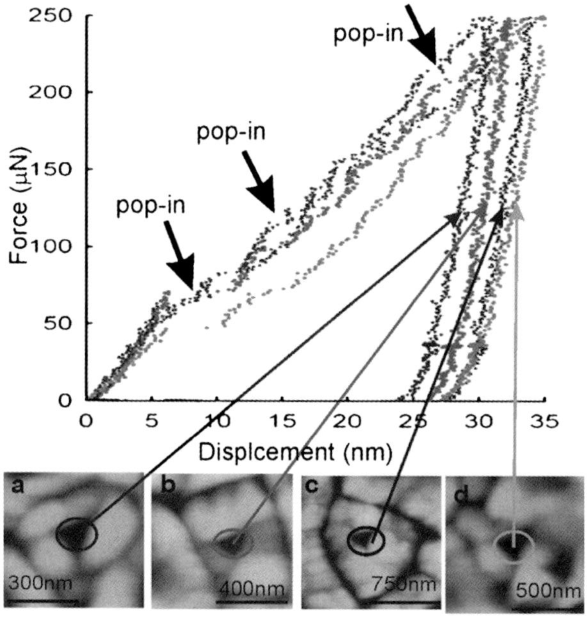

Fig. 8. Nanohardness *vs.* grain size (peak force 250 μN).

of sources in the adjacent grains. Especially, since no additional pop-ins were observed for grains larger than 1 μm, hence above the typical pile-up length.

The pile-up length L is related to the number of dislocation n, $L = D/2 = 2nA/\tau$, where A is a constant, D the grain size and τ the external stress. If we now assume that the stress for the activation of sources in adjacent grains is nearly constant as assumed by Hall and Petch [14], the pop-in load must be higher for smaller grains (beside of scatter due to orientation effects) and they must occur at lower strains as observed in the experiments. Therefore, we interpret our data as a method to measure for the first time the assumptions of the Hall–Petch law locally. These strong interactions between the dislocations nucleated below the indenter not only alter the pop-in width and hardness but also change the strain rate sensitivity of single grains as shown below.

5. Strain rate effects in nanomaterials

5.1 Strain rate sensitivity of Ni as a function of grain size

The strain rate sensitivity of typical fcc metals is usually benign. However, recently many groups observed a strong strain rate dependence in nanocrystalline

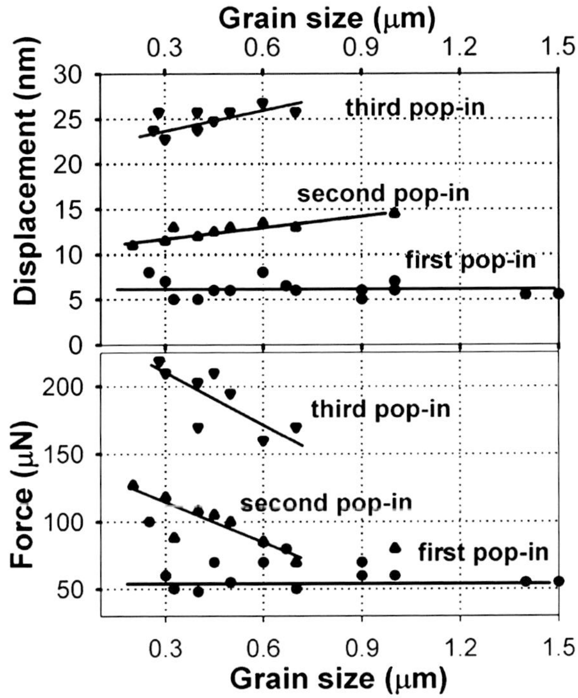

Fig. 9. Force and displacement at which pop-ins occurs.

fcc materials [15–22], even at low temperatures. Depending on grain size and temperature additional deformation mechanisms, such as emission of partial dislocations from grain boundaries [23, 24], micro twinning [16, 25], grain rotation [26–28] and conventional creep mechanisms [29, 30] which should operate even at room temperature are discussed, simulated and shown experimentally under special conditions.

One way to distinguish between these mechanisms is to conduct creep and tensile tests at different temperatures and loads. However, this only yields results averaged over the whole microstructure, whereas in simulations [*e.g.* [15, 16, 27, 32]], relatively small model systems consisting of typically one million of atoms can be considered. Therefore, in this paper we try to combine the results of global measurements (like tensile tests with massive nanocrystalline specimens [33]) with local measurements (like nanoindentation) at single grains. Incremental strain rate and creep tests were conducted in order to obtain an idea of the dominating mechanisms which were then tested locally by specifically designed nanoindentation tests.

The hardness measurements have shown that the mechanical properties start to change at a grain size of below 1 μm. Therefore we conducted our incremental strain rate and creep tests below this size [34, 35]. At a grain size of

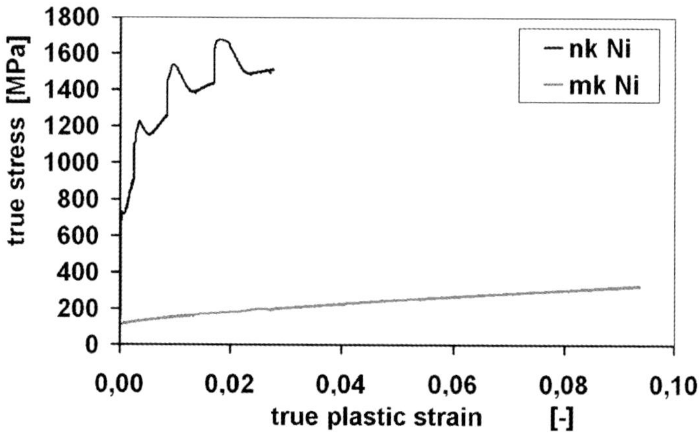

Fig. 10. Strain rate effects in nc and mc Ni shown in the same scale, the strain rates were increased by a factor 10 and 100 based on a reference strain rate of $10^{-4}\,s^{-1}$.

around 1 μm creep effects are not expected. However, below this grain size, it is known, that the typical cell structure does not develop [35]. In other words, dislocations more often interact with grain boundaries than with each other. This can result in rate effects. Therefore one goal of this work is to make a distinction between rate effects due to dislocation mechanisms and rate effects due to diffusion as a function of grain size.

Figure 10 compares the true stress–true plastic strain curve of nc and mc Ni. After given strain increments the strain rate was increased and reduced, respectively. The results clearly indicate, that the strain rate sensitivity in nc Ni (grain size of 30 nm) is much higher than in mc Ni.

Figure 11a and b show the strain rate sensitivity m

$$m = \left.\frac{\partial \ln \sigma}{\partial \ln (\dot{\varepsilon})}\right|_{T,\,\varepsilon} \tag{25}$$

and Fig. 12a and b the activation volume V

$$V = kT \left.\frac{\partial \ln (\dot{\varepsilon})}{\partial \tau_{\mathrm{Eff}}}\right|_{T,\,\varepsilon} \tag{26}$$

of mc and nc Ni, respectively. The Taylor factor M ($M = 3.06$ for statistically oriented polycrystalline fcc metals) relates the effective stress τ_{Eff} to the applied stress σ. The resulting activation volume V (Fig. 12a) in nc Ni is a factor of 100 smaller than in mc Ni (Fig. 12b) [22, 35]. The grain size was varied either by heat treatment (Fig. 11a) or by changing the pulse electrodeposition conditions. In both cases, the activation volume increased with increasing strain and with increasing grain size, independent of the processing conditions. The measured

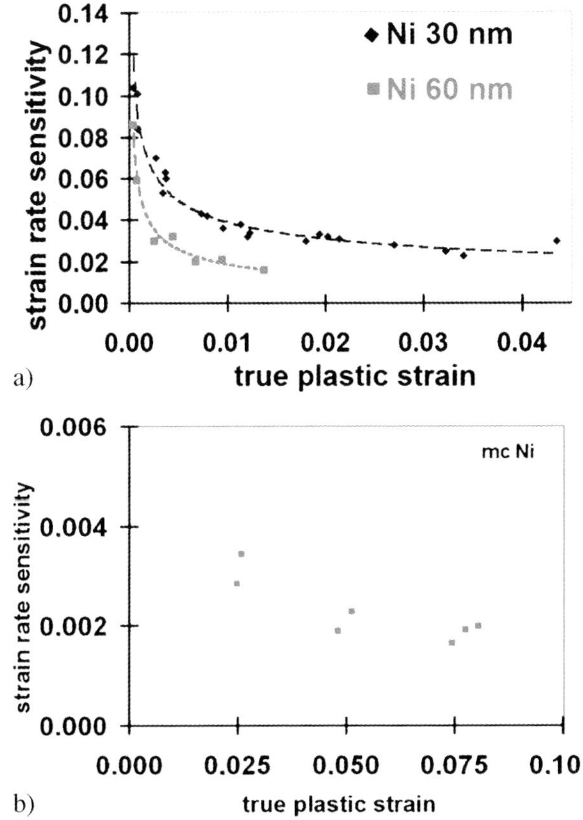

Fig. 11. The effect of grain size on strain rate sensitivity as a function of true plastic strain for nc Ni (a) and mc Ni (b).

activation volume was measured to be independent of temperature at and below room temperature. This was tested in incremental creep tests.

5.2 Incremental creep tests

The result of a typical incremental creep test is plotted in Fig. 13.

The grain size of 30 nm was systematically investigated at different temperatures giving the five sets of data in Fig. 14. The curves clearly indicate a mechanism change with increasing stress [37]. For the following, we only use the observation that in the measured range no creep effects were observed at temperatures below room temperature, hence the measured strain rate sensitivity is not due to creep effects but due to thermal activation only. This is also supported by the fact that the activation volume measured for different low temperatures is constant at the same stress and for the same

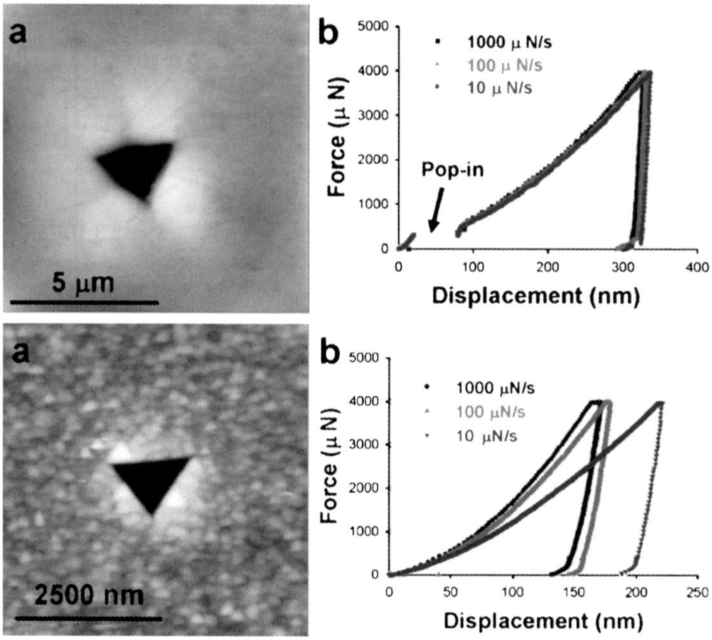

Fig. 15. Effect of loading rate on the load–penetration rate for Ni with different grain sizes (grain sizes: (a) 40 000 nm, (b) 250 nm.

sults with tensile tests and FEM simulations could serve to calibrate the strain rate effects measured during nanoindentation. This offers the possibility of measuring strain rate effects in thin films or even in nano materials which are only available in small quantities. Therefore, not only the tests where the grain size is in the order of the tip size are interesting for nanoindentation but also the strain rate effects in materials where the grain size is much smaller than the tip.

However, although these experiments were more difficult to evaluate quantitatively, they encouraged us to test the effect of grain boundaries on strain rate sensitivity for cases where only single boundaries interact with the displacement field under the indenter. For this purpose we choose the grain size in such a way (Fig. 1b), that the indent can be centred in the grain that means grain sizes above 200 nm.

Referring to the results that the pop-in width is limited by the grain size (Fig. 6), we studied the effect of grain size on strain rate sensitivity for grain sizes above and below this range. The maximal load was kept constant at 300 μN and the loading rates were changed from 10 $\mu N s^{-1}$ over 20 $\mu N s^{-1}$ up to 40 $\mu N s^{-1}$. An example of a typical measurement is given in Fig. 16. In Fig. 16a it can be seen that in the case where the pop-in width is independent of the grain size (above 900 nm in Fig. 7b) no rate effects occurred. However, in the second case when the boundaries restrict the pop-in width, strong rate

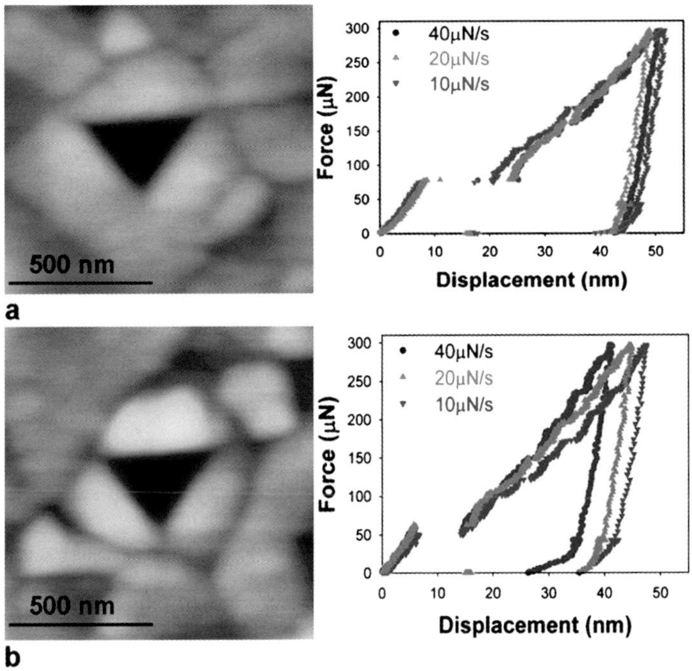

Fig. 16. Rate dependence of the force-penetration curve: (a) for larger grains (900 nm) no rate effects; (b) for smaller grains (450 nm): pop-in width dependent on grain size, strong rate effects were observed.

dependence and a decreasing penetration depth at a constant load were found (grain size below 900 nm; Fig. 7b).

This indicates that the interaction of dislocations with the grain boundaries as dislocation pinning at boundaries is the rate controlling mechanism at least in the ultra fine grained range tested by nanoindentation. However, for smaller grain sizes the nanoindentation tests (Fig. 15b) show similar strong rate dependences.

Combining these different results, mainly the effect of grain size on pop-in width (Fig. 7b), on the hardness of single grains (Fig. 7a) and on the strain rate sensitivity of single grains (Fig. 16b) lead us to the interpretation that dislocations coming from different sources below the indenter primarily interact with the grain boundaries. This behaviour results in strain rate sensitivity. Assuming that the dislocations are pinned at the boundary, with the structural elements of the boundary as pinning points, the activation volume should simply scale with the grain size $l = d$;

$$V \approx b^2 l \tag{27}$$

sketched in the simplified drawing (Fig. 17).

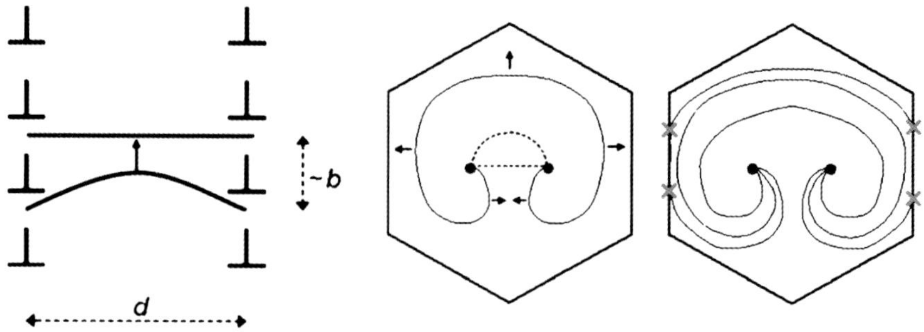

Fig. 17. Schematic drawing of dislocation pinning at grain boundaries ($V \approx b^2 d$ with $l \approx d$).

Figure 17 shows a typical Orowan source in a grain. If only one source is operating, we expect a strong hardening depending on grain size and a strong influence of the grain boundaries on dislocation multiplication, since the dislocation must glide along the boundary as indicated by the two crosses in Fig. 17. In our view this slipping along the interface leads to rate effects. For one source interacting with a free surface, rate dependence was observed in a thin single crystal during compressions tests by Dimiduk *et al.* [38]. With decreasing grain size from 900 nm (for nanoindentation tests) to 30 nm (for strain rate tests), we believe that the small activation volume for the small grains (30 nm) indicates, that the grain boundary operates as dislocation source [19].

We present the following idea of the deformation process: partial dislocations nucleate at boundaries (indicated by our findings, that l in Eq. (3) was found to be consistently smaller than the grain size d but scaled with the grain size (Fig. 12). During the propagation the dislocation is pinned at the boundary which then leads to thermally activated glide as indicated by the results obtained from the nanoindenter (Fig. 16).

Increasing the temperature above room temperature, creep was observed in the experiments at constant stresses. The deformation rate measured in the experiments changed with increasing stress and this transition shifted to lower stresses with higher temperatures (Fig. 12). Around room temperature the stress range at which this transition occurs, is in the same range where we found the first indication of thermally activated glide. Therefore we believe that the stresses are high enough to nucleate dislocations. Since now deformation occurs at constant stresses, it can be assumed that the dislocations, which pile up at the boundaries can be absorbed by the boundaries resulting in a creep rate which is controlled by the absorption of dislocations. This is not the classical Taylor–Weertman model, but still a balance between multiplication and annihilation. However, when the dislocations were absorbed by the boundaries orientation changes and grain rotations should occur as predicted by some molecular dynamic calculations. Therefore, we are currently constructing a testing

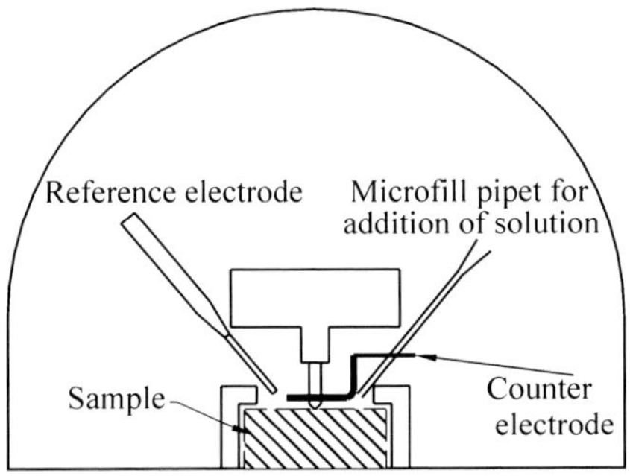

Fig. 18. Experimental setup.

machine to measure the lattice rotation during creep directly with a high resolution electron backscatter diffraction (EBSD)-system.

One feature remains, which is surprising in our view. In the incremental creep tests where a change in creep behaviour was observed with increasing stress (Fig. 14), the deformation rate was proportional to the stress indicating diffusional creep for lower stresses. This topic is still under examination. However, the following findings are consistent with our current interpretation that in this range grain boundary sliding occurs:

At room temperature the stresses applied in the creep tests were below the stresses at which plastic deformation was observed. The transition between the two types of creep behaviours shifted with increasing temperature to lower stresses. Even the low stresses in our experiments are high and fall in a stress range where grain boundary sliding is expected. Yang and Vehoff [12, 39] observed grain boundary sliding directly during *in-situ* crack growth in an AFM with nc Ni and in nanoindentation tests. However, besides the fact that different diffusion related deformation mechanisms show different stress exponents, the effect of grain size is even stronger, $1/d^3$ for grain boundary sliding. To test the assumption experiments at different grain sizes in the range from 20 nm to 900 nm are under way.

5.4 *In situ* electrochemical nanoindentation

During the SFB program the nanoindentation technique was further developped resulting in testing maschine which allows for testing in different corrosive environments. A schematical drawing of this maschine is shown in Fig. 18.

With this set up the effect of hydrogen on the nucleation of dislocations was studied on nickel(111) surfaces in a 0.05 M Na_2SO_4 electrolyte prepared from an analytical grade Na_2SO_4 and double distilled water. To reduce the air formed oxide film of nickel, the pH of the solution was set to 6 by adding trace amounts of analytical grade H_2SO_4 [40]. A new passive film, which is stable under cathodic potentials, forms after subsequent passivation *i.e.* this new

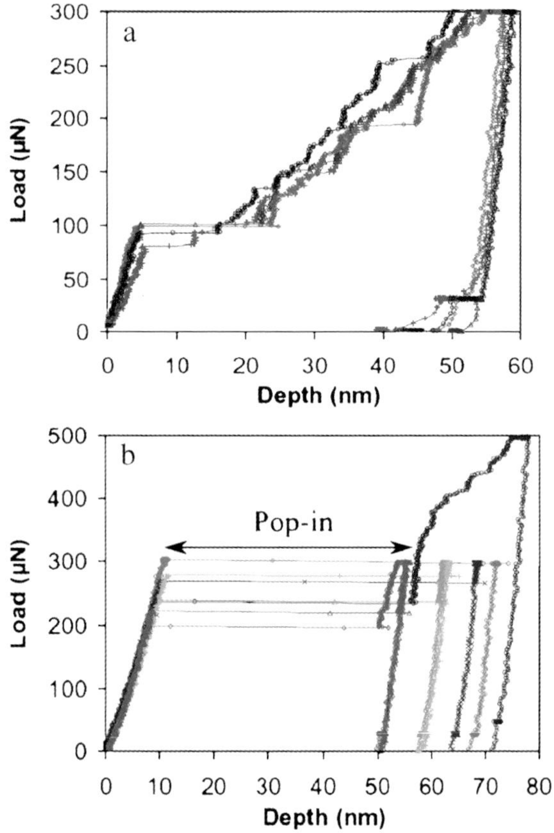

Fig. 19. Load displacement curves for nickel at cathodic (a) and anodic (b) potentials.

passive film is not vulnerable to cathodic reduction. This stable surface film removes discrepancies due to film effects [4, 41].

Typical load displacement curves obtained under cathodic potential are shown in Fig. 19a. After several indentations, the potential was again switched to anodic and maintained for several seconds before indentation. Figure 19b shows typical anodic load displacement curves with pop-in loads that are considerably higher than those at cathodic potential. This sequence was performed several times to check the reproducibility of the observations for each polarization condition.

Figure 20 shows the measured distribuion of pop-in loads under anodic and cathodic conditions. It can be clearly seen that hydrogen reduces the pop-in loads drastically. Combining these results with classical nucleation theory the effect of hydrogen on dislocation nucleation can be measured directly for the first time.

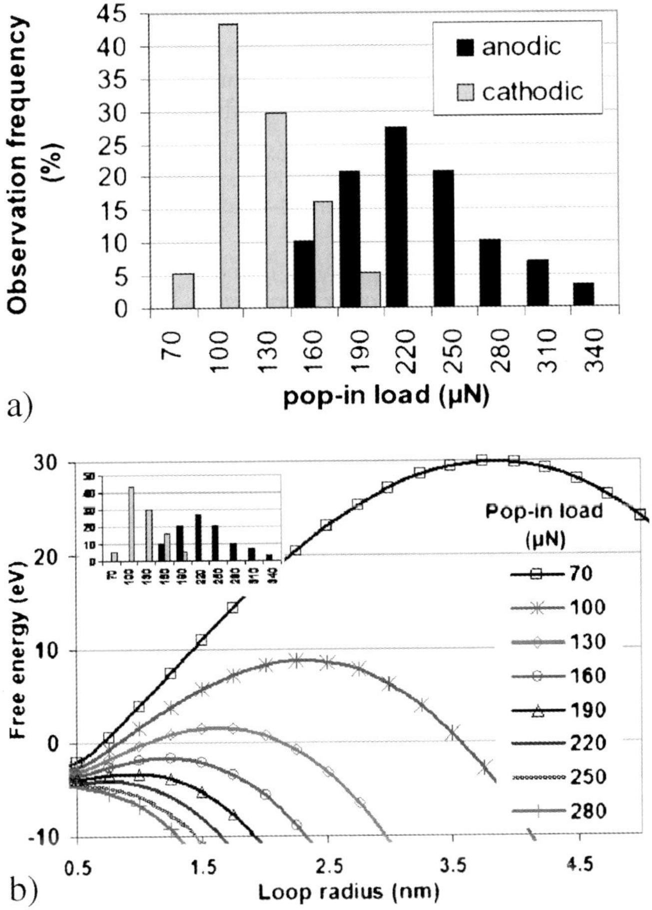

Fig. 20. a) Pop-in load distribution under various polarization conditions for nickel. b) Change in free energy of homogenous dislocation nucleation for nickel as a function of dislocation loop radius (r) for various experimentally observed pop-in loads in Fig. 20a (pop-in load distribution plot, upper left corner).

The free energy of a circular dislocation loop of radius r is given by

$$\Delta G = 2\pi r W + \pi r^2 \gamma - \pi r^2 b \tau \tag{28}$$

where W is the line energy of the dislocation loop, b is the Burgers vector for a partial dislocation in an fcc lattice, τ is the external shear stress acting on the loop, and γ is the stacking fault energy (SFE). The first term on the right-hand side of Eq. (28) describes the energy required to create a dislocation loop in a defect free lattice. The second term indicates the increase in energy due to the

Equilibrium and Nonequilibrium Behaviour of Ferrofluids – Experiments and Theory

By Jan Peter Embs[1,*], Björn Huke[2], Andreas Leschhorn[2], and Manfred Lücke[2]

[1] Laboratory for Neutron Scattering ETH Zurich, Paul Scherrer Institut, CH-5232 Villigen-PSI, Switzerland
[2] Universität des Saarlandes, D-66041 Saarbrücken, Germany

Magnetic Fluids / Transport Properties / Pattern Formation

Selected results on the spatiotemporal behaviour of equilibrium and nonequilibrium properties of ferrofluids in different magnetic fields are reviewed. They have been obtained in the project B13 *Transport, response and instability behaviour of ferrofluids* of the SFB 277 by experiments and by various theoretical methods ranging from purely analytical calculations to full numerical approaches.

1. Introduction

Ferrofluids are colloidal suspensions of permanent magnetic particles with magnetic moments of about $10^4 \mu_B$ in carrier liquids such as, *e.g.*, oil or water. The small size of the particles in the range of about 10 nm provides stability against segregation due to gravity and common magnetic field gradients. To prevent aggregation due to van-der-Waals forces one customarily coats the particles by suitable surfactants.

Macroscopically, ferrofluids appear as paramagnetic liquids with a susceptibility $\chi \approx 1$ and higher, which is very large compared to molecular paramagnetic liquids. For this reason they are called super-paramagnetic. The large magnetic moments easily allow to exert significant magnetic forces and magnetic torques on the particles. This in turn has been stimulating not only a wealth of interesting basic research activities but also the development of many technical and medical applications: The forces and torques on the particles are mediated *via* the surfactant coating to the carrier liquid and with it to the ferrofluid. Thus, one can easily modify and control – on macroscopic

* Corresponding author. E-mail: jan.embs@psi.ch

Z. Phys. Chem. **222** (2008) 527–586

scales – the spatiotemporal properties of the equilibrium, transport, response and instability behaviour of the ferrofluid by magnetic fields.

Here we review selected results on the equilibrium as well as on the nonequilibrium behaviour of ferrofluids. They were obtained in the project B13 of the SFB 277 by experiments and by various theoretical methods ranging from purely analytical calculations to full numerical methods.

In Sect. 2 we present theoretical results concerning the equilibrium magnetization of monodisperse and polydisperse ferrofluids as obtained by Born–Mayer cluster expansion methods for dipolar hard spheres. The third section deals with a ferrofluid in a driven torsional pendulum undergoing small amplitude rotary oscillations close to resonance. With this setup one can measure among others the rotational viscosity of the ferrofluid in the absence of any macroscopic shear by enforcing the flow to be like a rigid body. Section 4 deals with a related problem, namely the relationship between magnetization and magnetic field in a ferrofluid that is rotating stationarily like a rigid body. Also here experiments as well as theoretical results are discussed and compared with each other. In Sect. 5 we investigate waves on the ferrofluid surface in vertically shaken containers in the presence of a magnetic field by experimental and numerical methods. In this experiment Faraday and Rosensweig instabilities of the flat ferrofluid surface can compete with each other. The final Sect. 6 presents a theoretical/numerical analysis of convection of ferrofluids in a Rayleigh–Bénard setup of a fluid layer heated from below. In particular the influence of the strong Soret effect and of the very slow concentration diffusion of the magnetic particles on the spatiotemporal behavior of convection is investigated.

2. Equilibrium magnetization: Born–Mayer cluster expansion

The most simple models of the microstructure of ferrofluids treat them as a gas of noninteracting, identical particles carrying a magnetic moment. The equilibrium magnetization in a magnetic field can then easily be found by analytical means and is described by a Langevin function.

However, the energy of the particle–particle interaction in ferrofluids can easily reach the order of kT and therefore cannot simply be neglected anymore. Even if the particles can still be described as monodisperse, deviations of the magnetization curve from a Langevin curve will become apparent [1]. When particle–particle interaction is taken into account an analytical solution for the equilibrium magnetization is not possible anymore.

We used the Born–Mayer or cluster expansion technique to derive an approximate analytical expression describing the equilibrium magnetization of interacting ferrofluid particles [2, 3]. This expression is given by the leading terms of an expansion in two relevant dimensionless parameters, namely the volume fraction ϕ of the magnetic material and an interaction parameter ϵ describing the potential energy of a particle pair. The expansion is controlled. The calculated terms are exact for their respective orders in ϕ and ϵ within the model used, and deviations from the result for the equilibrium magnetization are thus of higher order in ϕ or ϵ than those calculated. We will describe the technique and the results in the next subsection.

Real ferrofluids are polydisperse, *i.e.* the particles have different sizes and magnetic moments. This fact alone leads to deviations from the Langevin behavior that can be as strong or stronger than the influence of the particle–particle interaction. A model of monodisperse interacting ferrofluid particles is thus in general not sufficient to explain the magnetization curves found in experiments and the effects of polydispersity [4, 5] have to be taken into account. We describe the necessary changes in the Born–Mayer expansion and the modifications of the theoretical results in the third subsection. In that section we also report applications of the theory to experimental magnetization curves.

2.1 Monodisperse ferrofluids

In monodisperse ferrofluids, calculating the equilibrium magnetization requires the evaluation of high-dimensional integrals of the form

$$\int \exp\left(\sum_i \frac{\boldsymbol{m}_i \boldsymbol{H}_e}{kT} - \sum_{i<j} \frac{V_{ij}}{kT}\right) \mathrm{d}\boldsymbol{m}_i \mathrm{d}\boldsymbol{x}_i\,, \tag{1}$$

that appear in the expression for the partition function. Here, $\boldsymbol{m}_i$ is the magnetic moment of the i-th particle at position $\boldsymbol{x}_i$ and V_{ij} the interaction energy of the two particles i and j. $\boldsymbol{H}_e$ is the applied magnetic field.

When V_{ij} is ignored, the integral factorizes into one-particle terms and the analytical evaluation does not pose a problem. Within the Born–Mayer expansion terms f_{ij} are introduced *via*

$$\exp\left(-\frac{V_{ij}}{kT}\right) = 1 + f_{ij}\,, \tag{2}$$

allowing to rewrite (1) in a form

$$\int \exp\left(\sum_i \frac{\boldsymbol{m}_i \boldsymbol{H}_e}{kT}\right) \prod_{i<j} (1 + f_{ij})\, \mathrm{d}\boldsymbol{m}_i \mathrm{d}\boldsymbol{x}_i\,. \tag{3}$$

When the typical interaction energy is small, the f_{ij} are also small in most configurations.

The exponential function is now expanded into the f_{ij}, such that in the n-th order the high-dimensional integral (3) still factorizes, mostly into one-particle contributions. Only integrals over n particles at maximum remain, allowing at least a numerical evaluation of the integrals.

We modeled the interaction between the ferrofluid particles as dipole–dipole-interaction V^{DD} plus a hard–sphere repulsion V^{HS}:

$$V_{ij} = -\frac{3(\boldsymbol{m}_i \boldsymbol{r}_{ij})(\boldsymbol{m}_j \boldsymbol{r}_{ij}) - r_{ij}^2 \boldsymbol{m}_i \boldsymbol{m}_j}{4\pi\mu_0 r_{ij}^5} + V_{ij}^{\mathrm{HS}}, \tag{4}$$

$\boldsymbol{r}_{ij}$ being the distance vector between particle i and j.

The carrier liquid is ignored in this treatment. That can safely be done if the interaction range of the carrier liquid's molecules is short compared to the typical distance of the ferro particles [6].

Making apart from the expansion in the f_{ij} an additional expansion of the $\exp(-V_{ij}/kT)$ in powers of V_{ij}/kT, the leading terms of the equilibrium magnetization can even be evaluated analytically. In the final result the two expansions translate into expansions in the volume fraction $\phi = N\pi D^3/6V$ of the magnetic material and the interaction parameter $\epsilon = m^2/4\pi\mu_0 kTD^3$, a typical interaction energy of two particles in contact (half the energy of a head-to-tail orientation). D is the hard sphere particle diameter, N/V the particle density. The expansion coefficients are functions of the dimensionless magnetic field $\alpha = mH/kT$:

$$M = M_{\mathrm{sat}} \left(\sum_{m,n} L_{m,n}(\alpha)\phi^m \epsilon^n \right). \tag{5}$$

$M_{\mathrm{sat}} = Nm/\mu_0 V$ the saturation magnetization. The leading term $L_{0,0}$ is the Langevin function. There are no other terms with $m = 0$ or $n = 0$. The terms of first order in ϕ can in principle be calculated analytically in every order of ϵ, only the integrands become more and more unwieldily. Using a computer algebra program we calculated the terms of order $\phi\epsilon^n$ up to $n = 13$, an expansion that is useful if the ferrofluid is dilute but the particles possess a large magnetic moment. We also derived an analytical expression for the term of order $\phi^2\epsilon^2$, (the term of order $\phi^2\epsilon$ vanishes) which yields important contributions when the ferrofluid is highly concentrated.

The equilibrium magnetization as function of an external magnetic field depends also on the probe geometry, a consequence of the fact that dipolar forces decrease only $\propto r^{-3}$ and are therefore of long-ranged nature. Applying the thermodynamic limit $N \to \infty$ thus requires special care. In Eq. (1) the term of the outer potential contains correctly the magnetic field in absence of all other particles, *i.e.* simply the external field $\boldsymbol{H}_e$. To make the resulting expression for $\boldsymbol{M}$ independent of the geometry one has to enter the macroscopic internal field $\boldsymbol{H}$ into the calculation. To do so, one divides the dipolar interaction between the particles into those within a cutoff radius R_s and those that are far away, *i.e.* beyond the cutoff. The long-ranging contributions within the sphere of radius R_s cancel out and the limit $R_s \to \infty$ can thus safely be taken. The far-away particles are treated as a continuum that contributes a field $H_s = H + M/3$ inside the sphere where M is the sought after equilibrium magnetization. H_s formally takes the role of H_e in (1). The magnetization is then calculated using the self-consistency condition $M = M(H_s) = M(H + M/3)$ which can be solved in the required order to yield an analytical expression for $M(H)$ in the form given in Eq. (5).

The leading term in the expression for the reduced equilibrium magnetization M/M_{sat} beyond the noninteracting approximation is of order $\phi\epsilon$. It has been explicitly considered in [7]. Furthermore, several other theories reproduce the correct $\phi\epsilon$-term when written in the form of a (ϕ, ϵ)-expansion [8–10]. However, no other theory also delivers the higher order terms correctly.

Taking the higher order functions $L_{m,n}(\alpha)$ into account significantly extends the range of validity of the formula (5). Deviations from the $\mathcal{O}(\phi\epsilon)$-approximation already become important when $\epsilon > 0$ or $\phi\epsilon = 0.04$. On the other hand, judging from the relative importance of the highest order terms we have calculated we expect our extended expression to fail at about $\epsilon = 3$ and $\phi\epsilon = 0.5$. This larger range is relevant for real ferrofluids: Volume fractions of about $\phi = 0.5$ have been reported as well as dipolar interaction strengths of $\epsilon = 2.7$ [11].

2.2 Polydisperse ferrofluids

Polydispersity has an influence on the magnetization curve already in the non-interacting case, and the influence on the interaction contributions is even stronger. To compare the results of the Born–Mayer expansion to experiments it is thus necessary to extend the theory to take polydispersity into account.

The necessary modifications are straightforward. In the integral (1) the absolute values m_i of the magnetic particles will now be different, as well as the hard sphere diameters that enter into the interaction term V_{ij}. Sums over all N particles are replaced by integrals $N \int P(m, D)\mathrm{d}m\,\mathrm{d}D$ for a given distribution function $P(m, D)$ that can be replaced by a function $P(D)$ if a fixed relationship $m = m(D)$ is assumed. The result (5) now takes the form

$$M = \overline{M}_{\text{sat}} \left(\sum_{m,n} L^{\text{poly}}_{m,n}(\overline{\alpha})\overline{\phi}^{m}\overline{\epsilon}^{n} \right), \tag{6}$$

where the overlined quantities are now defined *via* "typical" values $\overline{D}$ and $\overline{m}$ for the particle diameter and magnetic moment. The functions $L^{\text{poly}}_{m,n}(\overline{\alpha})$ are polydisperse generalizations given by m-fold integrals over the reduced diameters $D_i/\overline{D}$ and magnetic moments $m_i/\overline{m}$.

For reasons of simplicity we will assume the definitions $\overline{D}^3 = \langle D^3 \rangle$ and $\overline{m} = \langle m \rangle$ which unambiguously defines $L^{\text{poly}}_{m,n}$ if $P(m, D)$ is given. The reference values $\overline{D}$ and $\overline{m}$ can in fact be chosen freely however, which facilitates the application of the theory in cases where the actual distribution function might not be known beforehand, *e.g.* in magneto-granulometric measurements.

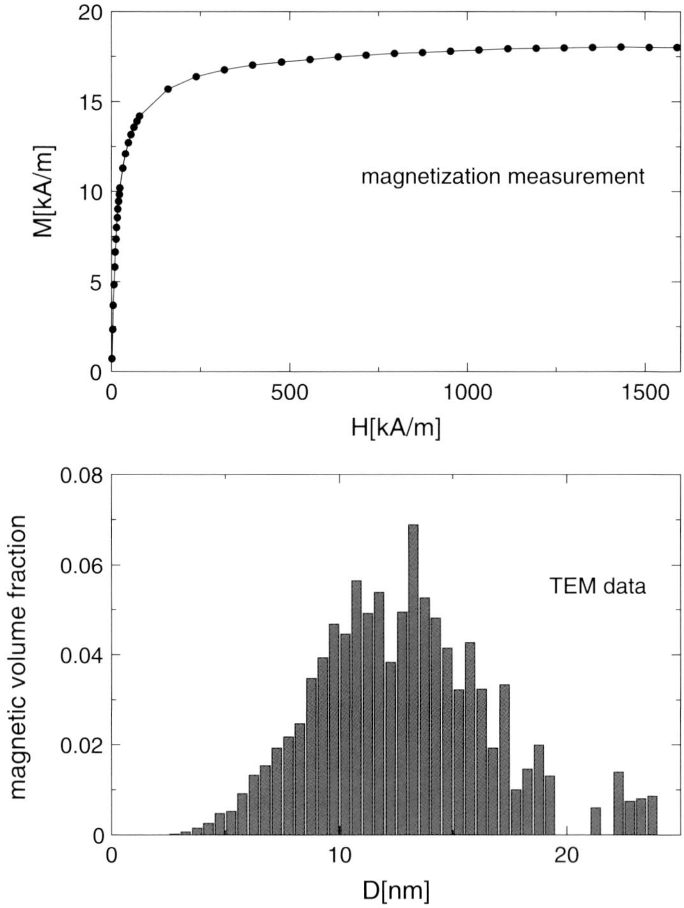

Fig. 3. Equilibrium magnetization of the ferrofluid APG 933 as measured in experiments (upper plot, points) *versus* the results of our theory (solid line) for the diameter distribution found in TEM measurements (lower plot).

the calculation. No fit parameters were required. Nevertheless the agreement between theoretically and experimentally found magnetization curves is excellent.

We also applied our theory to magnetization curves of a series of ferrofluids prepared by Kuznetsova and Ivanov *via* diluting a highly concentrating basic ferrofluid. In this case the diameter distribution $P(D)$ was a priori unknown. Assuming a Γ-distribution $P(D) \sim D^{\beta} e^{-D/D_0}$ the task of finding $P(D)$ can be reduced to determining two fit parameters β and D_0. Under the condition that the dilution process does not change $P(D)$, different models for the equilibrium magnetization can be judged by comparing the fit results for the different samples. The results for the volume weighted mean diameter (in nm) are shown in the following table:

Degree of dilution:	0.088	0.137	0.197	0.296	0.444	0.664	1
Langevin	7.1	7.2	7.0	6.4	6.3	6.3	6.0
Weiss [8]	7.3	7.6	7.7	7.6	8.0	8.5	8.9
MSM [9]	7.3	7.5	7.5	7.1	7.2	7.2	6.9
HTA [7]	7.3	7.4	7.4	6.9	6.7	6.5	5.6
MMFM-1 [10]	7.3	7.5	7.5	7.1	7.1	7.1	6.8
CET [5]	7.13	7.13	7.13	7.19	7.24	7.46	8.25
MMFM-2 [14]	7.30	7.32	7.32	7.31	7.30	7.30	7.29

Our cluster expansion theory (CET) showed the second most consistent results for the mean volume and only fails for the highest concentrations where the second order modified mean field model by Ivanov [14] turned out to be best. The results for the variance that are not shown here are similar. The shortcoming of the cluster expansion here is probably due to the strong contribution of the higher order $\phi\epsilon^n$-terms. In Ivanov's theory such terms are not taken into account at all. The small differences for the fit results of our theory compared to all others already at the smallest concentrations is due to the fact that we used a binned approximation to the Γ-distribution for $P(D)$ instead of a smooth curve. This also demonstrates that the calculated means and variances may depend sensitively on the model distribution used.

To conclude: Using the Born–Mayer or cluster expansion technique we have derived a formula for the equilibrium magnetization given by the leading terms of a controlled expansion in two parameters ϕ and ϵ modelling the ferrofluid as a system of dipolar hard spheres. We also derived a polydisperse generalization that allows to calculate the magnetization for arbitrary distributions of magnetic moments and hard sphere diameters. The formulas have been tested on both data from numerical simulations and experimental data.

3. Resonantly driven torsional pendulum in a magnetic field

A ferrofluid in solid-body rotation experiences a viscous dissipation if an external magnetic field is applied perpendicular to the rotation axis. This so-called magneto-viscous effect [15] is attributed to a field dependent "rotational viscosity" η_R. With a magnetic torque acting on each magnetic colloidal particle its rotation in phase with the local vorticity of the carrier liquid is impeded thus giving rise to dissipative losses. In a mesoscopic picture the latter are due to friction between the particles and the carrier liquid. Now, the velocity gradient tensor of an arbitrary macroscopic incompressible flow can be decomposed into antisymmetric contributions (vorticity) and a traceless symmetric tensor

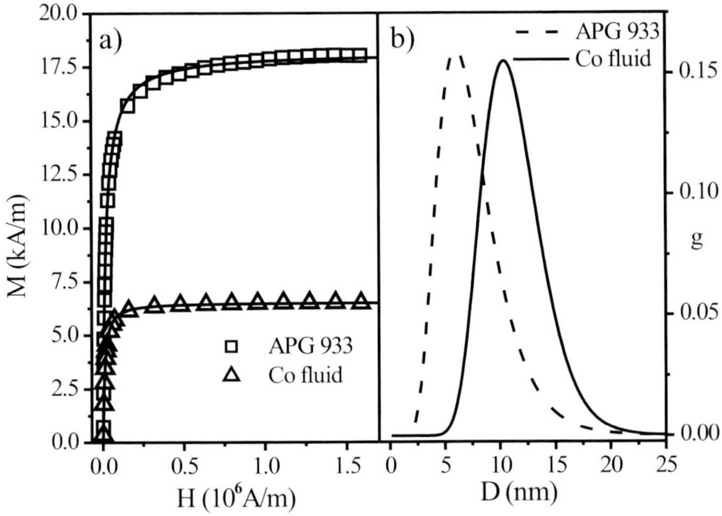

Fig. 4. Symbols denote the measured equilibrium magnetization for APG 933 as a function of the macroscopic magnetic field H within the sample. The solid line is a least-squares fit according to Eq. (8) with D_0, σ, and M_s as given in Table 2. The dashed line is a fit for a monodisperse ferrofluid (Langevin behavior) with $D = 11.1$ nm and $M_s = 17\,563$ A/m.

diameters, however, results in close agreement between measured and calculated magnetization curves (Fig. 4). The lognormal distribution parameters obtained from magneto-granulometry analyses of the ferrofluid samples are listed in Table 2. Additionally, Table 2 contains values for the volume-averaged diameter $D_v = \langle D^3 \rangle^{1/3} = D_0 \exp[(3/2)\ln^2 \sigma]$, which can be measured by the torsional-pendulum method (Sect. 3.2), and for the volume-weighted average diameter $D_{\mathrm{vol}} = \langle D^4 \rangle \langle D^3 \rangle = D_0 \exp[(7/2)\ln^2 \sigma]$, which is the size parameter customarily provided by X-ray diffractometry (Sect. 3.1.2).

To rule out a possible concentration dependence of these parameters, we investigated a dilution series with n-heptane ranging from the non-diluted volume concentration of $\varphi \simeq 3.3\%$ down to concentrations as low as 0.019%. From the scatter in the parameter values, we estimate the uncertainties associated with the magneto-granulometric size determination to be ± 1.0 nm for D_0 and ± 0.02 for σ.

3.1.2 X-ray diffraction (XRD)

Using XRD, we determined the volume-weighted average core size of the ferrofluid particles. A glass substrate holder was filled with undiluted ferrofluid and mounted in a Siemens D-5000 diffractometer with θ–θ geometry. Wide-angle diffraction scans were recorded from 10° to 140° (2θ) with a typical step

Table 2. The core-size distribution parameters D_0 and σ and the saturation magnetization M_s for the ferrofluid samples, as determined by magneto-granulometry using Eq. (8). Also listed are the volume-averaged core diameter $D_v = \langle D^3 \rangle^{1/3}$ and the volume-weighted core diameter $D_{vol} = \langle D^4 \rangle / \langle D^3 \rangle$.

Ferrofluid	M_s (A/m)	D_0 (nm)	σ	D_v (nm)	D_{vol} (nm)
APG 933	18149	7.0	1.47	8.7	11.7
APG 934	16400	7.1	1.42	8.5	10.9
APG 935	17886	6.3	1.44	7.6	9.7
APG 936	19013	7.0	1.42	8.4	10.8
Co	6512	11.0	1.27	12.0	13.5

size of 0.02° and count time of 20 s using Cu K_α radiation ($\lambda = 1.54$ Å). All Bragg peaks recorded in the diffraction measurements of the ferrofluid samples APG 933–936 could be indexed to magnetite (Fe_3O_4); no evidence for the presence of secondary phases was found. The Bragg peaks were broadened considerably with respect to the peak widths expected for diffraction from a coarse-grained (> 1 μm), strain-free sample of magnetite. This indicates the presence of a significant amount of inhomogeneous lattice strain and/or a broad size distribution of the magnetite cores, with an average size well below 50 nm [22]. By performing a standard peak-profile analysis according to the method of Warren and Averbach [23–26] or a full-pattern profile fit [27], it is possible to separate the lattice-strain effect from that of the crystallite size and thereby determine the inhomogeneous strain level e and the volume-weighted average crystallite size D_{vol}. Furthermore, one can also estimate the geometric standard deviation σ of the lognormal distribution of spherical crystallites that would give rise to Bragg peak profiles most nearly identical to those actually measured [27].

The result of a full-pattern profile fit to the diffraction scan of sample APG 936 is shown in Fig. 5. This fitting technique simultaneously refines D_{vol}, σ, e, the lattice parameter of magnetite and a polynomial background function in order to obtain a least-squares fit to the measured diffraction intensity.

In Table 3 the results of full-pattern profile fits to the four APG ferrofluid specimens are collected. Within the uncertainty range of this analysis technique (±10%), the crystallite sizes and lattice strains were found to be identical for all samples. Averaging over the indicated values, we find D_{vol} equal to (11.8 ± 0.7) nm and σ equal to 1.58 ± 0.04. Averaging over the inhomogeneous lattice-strain values yields $e = (0.29 \pm 0.05)\%$.

Comparing the size distributions obtained by XRD with those of MG, we observe that the D_0 values from XRD are systematically smaller (on average by 20%), and the width parameter σ is larger by about 8%. However, the results of the two methods are consistent within the ranges of experimental uncertainty for D_0 and σ. The agreement between the values obtained by each method for

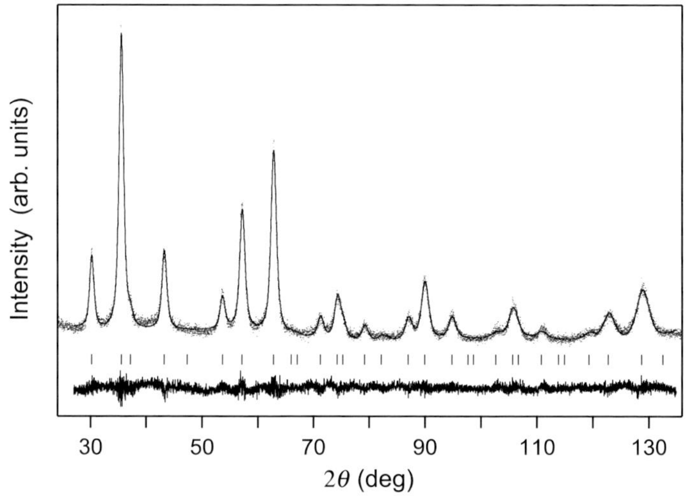

Fig. 5. X-ray diffraction scan of ferrofluid APG 936 (dots). The solid curve superimposed on the measured data is the result of a full-pattern profile fit between 27° and 135°; below that lies the difference curve between the measured and calculated diffraction intensities. Vertical lines indicate the expected positions of Bragg reflections from Fe_3O_4. The fit curve represents the calculated diffraction intensity arising from a lognormal distribution of spherical crystallites with a median diameter of $D_0 = 5.3$ nm and a geometric standard deviation of $\sigma = 1.60$.

Table 3. The parameters D_0 and σ of a lognormal distribution of crystallite sizes, as obtained from XRD analysis of the magnetite-based ferrofluids. Also included are the volume-averaged diameter $D_v = \langle D^3 \rangle^{1/3}$ and the volume-weighted average diameter $D_{vol} = \langle D^4 \rangle / \langle D^3 \rangle$.

Ferrofluid	D_0 (nm)	σ	D_v (nm)	D_{vol} (nm)
APG 933	5.4	1.64	7.8	12.7
APG 934	6.3	1.54	8.4	12.1
APG 935	5.7	1.53	7.5	10.7
APG 936	5.3	1.60	7.3	11.5

the volume-averaged diameter D_v is even better, with a mean difference of less than 7%.

3.1.3 Transmission electron microscopy (TEM)

Samples for TEM analysis were prepared by diluting the magnetite-based ferrofluid samples with n-heptane in an ultrasonic bath and then transferring a small amount to copper grids coated with amorphous carbon. Bright-field/dark-field image pairs and electron-diffraction patterns were recorded at 200 kV on a JEOL 2010 microscope. Cryo-TEM investigations were performed on APG 933, comparing an undiluted sample ($\varphi = 4.1\%$) with a diluted one ($\varphi = 0.007\%$). The cryo-TEM studies-performed by the base-grid method-entailed shock-freezing the sample holder carrying the ferrofluid before introducing it into the TEM. In addition, a high-resolution TEM analysis was carried out on individual magnetite particles in highly diluted specimens.

Unlike XRD scans, dark-field TEM images provide a direct measure for the spatial extent of the individual crystallites in a sample (see inset in Fig. 6). From such images one can determine the crystallite-size distribution without having to rely on an assumed distribution shape, as is the case with XRD or MG. Histograms approximating the crystallite-size distribution in the ferrofluid samples were generated by computer-assisted analysis of several dark-field images (Fig. 6). The resulting distribution can indeed be closely approximated by a lognormal function, as was assumed in the XRD and MG analyses. Furthermore, the corresponding values for D_0 and σ are very close to those obtained by XRD and MG.

To investigate the possible occurrence of particle agglomeration caused by diluting the sample with n-heptane or subsequent evaporation of heptane from the sample while in the microscope, we performed cryo-TEM investigations of a series of ferrofluids of varying concentration. Figure 6 shows the images of an undiluted (a) and highly diluted (b) sample of APG 936. The associated crystallite-size histograms shown in (c) do not differ significantly, indicating that there is no measurable particle agglomeration.

Figure 6a also sheds light on the hydrodynamic diameter of the particles in the APG 936 sample. Taking the undiluted volume fraction of $\varphi = 4.3\%$ and the measured particle core diameter of $D_v \simeq 8.4$ nm (Table 1 and Table 2), one obtains for the average light distance between the surfaces of adjacent particles 11 nm. However, the analysis of Fig. 6a (also performed on the undiluted sample) reveals that the particles are able to approach closer to each other. Here the evaluated average light distance is only (5.2 ± 1.7) nm. We take half of this value as an upper limit for the thickness of the particle coating.

3.1.4 Photon-correlation spectroscopy (PCS)

Dynamic light-scattering experiments performed on diluted ferrofluid samples yield a value for the diffusion coefficient of the particles, which in turn depends on the hydrodynamic particle diameter D^{hyd}. In our studies, PCS was carried out on an ALV 5000 spectrometer equipped with a krypton-ion laser (Spectra Physics). Dilution series were investigated with volume fractions ranging from $\varphi = 0.002\%$ to 0.02%. The intensity correlation function of the scattered light was recorded with an ALV-5000E correlator. Thereafter, the data were

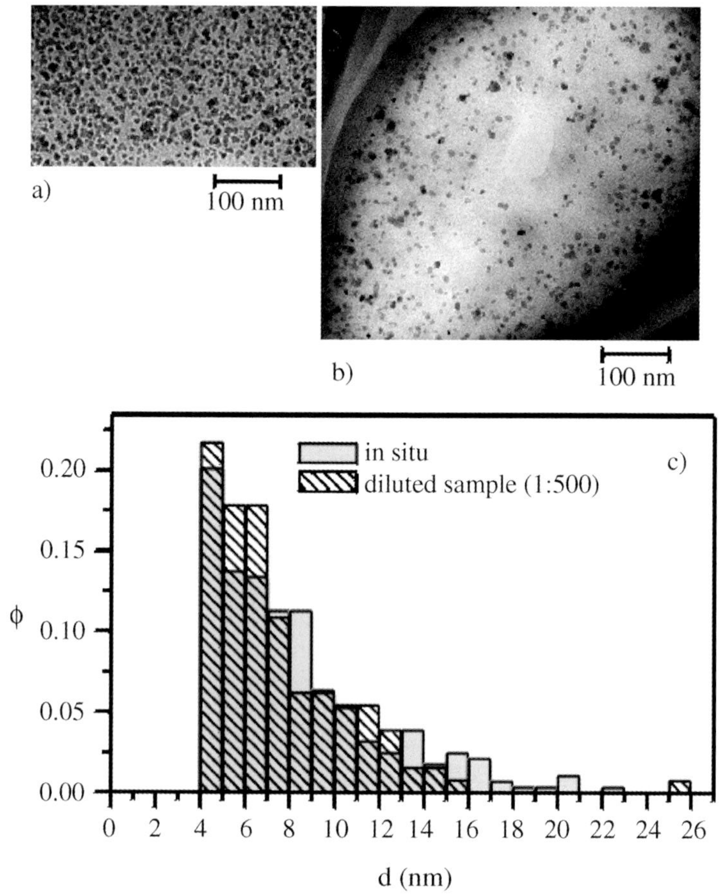

Fig. 6. Crystallite-size histograms (c) obtained by cryo-TEM studies of an undiluted sample (a) of APG 936 (grey histogram bars) and after strong dilution (b) with n-heptane to $\varphi = 0.0067\%$ (striped histogram bars).

fitted using the ALV-NonLin routine, which is based on the CONTIN regularization [28]. This analysis yields an intensity-weighted distribution function for the decay times. Under the assumption that the scattering particles behave like hard spheres in dilute solution, one can deduce the distribution of the hydrodynamic radii from the Stokes–Einstein relation. Applying an appropriate data weighting, one then obtains the number-weighted hydrodynamic-diameter distribution function $g(D^{\text{hyd}})$ from the intensity-weighted distribution.

Figure 7 illustrates the distribution $g(D^{\text{hyd}})$ for the magnetite-based ferrofluid APG 934 and for the Co-based sample. By fitting such CONTIN curves with a lognormal distribution (dashed lines), we obtain the size parameters D_0 and σ, which can be compared directly to the results of the other sample

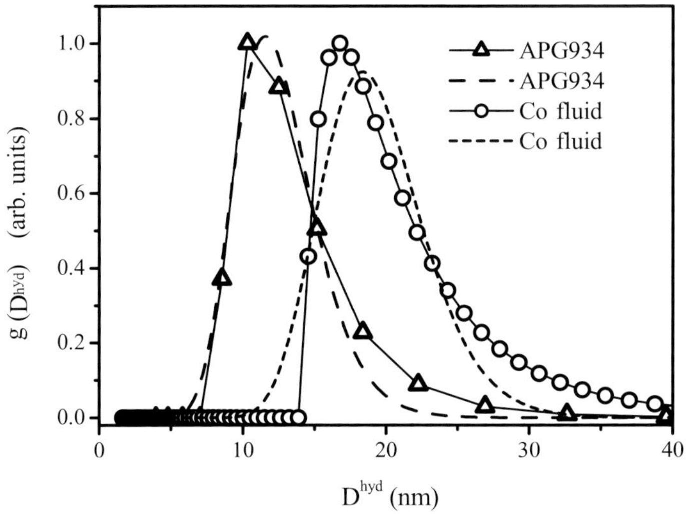

Fig. 7. Number-weighted distribution $g(D^{\mathrm{hyd}})$ for the hydrodynamic diameter, as obtained from the CONTIN regularization of the intensity correlation. Fitting the curves with a log-normal distribution (dashed lines) yields the parameters D_0 and σ listed in Table 4.

Table 4. The parameters D_0 and σ of a lognormal distribution of hydrodynamic diameters, as obtained from PCS measurements. Values for the volume-averaged hydrodynamic diameter $D_v = \langle D^3 \rangle^{1/3}$ and the volume-weighted average hydrodynamic diameter $D_{\mathrm{vol}} = \langle D^4 \rangle / \langle D^3 \rangle$ are also listed.

Ferrofluid	D_0 (nm)	σ	D_v^{hyd} (nm)	$D_{\mathrm{vol}}^{\mathrm{hyd}}$ (nm)
APG934	12.1	1.2	12.7	14.0
APG935	14.5	1.2	15.2	16.8
APG936	18.6	1.2	19.5	21.6
Cobalt	19.0	1.2	20.0	22.0

characterizations. The resulting size parameters are listed in Table 4 together with the mean values D_v^{hyd} and $D_{\mathrm{vol}}^{\mathrm{hyd}}$. From the scatter in measurements made at different angles, we estimate the uncertainty of the mean particle size to be ± 1.5 nm.

3.2 Measurements of the rotational viscosity

One of the many fascinating properties of ferrofluids [15] is their tunable viscosity which can be controlled by an external magnetic field. This so called magneto-viscous effect has first been observed by McTague [29]. Qualitatively it can be accounted for [30] by magnetic torques acting upon the suspended

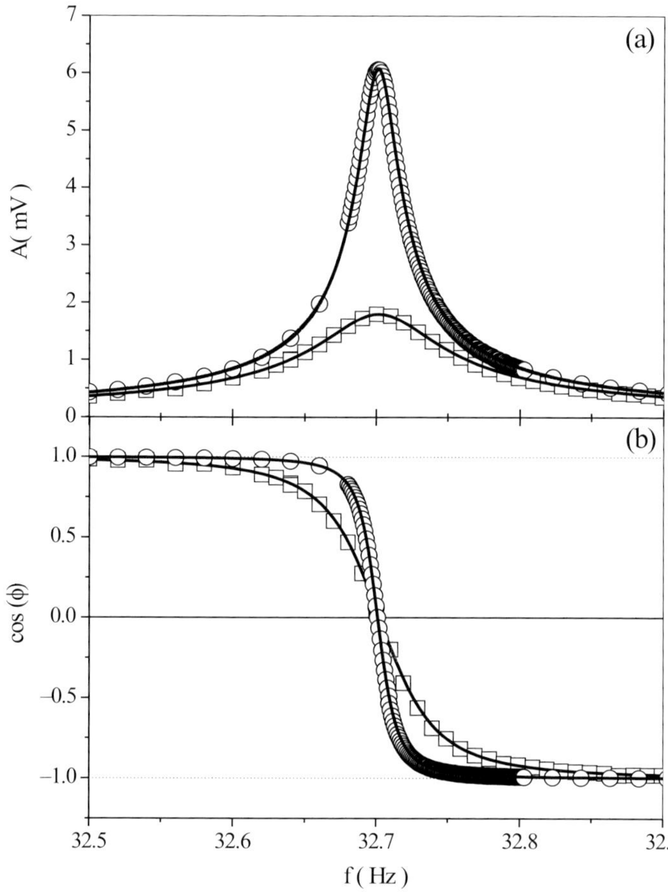

Fig. 9. Experimental raw data for the deflection amplitude (a) (measured in mV at the capacitor detector) and for the phase lag (b) relative to the external drive signal. Circles denote data taken for $H_0 = 0$, squares are for $H_0 = 14.2 \times 10^4$ A/m. Solid lines are fits to the harmonically driven linear oscillator. For the damping coefficient we obtain $\Gamma = 0.521$ Hz ($\Gamma_P = 0.178$ Hz) in the presence (absence) of the magnetic field.

field. This generates the torque [31]

$$N = \int_V \mu_0 (\boldsymbol{M} \times \boldsymbol{H})_z \mathrm{d}V \tag{9}$$

acting in z-direction on the ferrofluid volume V. Here $\boldsymbol{H}$ denotes the macroscopic magnetic field inside the ferrofluid probe. The magnetization $\boldsymbol{M}(\boldsymbol{r}, t)$ is an average over a large number of ferromagnetic particles, each carrying a magnetic moment $\boldsymbol{m}$, say. Their statistics is largely governed by the Langevin parameter $\alpha_{\text{local}} = \boldsymbol{m}\boldsymbol{H}_{\text{local}}/(k_B T)$, where k_B is the Boltzmann constant and T the absolute temperature. The local field $\boldsymbol{H}_{\text{local}} = \boldsymbol{H}_0 - \bar{N}\boldsymbol{M} + \frac{1}{3}\boldsymbol{M}$ seen by an individual dipole differs from the local macroscopic field $\boldsymbol{H} = \boldsymbol{H}_0 - \bar{N}\boldsymbol{M}$ by the Lorentz field of the cavity which encloses the particle [20, 21]. The demagnetizing field $\bar{N}\boldsymbol{M}$ is geometry dependent; for our cylindrical container of aspect ratio $\simeq 1$ we get [38] $\bar{N} \simeq 0.3485$ so that $\boldsymbol{H}_{\text{local}} \simeq \boldsymbol{H}_0$ and $\alpha_{\text{local}} \simeq \alpha_0$. Furthermore, the tiny oscillation amplitude of our apparatus guarantees that $\boldsymbol{M}$ deviates only slightly from the equilibrium magnetization $\boldsymbol{M}_{\text{eq}} = M_{\text{eq}}\boldsymbol{H}_0/H_0$ in the absence of rotation, $\dot{\varphi} = 0$.

We model the dynamics of the nonequilibrium magnetization $\boldsymbol{M} - \boldsymbol{M}_{\text{eq}}$ by the relaxation approximation

$$\left[\partial_t + (\boldsymbol{v}\nabla) - \frac{\nabla \times \boldsymbol{v}}{2} \times\right] \boldsymbol{M} = -\frac{1}{\tau_\parallel}\left(\boldsymbol{M}_\parallel - \boldsymbol{M}_{\text{eq}}\right) - \frac{1}{\tau_\perp}\boldsymbol{M}_\perp , \tag{10}$$

which can also be motivated by studying single-particle rotary Brownian motion [39]. Here $\boldsymbol{v}(\boldsymbol{r}, t) = \dot{\varphi}(\boldsymbol{e}_z \times \boldsymbol{r})$ is the velocity field and $\parallel$ and $\perp$ denote the respective directions parallel and perpendicular to $\boldsymbol{H}_0$. The quantities $\tau_\parallel$, $\tau_\perp$ are the associated field dependent relaxation times. In the limit $H_0 = 0$ they reduce to the Brownian rotational diffusion time, τ_B. Typically τ_B lies between 10^{-3}–10^{-6} s depending on the shape of the particle and the viscosity of the carrier liquid. With increasing field strength the relaxation times decrease according to complicated analytical expressions [39]. Ignoring spatial inhomogeneities and higher order contributions $\propto \dot{\varphi}(M_\parallel - M_{\text{eq}})$ in Eq. (10) the magnetization M_y in the relevant direction perpendicular to $\boldsymbol{H}_0$ and the rotation axis is given by

$$\partial_t M_y = \dot{\varphi} M_{\text{eq}} - \frac{1}{\tau_\perp} M_y . \tag{11}$$

Thus, in the low-frequency limit, $\tau_\perp \Omega_0 \ll 1$, which prevails in our experiment

$$M_y(t) = \tau_\perp \dot{\varphi}(t) M_{\text{eq}} \tag{12}$$

follows the rotation frequency $\dot{\varphi}(t)$ without phase lag [40]. In this limit the torque (9) is purely dissipative

$$N \simeq N_{\text{d}} = -V\mu_0 H_0 M_{\text{eq}} \tau_\perp \dot{\varphi} , \tag{13}$$

whereas in general the magneto-viscous torque resulting from (11) also acquires a reactive contribution. It gives rise to an $O(\Omega_0 \tau_\perp \Gamma_{\text{FF}})$-increase in the eigenfrequency of the pendulum which, however, cannot be resolved with our present apparatus.

The overall damping rate $\Gamma(H_0) = \Gamma_{\text{P}} + \Gamma_{\text{FF}}(H_0)$ measured with our pendulum (see Fig. 8) is due to parasitic effects like air friction and mechanical detected deviations are due to particle interactions (*e.g.* chain formation) not covered by the theory.

For a *quantitative* comparison between theory and experiment the above listed experiments have to cope with (some of) the following problems: (i) How to ensure that the assumed flow profiles (Poiseuille pipe flow, circular Couette flow, etc.) remain undeformed when the magnetic field is applied? (ii) The reported flows exhibit shear as well as rotational friction giving rise to a *total* effective viscosity, from which the small rotational contribution has to be separated. (iii) The important problem of identifying the proper magnetic field entering into the theoretical analysis of the experiments (*cf.* our discussion of local *vs.* macroscopic field) has largely been ignored – Refs. [21, 34] are an exception. (iv) For magnetizations $\boldsymbol{M}$ that deviate strongly from (local) equilibrium $\boldsymbol{M}_{\text{eq}}$ a reliable constitutive equation as required in the analysis is not available.

With the low-frequency, small-angle, rigid-body rotation of the fluid in our torsional viscometer we avoid or solve these problems and we obtain a direct and sensitive measurement of the rotational viscosity in a novel and straightforward way. By an appropriate choice of the probe geometry it was possible to evaluate the dissipative torque in terms of the (easily accessible) external magnetic field strength. For the moderately concentrated ferrofluid investigated here the results compare reasonably well with theoretical predictions based upon single-particle relaxation behavior and monodisperse size distribution.

3.3 Effect of polydispersity

Polydispersity strongly influences the macroscopic magnetic properties of the ferrofluid. We investigate here itseffect on the dynamic response of a ferrofluid pendulum [43].

A torsional pendulum containing a ferrofluid is forced periodically in a homogeneous magnetic field $\boldsymbol{H}^{\text{ext}} = H^{\text{ext}}\boldsymbol{e}_x$ that is applied perpendicular to the pendulum axis $\boldsymbol{e}_z$ (see Fig. 11). Such a ferrofluid pendulum is used for measuring the rotational viscosity [16, 32]. The cylindrical ferrofluid container is here of sufficiently large length to be approximated as an infinite long cylinder. We consider rigid-body rotation of the ferrofluid with the time dependent angular velocity $\boldsymbol{\Omega} = \dot{\varphi}\boldsymbol{e}_z$ as can be realized with the set-up of [16, 32]. The fields $\boldsymbol{H}$ and $\boldsymbol{M}$ inside the cylinder are spatially homogeneous and oscillating in time.

3.3.1 Equations

First, the Maxwell equations demand that the fields $\boldsymbol{H}$ and $\boldsymbol{M}$ within the ferrofluid are related to each other *via*

$$\boldsymbol{H} + \bar{N}\boldsymbol{M} = \boldsymbol{H}^{\text{ext}}\,, \tag{18}$$

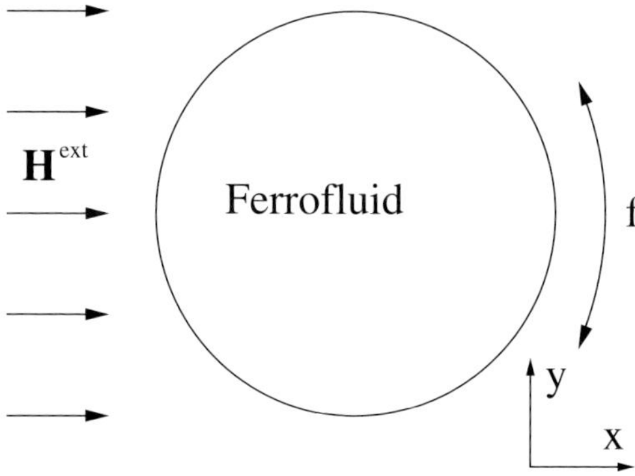

Fig. 11. Schematic plot of the FF-pendulum.

with $\bar{N} = 1/2$ for the infinitely long cylinder. Then we have the torque balance of the torsional pendulum

$$\ddot{\varphi} = -\omega_0^2 \varphi - \Gamma_0 \dot{\varphi} + \frac{N}{\Theta} + f(t) \,, \tag{19}$$

with the eigenfrequency ω_0 and the damping rate Γ_0 of the pendulum without field and the total moment of inertia Θ. The magnetic field $\boldsymbol{H}$ exerts on the magnetic moments of the ferrofluid, *i.e.*, on the pendulum the torque

$$N = \mu_0 \int dV (\boldsymbol{M} \times \boldsymbol{H})_z = -\mu_0 V M_y H^{\text{ext}} \,, \tag{20}$$

and $f(t)$ is the external mechanical forcing.

Finally, we need an equation describing the magnetization dynamics. Here, we consider the polydisperse ferrofluid as a mixture of ideal monodisperse paramagnetic fluids. Then the resulting magnetization is given by $\boldsymbol{M} = \sum \boldsymbol{M}_j$, where $\boldsymbol{M}_j$ denotes the magnetization of the particles with diameter d_j. We assume that each $\boldsymbol{M}_j$ obeys a simple Debye relaxation dynamics described by

$$d_t \boldsymbol{M}_j - \boldsymbol{\Omega} \times \boldsymbol{M}_j = -\frac{1}{\tau_j} \left[\boldsymbol{M}_j - \boldsymbol{M}_j^{\text{eq}}(\boldsymbol{H}) \right] . \tag{21}$$

We take the equilibrium magnetization to be given by a Langevin function

$$\boldsymbol{M}_j^{\text{eq}}(\boldsymbol{H}) = \chi_j(H) \boldsymbol{H} = w_j \mathcal{L} \left(\frac{\mu_0 \pi M_{\text{mat}}}{6 k_{\text{B}} T} d_j^3 H \right) \frac{\boldsymbol{H}}{H} \,, \tag{22}$$

with the saturation magnetization of the material M_{mat} and the magnetization distribution $w_j(d_j)$. Note that the magnetization equations (21) for the different particle sizes are coupled by the internal field $\boldsymbol{H} = \boldsymbol{H}^{\text{ext}} - \bar{N} \boldsymbol{M}$. As relaxation rate we combine Brownian and Néel relaxation $1/\tau_j = 1/\tau_{\text{B}}^j + 1/\tau_{\text{N}}^j$. The relaxation times depend on the particle size by $\tau_{\mathrm{B}}^{j} = \pi\eta/2k_{\mathrm{B}}T(d_j + 2s)^3$ and $\tau_{\mathrm{N}}^{j} = f_0^{-1}\exp(\pi K d_j^3/6k_{\mathrm{B}}T)$ with η the viscosity, s the thickness of the nonmagnetic particle layer and K the anisotropy constant.

Altogether we use the following system of equations:

$$\dot{\varphi} = \Omega \tag{23}$$

$$\dot{\Omega} = -\omega_0^2\varphi - \Gamma_0\Omega - \mu_0\frac{V}{\Theta}H^{\mathrm{ext}}M_y + f(t) \tag{24}$$

$$\dot{M}_x^j = -\Omega M_y^j - \frac{1}{\tau_j}\left[M_x^j - \chi_j(H)(H^{\mathrm{ext}} - \bar{N}M_x)\right] \tag{25}$$

$$\dot{M}_y^j = \Omega M_x^j - \frac{1}{\tau_j}M_y^j - \frac{1}{\tau_j}\bar{N}\chi_j(H)M_y\,. \tag{26}$$

3.3.2 Linear response analysis

First we should like to emphasize that we use in this paragraph the index 0 to identify the equilibrium situation of the unforced pendulum at rest: $\varphi_0 = \Omega_0 = M_y^{j0} = 0$ and $M_x^{j0} = M_{\mathrm{eq}}^j(H_0)$. Furthermore, $M_0 = \sum M_{\mathrm{eq}}^j(H_0)$ with H_0 being the internal field in the pendulum at rest, *i.e.*, $H_0 = H^{\mathrm{ext}} - \bar{N}M_0(H_0)$.

External forcing with small $|f|$ leads to small deviations of φ, of Ω, and of $\delta\boldsymbol{H} = \boldsymbol{H} - \boldsymbol{H}_0 = -\bar{N}(\boldsymbol{M} - \boldsymbol{M}_0) = -\bar{N}\delta\boldsymbol{M}$ from the above described equilibrium state. We expand each $\chi_j(H)$ up to linear order in $\delta\boldsymbol{H}$

$$\chi_j(|\boldsymbol{H}_0 + \delta\boldsymbol{H}|) = \chi_{j0} - \chi_{j0}'\bar{N}\delta M_x + \mathcal{O}(\delta\boldsymbol{H})^2\,. \tag{27}$$

Here, $\chi_{j0} = \chi_j(H_0)$ and χ_{j0}' is the derivative of χ_{j0}. Then we get the linearized equations

$$\dot{\varphi} = \Omega \tag{28}$$

$$\dot{\Omega} = -\omega_0^2\varphi - \Gamma_0\Omega - \kappa y + f(t) \tag{29}$$

$$\dot{x}_j = -\frac{1}{\tau_j}x_j - \frac{1}{\tau_j}\bar{N}\left(\chi_{j0} + \chi_{j0}'H_0\right)x \tag{30}$$

$$\dot{y}_j = \Omega x_j^0 - \frac{1}{\tau_j}y_j - \frac{1}{\tau_j}\bar{N}\chi_{j0}y\,. \tag{31}$$

We have introduced the abbreviations $x_j = \delta M_x/M_0$, $x_j^0 = M_x^{j,0}/M_0$, $y_j = \delta M_y/M_0$ and $x = \sum_j x_j$, $y = \sum_j y_j$. The strength of the coupling constant between the mechanical degrees of freedom φ, Ω and the magnetic ones is $\kappa = \mu_0 H^{\mathrm{ext}}M_0V/\Theta$.

For periodic forcing $f(t) = \hat{f}\mathrm{e}^{-\mathrm{i}\omega t}$ we look for solutions in the form

$$\begin{pmatrix}\varphi(t)\\ \Omega(t)\\ x_j(t)\\ y_j(t)\end{pmatrix} = \begin{pmatrix}\hat{\varphi}\\ \hat{\Omega}\\ \hat{x}_j\\ \hat{y}_j\end{pmatrix}\mathrm{e}^{-\mathrm{i}\omega t}\,. \tag{32}$$

Inserting the ansatz (32) into the linearized equations (28)–(31) yields

$$\hat{\Omega} = -\,\mathrm{i}\omega\hat{\varphi} \tag{33}$$

$$\hat{x} = 0 = \hat{x}_j \tag{34}$$

$$\hat{y}_j = -\left[\frac{\mathrm{i}\omega\tau_j}{1-\mathrm{i}\omega\tau_j}x_j^0 - \frac{\bar{N}\chi_{j0}}{1-\mathrm{i}\omega\tau_j}\frac{\omega}{\kappa}\Sigma\right]\hat{\varphi} \tag{35}$$

$$\hat{y} = -\frac{\omega}{\kappa}\Sigma\hat{\varphi} \tag{36}$$

and

$$\hat{\varphi} = G\hat{f} = \left[\omega_0^2 - \omega^2 - \mathrm{i}\omega\Gamma_0 - \omega\Sigma\right]^{-1}\hat{f}\,. \tag{37}$$

The ferrofluid-induced 'selfenergy' $\Sigma(\omega)$ in the expression for the dynamical response function $G(\omega)$ of the torsional pendulum is

$$\Sigma(\omega) = \mathrm{i}\kappa\left(1+\bar{N}\sum_j\frac{\chi_{j0}}{1-\mathrm{i}\omega\tau_j}\right)^{-1}\sum_j\frac{\tau_j x_j^0}{1-\mathrm{i}\omega\tau_j}\,. \tag{38}$$

Its imaginary part changes the damping rate Γ_0 of the pendulum for $\kappa = 0$, *i.e.*, in zero field. The real part shifts the resonance frequency of the pendulum. In the special case of a monodisperse ferrofluid one has

$$\Sigma(\omega) = \frac{\mathrm{i}\kappa\tau}{1-\mathrm{i}\omega\tau+\bar{N}\chi_0}\,. \tag{39}$$

3.3.3 Results

We evaluated the linear response function $G(\omega) = \hat{\varphi}(\omega)/\hat{f}$ of the pendulum's angular deviation amplitude $\hat{\varphi}(\omega)$ to the applied forcing amplitude $\hat{f}$ and the selfenergy $\Sigma(\omega)$ for some experimental parameters from [16, 32]: $\omega_0/2\pi = 32.7$ Hz, $\Gamma_0 = 0.178$ Hz, $V/\Theta = 20$ m/kg. The cylinder is filled with the ferrofluid APG 933 of FERROTEC. Therefore, we used in Eq. (39) an experimental $\tau = 0.6$ ms and the experimental $M_{\mathrm{eq}}(H)$. These monodisperse results were compared with the expression (38) for the polydisperse case for the typical parameter values $M_{\mathrm{mat}} = 456$ kA/m, $\eta = 0.5$ Pa s, $s = 2$ nm, $K = 44$ kJ/m^3 and $f_0 = 10^9$ Hz. The contributions $w(d_j)$ that enter into the formulas (22) for the susceptibilities χ_j are given by a lognormal distribution [17]:

$$w(d_j) = M_{\mathrm{sat}}\frac{g(d_j)d_j}{\sum_{k=1}^{30} g(d_k)d_k}$$

$$\text{with } g(d_j) = \frac{1}{\sqrt{2\pi}d_j\ln\sigma}\exp\left(-\frac{\ln^2(d_j/d_0)}{2\ln^2\sigma}\right). \tag{40}$$

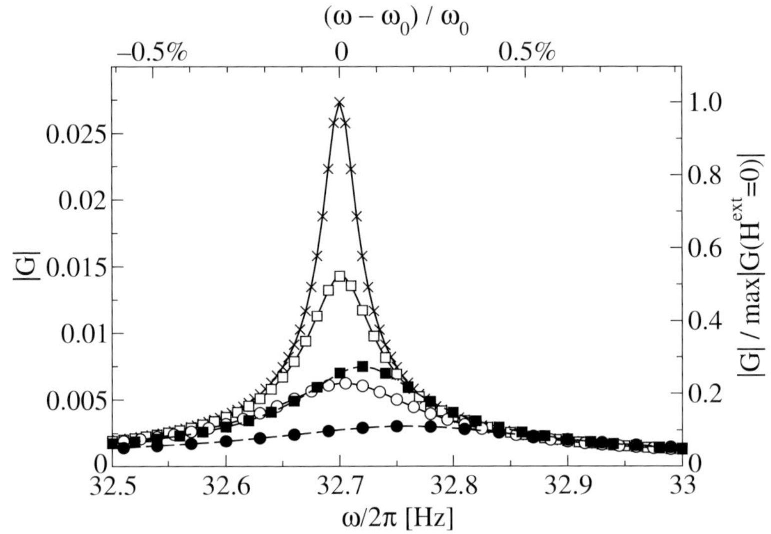

Fig. 12. $|G|$ near the resonance ω_0; $\times$ $H^{\text{ext}} = 0\,\text{kA/m}$, squares $H^{\text{ext}} = 5\,\text{kA/m}$, circles $H^{\text{ext}} = 10\,\text{kA/m}$; filled symbols: polydisperse.

Fitting the experimental $M_{\text{eq}}(H)$ with a sum of Langevin functions (22) yields $M_{\text{sat}} = 18\,149\,\text{A/m}$, $d_0 = 7\,\text{nm}$ and $\sigma = 1.47$. We used here 30 different particle sizes from $d_1 = 1\,\text{nm}$ to $d_{30} = 30\,\text{nm}$.

The calculations show the additional damping rate caused by the interaction between ferrofluid and external field. An increasing magnetic field leads to smaller amplitudes; in polydisperse ferrofluids the amplitude decreases faster (Figs. 12 and 13a). Furthermore, one can see a shift of the peak position to higher frequencies $\omega_{\max}$, which is stronger in polydisperse ferrofluids (Figs. 12, 13b and 14a).

4. Rotating ferrofluid cylinder in a magnetic field

The prospect of influencing the rotational dynamics of the nanoscaled magnetic particles in a ferrofluid by a macroscopic flow and/or by a magnetic field in order to then observe the resulting response *via* the magnetization and/or *via* changes in the flow has been stimulating many research activities [15, 44–46] ever since McTague measured [29] the so-called magneto-viscous effect. Of particular interest are in this context flows that are shear free on the macroscopic scale as in a fluid that is rotating like a rigid body with a rotation frequency, say, $\boldsymbol{\Omega} = \Omega \boldsymbol{e}_z$.

While the colloidal magnetic particles then undergo thermally sustained rotational and translational Brownian motion on the microscopic scale they

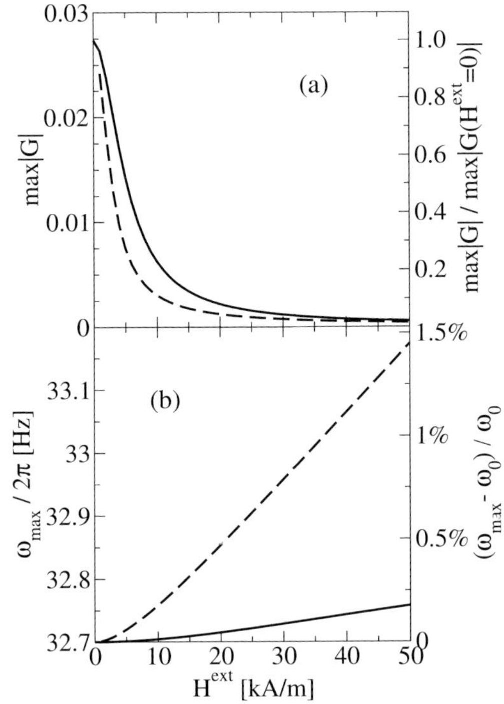

Fig. 13. Maximum value max $|G|$ (a) and peak position $\omega_{\max}$ (b) as a function of external field H^{ext}; full line monodisperse, dashed line polydisperse.

co-rotate in the mean with the deterministic macroscopic rigid body flow. However, this mean co-rotation can be hindered by magnetic torques on their moments when a magnetic field, say, $\boldsymbol{H}^{\text{ext}} = \boldsymbol{H}_0 = H_0\boldsymbol{e}_x$ is applied perpendicular to the rotation axis $\boldsymbol{e}_z$ of the flow. The combination of the externally imposed forcing of the particle motion by (i) the rigid body flow in which they are floating and by (ii) the magnetic torques on their magnetic moments drives the colloidal suspension out of equilibrium. Concerning the magnetic moments, this forcing causes the mean orientation of the moments, *i.e.*, of the magnetization $\boldsymbol{M}$ of the ferrofluid to be no longer parallel to the internal magnetic field $\boldsymbol{H}$. Instead, $\boldsymbol{M}$ is pushed out of the direction of $\boldsymbol{H}$ as well as of that of $\boldsymbol{H}_0$ thereby acquiring a nonzero transverse component M_y.

This transverse magnetization M_y of a slender cylindrical column of ferrofluid that was enforced to rotate like a rigid body with constant frequency $\Omega\boldsymbol{e}_z$ in an applied homogeneous magnetic field $H_0\boldsymbol{e}_x$ was measured as a function of Ω and H_0 [47]. These measurements are compared with the predictions based on both single–relaxation time models [31, 48–52] and polydisperse models with multi-relaxation times [53] for the magnetization dynamics.

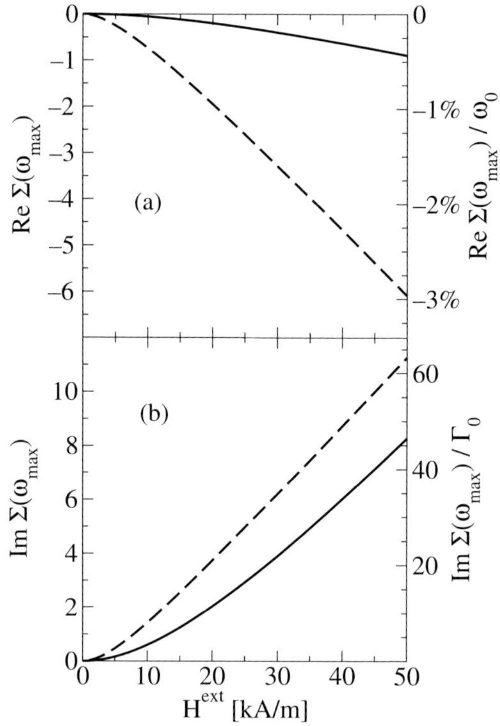

Fig. 14. Re(Σ) (a) and Im(Σ) (b) at $\omega = \omega_{\text{max}}$; full line monodisperse, dashed line polydisperse.

4.1 Experimental system

The experimental setup for measuring the magnetization of a rotating cylindrical column of ferrofluid [47] is sketched in Fig. 15. The ferrofluid is filled into a cylindrical plexiglass sample holder with inside radius $R = 3.2$ mm. This radius is so small that for our rotation frequencies the ferrofluid rotates as a rigid body with a flow field $\boldsymbol{u}(\boldsymbol{r}) = \boldsymbol{\Omega} \times \boldsymbol{r} = \Omega r \boldsymbol{e}_{\varphi}$. Here Ω is the externally enforced constant rotation rate of the sample and $\boldsymbol{e}_{\varphi}$ is the unit vector in azimuthal direction.

The sample is placed centrally between the poles of an electromagnet (BRUKER B-E 10V) providing the homogeneous and temporally constant magnetic field $\boldsymbol{H}_0 = H_0 \boldsymbol{e}_x$ perpendicular to the cylinder axis $\boldsymbol{e}_z$.

For such a combination of enforced rotation and applied field theoretical models allow for a spatially and temporally constant nonequilibrium magnetization $\boldsymbol{M}$ that is pushed out of the directions of $\boldsymbol{H}_0$ and $\boldsymbol{H}$ by the flow. However, in equilibrium, $\boldsymbol{\Omega} = 0$, the three fields are indeed collinear: the equilibrium magnetization $\boldsymbol{M}_{\text{eq}}(\boldsymbol{H}) = M_{\text{eq}}(H)\boldsymbol{H}/H$ is parallel to the internal field $\boldsymbol{H}$.

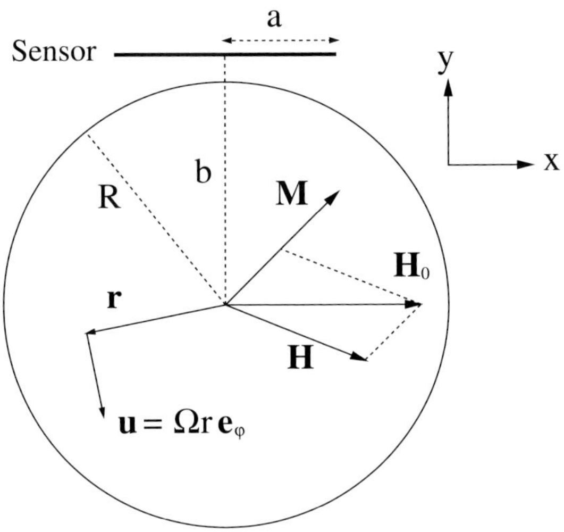

Fig. 15. Schematic plot of the rotating FF-cylinder. The cylindrical sample holder with inner radius R rotates with angular velocity $\boldsymbol{\Omega}$ in the applied static magnetic field $\boldsymbol{H}_0$ perpendicular to $\boldsymbol{\Omega}$. The magnetization M_y is measured with a Hall sensor. $\boldsymbol{M}$ and $\boldsymbol{H}$ denote the magnetization and internal magnetic field of the ferrofluid. Both are constant in space and time.

According to the Maxwell equations the fields $\boldsymbol{H}$ and $\boldsymbol{M}$ within the ferrofluid are related to each other *via*

$$\boldsymbol{H} = \boldsymbol{H}_0 - \frac{1}{2}\boldsymbol{M}. \tag{41}$$

for our long cylindrical sample and in particular $H_y = -M_y/2$ as indicated schematically in Fig. 15. In addition they demand that the magnetic field outside the ferrofluid cylinder

$$\boldsymbol{H}^{\text{out}} = \boldsymbol{H}_0 + \frac{1}{2}\frac{R^2}{r^2}\left(2\frac{\boldsymbol{r}}{r}\frac{\boldsymbol{M}\boldsymbol{r}}{r} - \boldsymbol{M}\right) \tag{42}$$

is a superposition of the applied field $\boldsymbol{H}_0$ and the dipolar contribution from $\boldsymbol{M}$. This result yields a relation between the perpendicular component of the magnetization M_y resp. of the internal field $H_y = -M_y/2$ and the field H_y^{sensor} measured by the Hall-sensor outside the sample as indicated in Fig. 15. Considering the finite size of the Hall-sensor, H_y^{sensor} is given by

$$H_y^{\text{sensor}} = \frac{1}{2a}\int_{-a}^{a} H_y^{\text{out}} \mathrm{d}x = -\frac{R^2}{a^2+b^2} H_y. \tag{43}$$

In our experimental setup $b = 4.75$ mm, $R = 3.2$ mm, and the horizontal extension of the Hall sensor $a = 2$ mm. So, we get $H_y^{\text{sensor}} = -0.386\, H_y$.

In our experiments we used the ferrofluid APG933 of FerroTec. We have measured the equilibrium magnetization of the ferrofluids (*cf.* Sect. 3.1.1). In order to get information on the particle size distribution of the ferrofluid under investigation, we used fits with a lognormal form of the distribution [17] as well as with a regularization procedure based on Tichonovs method [34, 47, 54].

The saturation magnetization of the ferrofluid sample was $M_{sat}^{FF} = 19108.6\,\mathrm{A/m}$. From the saturation magnetization the volume concentration of the magnetite particles was found to be $\phi = M_{sat}^{FF}/M_{sat}^{bulk} = 4.1\%$. For the initial susceptibility we found the value $\chi_0 = 1.09$.

4.2 Theoretical models

4.2.1 Monodisperse case

The model equations that we compare here imply a relaxational dynamics with a single relaxation time either of $\boldsymbol{M}$ towards the equilibrium magnetization

$$\boldsymbol{M}_{eq}(\boldsymbol{H}) = \frac{M_{eq}(H)}{H}\boldsymbol{H} = \chi(H)\boldsymbol{H} \tag{44}$$

or of the "local equilibrium" or "effective" field

$$\boldsymbol{H}_{eff}(\boldsymbol{M}) = \frac{M_{eq}^{-1}(M)}{M}\boldsymbol{M} = F(M)\boldsymbol{M} \tag{45}$$

towards the internal field $\boldsymbol{H}$. The equilibrium magnetization $M_{eq}(H)$ referring to the functional relation between internal field H and magnetization in the case of $\Omega = 0$ is a thermodynamic material property of the ferrofluid. The effective field $\boldsymbol{H}_{eff}$ lies parallel to $\boldsymbol{M}$ and can be seen as the inverse of the defining requirement

$$\boldsymbol{M} = \boldsymbol{M}_{eq}(\boldsymbol{H}_{eff})\,. \tag{46}$$

In equilibrium, $\Omega = 0$, one has $\boldsymbol{H}_{eff} = \boldsymbol{H}$ and $\boldsymbol{M} = \boldsymbol{M}_{eq}$.

We consider here the relations [48]

$$\text{Debye:}\qquad \boldsymbol{\Omega} \times \boldsymbol{M} = \frac{1}{\tau}\left(\boldsymbol{M} - \boldsymbol{M}_{eq}\right) \tag{47}$$

$$\text{S'72 [31]:}\qquad \boldsymbol{\Omega} \times \boldsymbol{M} = \frac{1}{\tau}\left(\boldsymbol{M} - \boldsymbol{M}_{eq}\right) + \frac{\mu_0}{4\zeta}\boldsymbol{M} \times (\boldsymbol{M} \times \boldsymbol{H}) \tag{48}$$

$$\text{FK [49, 50, 55]:}\qquad \boldsymbol{\Omega} \times \boldsymbol{M} = \gamma_H\left(\boldsymbol{H}_{eff} - \boldsymbol{H}\right) + \frac{\mu_0}{4\zeta}\boldsymbol{M} \times (\boldsymbol{M} \times \boldsymbol{H}) \tag{49}$$

$$\text{S'01 [51, 56]:}\qquad \boldsymbol{\Omega} \times \boldsymbol{H}_{eff} = \frac{1}{\tau}\left(\boldsymbol{H}_{eff} - \boldsymbol{H}\right) + \frac{\mu_0}{4\zeta}\boldsymbol{H}_{eff} \times (\boldsymbol{M} \times \boldsymbol{H}) \tag{50}$$

$$\text{ML [52]:}\qquad \boldsymbol{\Omega} \times \boldsymbol{M} = \xi(\boldsymbol{H}_{eff} - \boldsymbol{H}) \tag{51}$$

resulting for the rotating cylinder from the above 5 models. Here τ denotes a magnetic relaxation time, γ_H and ξ relaxation rates, ζ the vortex viscosity and μ_0 the vacuum permeability. In our calculations, we use $\gamma_H = \chi_0/\tau$ [57]. In ML we use the weak field variant of Ref. [52] with two different choices for the parameter ξ: Either the $\xi = \chi_0/\tau$ as in FK that is denoted here by ML(F). Or the variant $\xi = 1/[F(M)\tau]$ as in S'01 that is denoted here by ML(S).

Especially the perpendicular component $H_y = -\frac{1}{2}M_y$ of the magnetic field is suited for a comparison of the different models with each other and with experiments. Before doing the former we should like to draw the attention to the frequency behavior of $M_y(H_0, \Omega)$. We mentioned already that M_y vanishes for zero vorticity, $\Omega = 0$. Furthermore, one finds that M_y as well as M_x vanishes also in the limit $\Omega \to \infty$. Because of $M_y \geq 0$ one sees, that $M_y(\Omega)$ has a maximum as a function of Ω [48].

The differences in the results for the different models are easily captured by comparing their predictions for the maximum values of $|H_y| = \frac{1}{2}M_y$ and the locations of these maxima at $\Omega^{\max}$, each as a function of applied field H_0.

As an aside we mention that the maximal values of $|H_y|$ of Debye and S'72 are the same while their locations, $\Omega^{\max}$, differ. The models S'01, FK, and ML formulated in terms of the effective field also share a common maximal value of $|H_y|$ being larger than that of Debye and S'72 while the location, $\Omega^{\max}$, differ partly substantially. Hence,the magnetic torque, $\boldsymbol{M} \times \boldsymbol{H}$, entering into S'72, FK, and S'01 only shifts the frequency $\Omega^{\max}$ [48].

4.2.2 *Polydisperse case*

Here comparisons are made of the above single-relaxation-time approach with polydisperse models with multi-relaxation times [53]. We consider here as a first step polydisperse models with single-particle Brownian and Néel relaxation dynamics for the different particle sizes. Such models have been used to determine within a linear response analysis the effect of polydispersity on the dynamics of a torsional ferrofluid pendulum that was periodically forced close to resonance to undergo small amplitude oscillations in a rigid body flow [16, 32, 43].

We ignore the influence of any dipolar magnetic interaction and of any flow induced interaction on the (rotation) dynamics of the particles. Thereby collective, collision dominated long-range and long-time hydrodynamic relaxation dynamics of the ensemble of magnetic moments are discarded since only the individual relaxation of each magnetic moment is considered – albeit in the collectively generated internal magnetic field $\boldsymbol{H}$.

For numerical reasons we use a discrete partition of the particle size distribution. Then, without interaction, the magnetization of the resulting mixture of mono-disperse ideal paramagnetic gases is given by $\boldsymbol{M} = \sum \boldsymbol{M}_j$, where $\boldsymbol{M}_j$ denotes the magnetization of the particles with diameter d_j. We assume that each magnetic moment and with it each $\boldsymbol{M}_j$ obeys a simple Debye relaxation dynamics that drives them in the absence of flow towards their respective equilibrium value $\boldsymbol{M}_j^{\text{eq}}(\boldsymbol{H})$. Then, in the stationary situation resulting from the rigid body rotation with constant $\boldsymbol{\Omega}$ the Debye relaxation equation for each sub magnetization (21) reduces to

$$\boldsymbol{\Omega} \times \boldsymbol{M}_j = \frac{1}{\tau_j}\left[\boldsymbol{M}_j - \boldsymbol{M}_j^{\text{eq}}(\boldsymbol{H})\right]. \tag{52}$$

In the absence of interactions the equilibrium magnetization $\boldsymbol{M}_j^{\text{eq}}(\boldsymbol{H})$ of each species is determined by a Langevin function (22). For the magnetic weight $w_j(d_j)$ of species j appearing in (22) we take the experimentally determined values from the representation $\boldsymbol{M}^{\text{eq}} = \sum \boldsymbol{M}_j^{\text{eq}}$. Note that the magnetization equations (52) for the different particle sizes are coupled *via* the internal field $\boldsymbol{H}$: according to Maxwell's equations $\boldsymbol{H} = \boldsymbol{H}_0 - \frac{1}{2}\boldsymbol{M}$ is given in terms of the total $\boldsymbol{M} = \sum \boldsymbol{M}_j$.

In the relaxation rate $1/\tau_j$ we take into account Brownian and Néel relaxation processes by adding their rates with equal weight

$$\frac{1}{\tau_j} = \frac{1}{\tau_{\text{B}}^j} + \frac{1}{\tau_{\text{N}}^j}. \tag{53}$$

The relaxation times depend on the particle size according to $\tau_{\text{B}}^j = \pi\eta/2k_{\text{B}}T\,(d_j + 2s)^3$ and $\tau_{\text{N}}^j = f_0^{-1}\exp(\pi K d_j^3/6k_{\text{B}}T)$. Her η is the viscosity, s the thickness of the nonmagnetic particle layer, K the anisotropy constant and f_0 an attempt frequency of $\sim 10^9$ Hz [15, 58]. The combined relaxation rate (53) is dominated by the faster of the two processes. Thus, large particles relax in a Brownian manner with relaxation times of about some 10^{-3} s, while small particles have the much smaller Néel relaxation times. The typical range of K for magnetite-based ferrofluids is about (10–50) kJ/m^3 [59, 60]. The boundary between Néel and Brownian dominated relaxation as a function of particle size d depends sensitively on the anisotropy constant K. This is documented in Fig. 16 for the two values $K = 15\,\text{kJ/m}^3$ and $K = 50\,\text{kJ/m}^3$. For these specific examples the boundaries between Néel and Brownian dominated relaxation lie at about $d \simeq 20$ nm and $d \simeq 13$ nm, respectively.

4.3 Results

We measured H_y as a function of Ω ranging from 0 up to 3000 rad/s for a fixed value of H_0. After completing an experiment we increased H_0 and repeated the experiment. In this way we got $H_y(\Omega)$ for seven different values of H_0. One observes that the maximum amplitude of H_y increases with increasing H_0 and that the peak position $\Omega^{\max}$ is shifted towards higher Ω-values with increasing H_0.

In a first attempt we compared the experimental data for APG 933 with numerical calculations of $H_y^{\text{sensor}}(\Omega)$ with the single-relaxation time models (Sect. 4.2.1). Therefore we used τ as fit parameter and in addition an *ad-hoc*

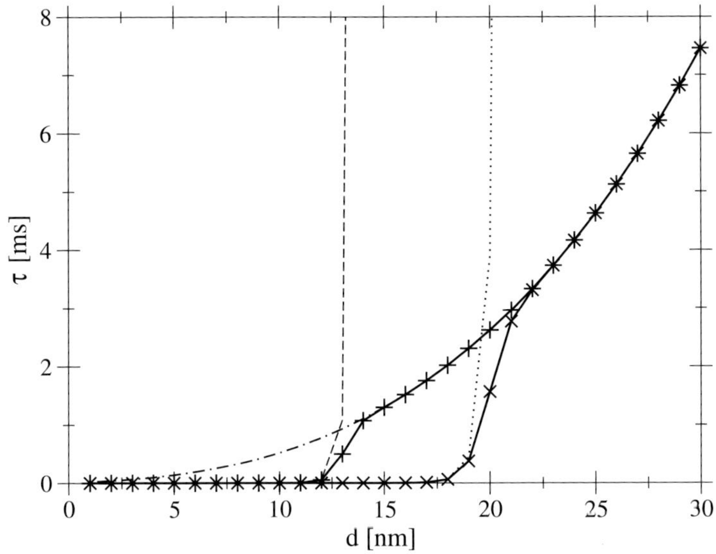

Fig. 16. Relaxation times as a function of particle diameter d: Brownian (dot-dashed line), Néel (dotted line for $K = 15\,\mathrm{kJ/m^3}$, dashed line for $K = 50\,\mathrm{kJ/m^3}$), and the combination (53) (solid line with crosses for $K = 15\,\mathrm{kJ/m^3}$, solid line with plusses for $K = 50\,\mathrm{kJ/m^3}$).

amplitude correction factor f discussed below that forces the model prediction for the maximal H_y^{sensor} to coincide with the measured one. Figure 17 shows the comparisons of experimental data with the fits for the case of the medium sized field $H_0 = 30\,\mathrm{kA/m}$. The parameters used to calculate $H_y(\Omega)$ are summarized in Table 5. We furthermore used as input the experimental equilibrium magnetization $M_{\text{eq}}(H)$ of APG 933, the experimental initial susceptibility $\chi_0 = 1.09$ and the vortex viscosity $\zeta = \frac{3}{2}\Phi\eta$ with viscosity $\eta = 0.5\,\mathrm{Pa\,s}$ and volume concentration $\Phi = 0.041$.

The necessity of an amplitude reduction factor $f < 1$ reflects in our opinion the fact that only a fraction of the particles contribute to the experimental signal: only those particles contribute that have a volume large enough to ensure the magnetic moment to be fixed within the particle and therefore to relax in a Brownian manner.

In the experiments done with polydisperse ferrofluids for frequencies up to $\Omega \simeq 3000\,\mathrm{rad/s}$ mainly the large particles contribute to M_y resp. to H_y^{sensor} since their magnetic moments are effectively frozen in the particle's crystal lattice. Only these magnetic moments can be pushed out of the direction of the magnetic field by the combined action of thermally induced rotary Brownian motion and deterministic macroscopic flow in the rotating cylinder. Smaller particles can keep their moment aligned with the magnetic field by the Néel process when these particles undergo rotational motion. Finally, the particle

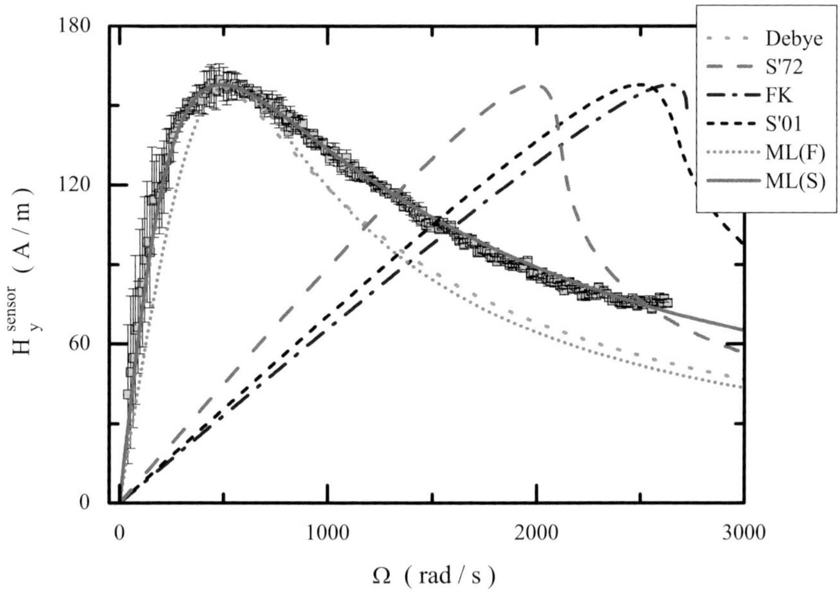

Fig. 17. Comparison of the experimental data (symbols) obtained at $H_0 = 30\,\text{kA/m}$ with the numerical results according to different theoretical models. See Table 5 for an identification of the curves and the fit parameters used.

Table 5. Models described in Sect. 4.2.1 are used in Fig. 17 in order to get the presented curves. *f-factor* refers to the amplitude factor.

model	color	f-factor	relaxation time τ [ms]
Debye	(dotted)	0.164	2.5
S'72	(dashed)	0.164	2.5
FK	(dashed dotted)	0.125	5
S'01	(short dashed)	0.125	3.5
ML(F)	(short dotted)	0.125	6
ML(S)	(solid)	0.125	3.5

diameter that separates Néel behavior from Brownian behavior in the size distribution and that thereby determines how many particles contribute to M_y resp. to the experimental signal H_y^{sensor} depends sensitively on the anisotropy constant K: The smaller K, the smaller is the number of Brownian particles according to Fig. 16, and the smaller is M_y resp. H_y^{sensor}.

The above sketched physical picture is corroborated by Fig. 18. There we compare the experimentally obtained $H_y^{\text{sensor}}(\Omega)$ (stars) as a function of Ω for a representative externally applied field $H_0 = 30\,\text{kA/m}$ with various

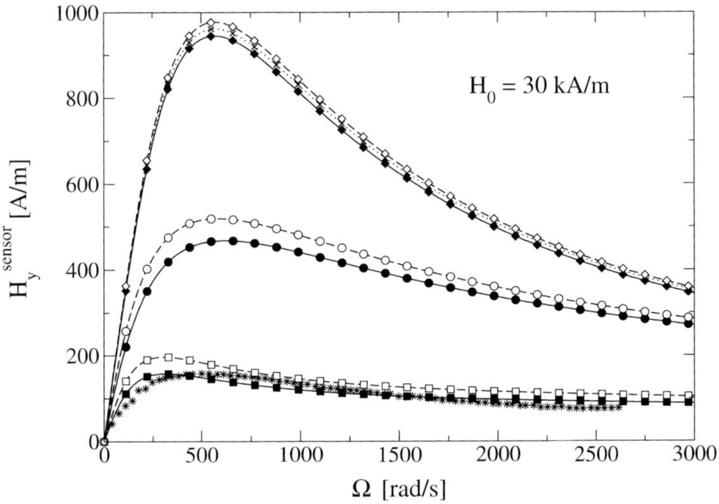

Fig. 18. H_y^{sensor} as function of Ω for $H_0 = 30\,\text{kA/m}$. Experimental data for APG 933 (stars) are compared with the results of different theoretical models. The three uppermost curves refer to single-time relaxation approximations, each with $\tau = 2\,\text{ms}$: monodisperse Debye model (dotted line with crosses), polydisperse models with common $\tau_j = \tau$ and magnetic weights w_j obtained with a lognormal distribution (full line with full diamonds) and with a regularization method (dashed line with open diamonds). Curves with circles and squares refer to truly polydisperse models (Eqs. (52–53)) with $K = 50\,\text{kJ/m}^3$ (circles) and $K = 15\,\text{kJ/m}^3$ (squares). Full lines with full symbols were obtained with a lognormal distribution. Dashed lines with open symbols refer to a distribution resulting from the regularization method. In all models the equilibrium magnetization $M_{\text{eq}}(H)$ was taken to be the experimental one.

model variants that take into account the polydispersity to different degrees. This is done for two different anisotropy constants, namely, $K = 15\,\text{kJ/m}^3$ and $K = 50\,\text{kJ/m}^3$ as representative examples. However, the three uppermost curves refer to single time relaxation approximations, each with $\tau = 2\,\text{ms}$: the dotted line with crosses is the result of the strictly monodisperse Debye model used above, however without amplitude correction factor ($f = 1$). The lines with diamonds refer to polydisperse models, however, with common $\tau_j = \tau = 2\,\text{ms}$ taken in Eq. (52) but different magnetic weights w_j obtained either from a lognormal distribution (full line with full diamonds) or from the regularization method (dashed line with open diamonds). The equilibrium magnetization $M_{\text{eq}}(H)$ was taken to be the experimental one, the distributions were obtained from this experimental $M_{\text{eq}}(H)$ by the lognormal ansatz resp. the regularization method. This, first of all, shows that models with only one relaxation time show roughly the same behavior of $M_y(\Omega)$ irrespective of whether the particle size and magnetic moment distributions are polydisperse or not.

The set of curves with circles and squares in Fig. 18 refer to truly polydisperse models, Eqs. (52–53), but different anisotropy constants of the magnetic material ($K = 50\,\mathrm{kJ/m^3}$ (circles), $K = 15\,\mathrm{kJ/m^3}$ (squares)). For the numerical calculations we used typical values for the ferrofluid APG 933: $M_{\mathrm{sat}}^{\mathrm{bulk}} = 456\,\mathrm{kA/m}$, $\eta = 0.5\,\mathrm{Pa\,s}$, $s = 2\,\mathrm{nm}$, and $f_0 = 10^9\,\mathrm{Hz}$. Again, full lines with full symbols were obtained with a lognormal distribution while dashed lines with open symbols refer to a distribution resulting from the regularization method.

The largest and most important difference between the curves with diamonds (*i.e.*, the single-time models) and the curves with circles and squares (*i.e.*, the genuine polydisperse models) come from the difference in the anisotropy constants of the magnetic material that governs how many particles contribute efficiently as Brownian ones to the transverse magnetization M_y resp. to H_y^{sensor}: for smaller K the magnetic moments of fewer particles being Brownian ones may be rotated out of the direction of the magnetic field by the flow in the cylinder. The curves for $K = 15\,\mathrm{kJ/m^3}$ yield roughly the same maximal size H_y^{sensor} as the experiments – they could be fine-tuned even further.

Furthermore we found that the models FK, S'01, and S'72 do not properly reproduce the position $\Omega^{\max}$ of the maximum of the experimental curve in Fig. 17. In contrast, using the other models with an appropriate choice of the fit-parameter τ we can reproduce the peak position $\Omega^{\max}$. However, only the model ML(S) reproduces the experimental curve in Fig. 17 for larger values of Ω.

From the measured curves of $H_y(\Omega, H_0)$ *versus* H_0 we extracted the position of the maximum (denoted as $\Omega^{\max}(H_0)$) and the maximum value $\max(H_y^{\mathrm{sensor}}) = H_y^{\mathrm{sensor}}(\Omega^{\max}, H_0)$. The latter is shown in Fig. 19 in comparison with the results of the models. In order to fit the maximum amplitude we had to introduce an *ad-hoc* additional amplitude reduction factor f (*cf.* Table 5). Note that the models Debye and S'72 yield the same maximum (Sect. 4.2.1). The other models [FK, S'01, ML(F), and ML(S)] also have a common but different maximal H_y. Thus, Table 5 contains only two different amplitude reduction factors f. Fig. 19 shows that already the simple Debye model reasonably well reproduces the variation of $\max(H_y^{\mathrm{sensor}})$ with H_0 if one ignores the *ad-hoc* amplitude reduction factor.

We have seen, that only the polydisperse model can reproduce the amplitude $\max(H_y^{\mathrm{sensor}})$ without an amplitude reduction factor. But also above polydisperse model yields locations, $\Omega^{\max}(H_0)$, of the maxima for different H_0 which are still off from the experimental ones as shown in Fig. 20a and 21b. However, the agreement between the predictions of the polydisperse models of Eqs. (52–53) and the experiments concerning the location $\Omega^{\max}(H_0)$ can be improved by allowing the relaxation rates τ_j of the differently sized particles to depend also on the internal field H. We used the ansatz [42]

$$\tau_j^{\gamma}(H) = \tau_j \, \frac{2\mathcal{L}(\gamma H)}{\gamma H - \mathcal{L}(\gamma H)}\,, \tag{54}$$

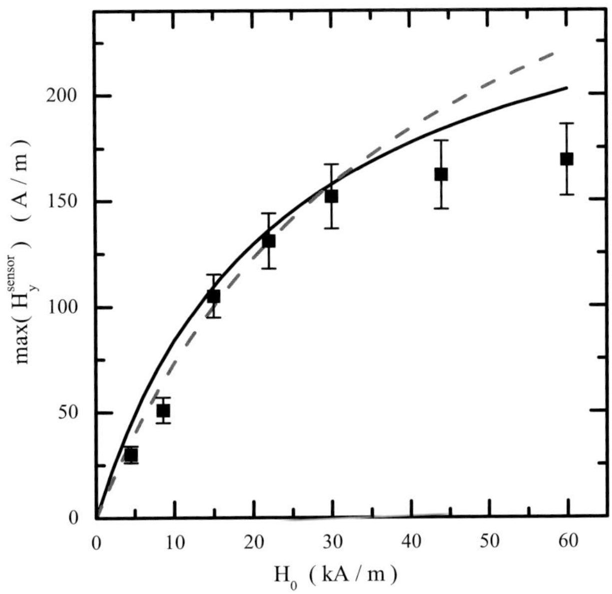

Fig. 19. Maxima of $H_y^{\text{sensor}} = H_y^{\text{sensor}}(\Omega^{\max})$ as a function of the external field H_0. Squares with error bars refer to experimental data. Solid line shows the result of the Debye and S'72 model. Dashed line is common to the models denoted as FK, S'01, and ML. Parameters are listed in Table 5.

with one additional fit parameter γ. Values of about $\gamma \sim 10^{-4}$ m/A yield maximum locations $\Omega^{\max}(H_0)$ that agree well with the experiments as can be seen in Fig. 20b and 21b. This generalization of the model (52–53) leaves the peak value $H_y^{\text{sensor}}(\Omega^{\max})$ almost unchanged, *cf.* Fig. 21a.

However, also this augmented polydispersive model reproduces with fixed values of K and γ the experimental data only in a small range of Ω and H_0, *cf.* Figs. 20 and 21.

4.4 Conclusion

We have compared the predictions of polydisperse models of the magnetization dynamics of ferrofluids with recent experiments measuring the transverse magnetization component M_y of a rotating ferrofluid cylinder. The models use mixtures of mono-disperse ideal paramagnetic gases of differently sized particles. The magnetization dynamics of the models take into account the rigid body rotation of the fluid combined with a simple Debye relaxation of the magnetic moments of each particle with size dependent Brownian and Néel magnetic relaxation times. Thus, in the absence of flow, each magnetic moment and with it each sub-magnetization would be driven independently of the others towards its respective mean equilibrium value that, however, depends

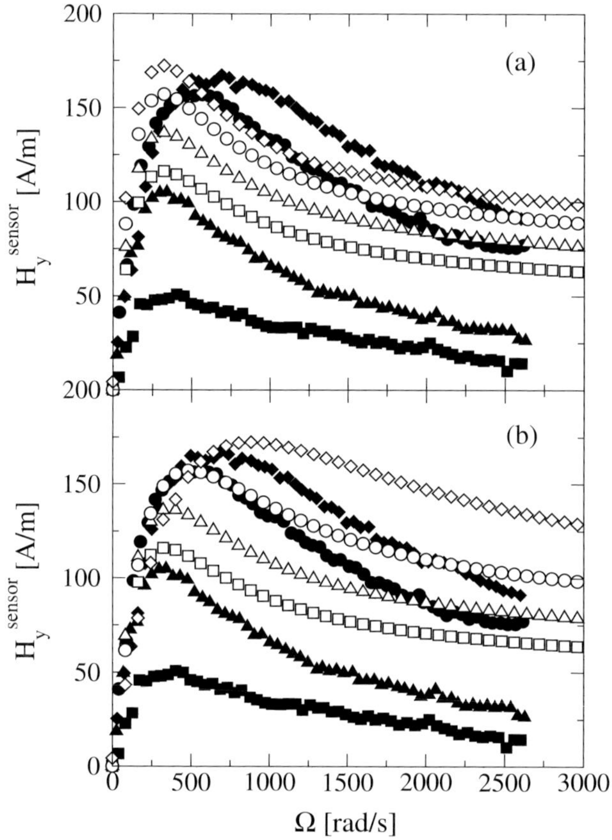

Fig. 20. H_y^{sensor} as function of Ω for $H_0 = 8.6\,\mathrm{kA/m}$ (squares), $15\,\mathrm{kA/m}$ (triangles), $30\,\mathrm{kA/m}$ (circles), and $60\,\mathrm{kA/m}$ (diamonds). Full symbols denote experimental data. Open symbols refer to the polydisperse model (Eqs. (52–53)) with lognormal distribution and $K = 15\,\mathrm{kJ/m^3}$: in (a) the relaxation times τ_j are independent of H, in (b) they are replaced by $\tau_j^{\gamma}(H)$ (54) with $\gamma = 10^{-4}\,\mathrm{m/A}$.

on the internal magnetic field $\boldsymbol{H}$ being collectively generated by all magnetic moments.

The comparison suggests that mainly the large particles contribute to M_y since their magnetic moments are effectively frozen in the particle's crystal lattice. Smaller particles can keep their moment aligned with the magnetic field by the Néel process when these particles undergo rotational motion.

Finally, the particle diameter that separates Néel behavior from Brownian behavior in the size distribution and that thereby determines how many particles contribute to M_y resp. to the experimental signal H_y^{sensor} depends quite sensitively on the anisotropy constant K of the magnetic material. K deter-

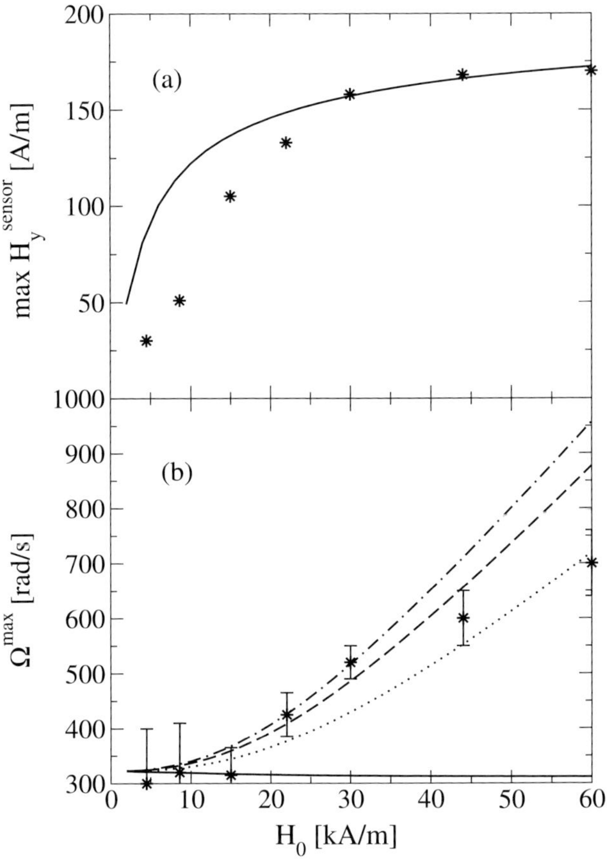

Fig. 21. Maximum magnitude $\max(H_y^{\text{sensor}}) = H_y^{\text{sensor}}(\Omega^{\max})$ (a) and location of the maximum $\Omega^{\max}$ (b) as functions of the external field H_0. Stars show experimental data. Lines refer to the results of polydisperse models with a lognormal distribution and $K = 15\,\text{kJ/m}^3$: H-independent relaxation times (solid); H-dependencies with $\gamma = 0.8 \times 10^{-4}$ m/A (dotted), $\gamma = 10^{-4}$ m/A (dashed), and $\gamma = 1.1 \times 10^{-4}$ m/A (dot-dashed). In (a) the differences between the lines are too small to be seen.

mines how many magnetic moments are "frozen" or "blocked" in particles and thus can be rotated by the rigid body flow: The smaller K, the smaller is the number of Brownian particles with frozen moments, and the smaller is the resulting M_y. Or, *vice versa*, a large transverse magnetization can be expected in ferrofluids with large anisotropy constants.

An analysis of the rotation rates $\Omega^{\max}(H_0)$ for which M_y is largest indicates that the agreement between experiments and model predictions can be improved by allowing the relaxation rates of the differently sized particles to depend also on the magnetic field H.

5. Surface instabilities

Here we discuss and compare results from experimental investigations of the anomalous dispersion branch of surface waves on ferrofluids in a magnetic field and numerical analyses [61]. In our experiments we used the well known Faraday instability of parametrically excited surface waves on a ferrofluid in the presence of a magnetic field oriented perpendicular to the fluid's surface.

5.1 Earlier work

Cowley and Rosensweig [62] have investigated experimentally the instability of a flat ferrofluid surface in a static magnetic field H oriented perpendicular to the surface: above a certain critical value, H_C^R, the surface spontaneously deforms into regularly spaced peaks, that are usually arranged in a hexagonal pattern. At the critical field H_C^R of this so-called Rosensweig instability, the dispersion curve of surface waves touches the zero frequency axis at a finite wave number k_C^R, analogous *e.g.* to the soft mode behavior of phonons at structural phase transitions of solids. For magnetic fields slightly below H_C^R, the dispersion is non-monotonous with a section of negative slope (anomalous dispersion). There exists a number of papers dealing with theoretical modelling of the dispersion relation [63–66]. But up to now the anomalous dispersion has not been detected experimentally, although there is experimental evidence for a jump in wave number [67] that is expected to occur in the presence of anomalous dispersion when ramping the frequency up and down [68].

The excitation of surface waves has been realized by different mechanisms. Pétrélis *et al.* [69] have studied the influence of the Faraday instability mechanism of parametrically excited surface waves that arise in vertically shaken containers on the Rosensweig instability for driving frequencies $f \geq 30$ Hz. They have demonstrated experimentally that the Faraday instability inhibits the Rosensweig instability, a fact that was predicted by Müller in [68]. In [67] Mahr *et al.* have reported the existence of two different wave numbers occurring at the same excitation frequency in a V-shaped annular teflon channel. In order to generate surface waves they have directed an air jet towards the magnetic fluid's surface. Results of measurements at a magnetic field 3% below the threshold of the Rosensweig instability have been presented but the anomalous region of the dispersion has not been studied. Due to the finite geometry of the vessel a quantitative comparison of their experimental results with calculation was not possible. In Ref. [70] a mechanical excitation of surface waves in the presence of a dc magnetic field was used. The authors have found so-called twin-peak patterns in the non-monotonic regime of the dispersion relation 2% below the critical field H_C^R and furthermore domain structures, oscillating defects and relaxational phase oscillations. These findings have been attributed to the non-monotonic dispersion relation of ferrofluids. The same group has presented in [71] experimental results on surface waves, that have been excited by an oscillating magnetic field. Both, subharmonic and harmonic surface response have been detected. Higher order resonance tongues were generated by tuning the viscous damping with an external magnetic field. This was made possible by a specific peculiarity of ferrofluids, namely a very weak damping of surface waves when the wave number is close to k_C^R. Browaeys *et al.* [72] have used a small coil placed beneath the ferrofluid container in order to generate surface waves. With this method it was very difficult to excite surface waves with wavelengths small compared to the coils diameter. They limited the ratio H/H_C^R to 0.79 because above this ratio the peaks started to form at the boundary of the vessel.

5.2 Generating surface waves with anomalous dispersion

A drawback of generating surface waves by modulating the magnetic field is that the time variation of the magnetization $M(t) = \chi H(t)$ enters quadratically into the dispersion relation so that the effective excitation signal is composed of two frequencies. Furthermore, the effective magnetic drive is proportional to k^2 while it is linear in k for the gravity modulation. Thus, shorter wavelength are favored by the magnetic drive. To avoid these complications the Faraday forcing is used, where surface waves are generated only beyond a critical excitation amplitude [73]. In this type of experiment the earth's gravity acceleration g_0 is modulated in the co-moving frame according to $g(t) = g_0 + a\cos(2\Omega t)$. As long as the amplitude a is smaller than a critical one, a_C, the surface of the ferrofluid remains flat. By increasing a beyond a_C, the flat surface becomes unstable and the system forms standing waves. The wave pattern depends on the mass density ρ, surface tension σ, viscosity η of the fluid and its filling level h. For small values of h, an experimental investigation of the anomalous part of the dispersion relation was predicted to be feasible [68].

To get access to the anomalous dispersion branch we took advantage of the dissipation in the bottom boundary layer [68] that is responsible for the suppression of the growth of waves with wave numbers below $k \sim 1/h = 200\,\mathrm{cm}^{-1}$ for $h = 5$ mm. Being strongly damped the onset of such waves in the region of linear dispersion is delayed. Furthermore, by choosing the viscosity appropriately the marginal stability curve $a(k)$ of the flat surface against the growth of standing waves can be tuned such that its absolute minimum k_c lies in the wave number region of anomalous dispersion. Then the growth rates of waves in this region are largest and get selected when increasing the acceleration above threshold.

Besides the filling level h, the fluid's viscosity η influences the marginal curves in a way, that can be used to get access to the anomalous dispersion branch. The influence of the viscosity on the marginal curves is shown in Fig. 22. The upper panel of Fig. 22 shows the marginal curves for a ferrofluid of low viscosity ($\eta = 6.66$ mPa s), while in the lower panel the high-viscosity case ($\eta = 53.3$ mPa s) is presented. In the latter, the high viscosity clearly favors the

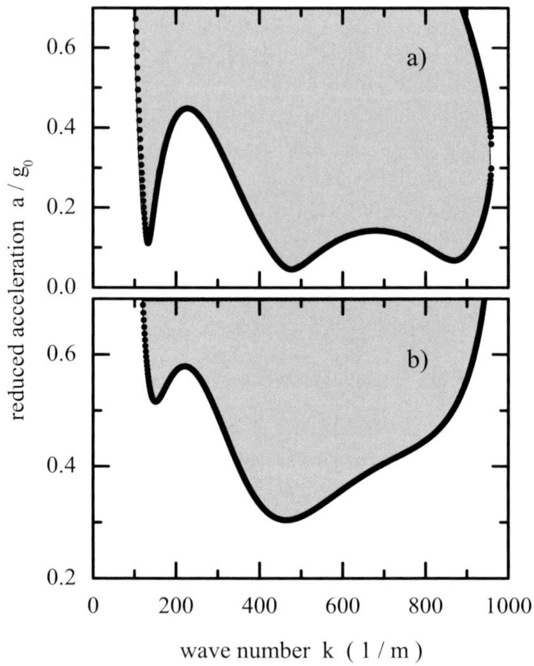

Fig. 22. Marginal stability curves of the flat surface against growth of standing waves for different values of the viscosity η (a) $\eta = 6.66\,\mathrm{mPa\,s}$, (b) $\eta = 53.3\,\mathrm{mPa\,s}$) for the following parameters: $f = 7\,\mathrm{Hz}$, $h = 5\,\mathrm{mm}$, $\sigma = 0.022\,\mathrm{N/m}$, $\rho = 1110\,\mathrm{kg/m^3}$, $\chi = 0.8$ and $M = 0.999\,M_C^R$. Within the gray regions, the plane surface is unstable.

local minimum belonging to the anomalous dispersion branch. Thus, choosing an appropriate combination of filling level and fluid's viscosity enabled us to see in our experiment only the dispersion branch of interest, *i.e.* the anomalous one.

5.3 Experimental setup

The experimental setup is sketched in Fig. 23. The cylindrical container filled with the ferrofluid is attached to a mechanical vibrator (Large Displacement Shaker V617, Gearing & Watson Electronics Ltd.) connected to a 4 kW power amplifier (DSA-4k, Gearing & Watson Electronics Ltd.). The container is placed in the center of a pair of Helmholtz coils providing a magnetic field oriented perpendicular to the fluid surface. The homogeneity of the field across the liquid level is better than 5%. The sinusoidal signal for the power amplifier was synthesized by a digital–analog card installed in a Pentium PC. The vertical acceleration of the container is measured by a piezo-electric device (Brüel & Kjaer 4393). The signal of this device is amplified (Brüel & Kjaer 2525)

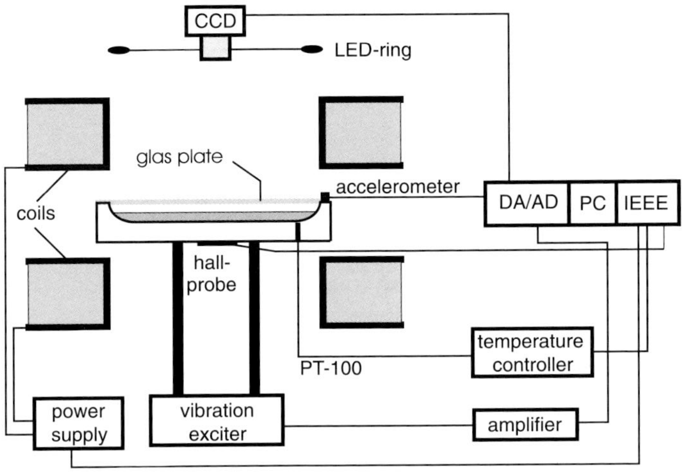

Fig. 23. Sketch of the experimental setup. For further details see text.

and routed to the PC for data acquisition. The temperature was regulated to $T = (24.5 \pm 0.1)$ K to hold the viscosity and the surface tension of the liquid constant. In order to visualize the surface structure a full frame CCD camera (Hitachi KPF-1) was placed about 1 m above the interface ferrofluid/air in the center of a ring (with radius $r = 0.3$ m) of 120 LEDs illuminating the surface. Using this visualization technique only surface elements of appropriate slope reflect light into the camera, while the flat regions of the surface appear black. The spatial symmetry of the surface deformation $\zeta(x, y)$ was analyzed by using two-dimensional Fourier transformation. For the measurements we used a commercial ferrofluid (APG J12, FerroTec) with the following specifications: dynamic viscosity $\eta = 45 \times 10^{-3}$ Pa s, mass density $\rho = 1110\,\mathrm{kg/m^3}$ and initial magnetic susceptibility $\chi = 0.8$.

5.4 Dispersion relation

The dispersion relation of surface waves for an ideal inviscid fluid in a perpendicular magnetic field,

$$\omega_0^2(k) = \tanh(kh)\left\{g_0 k + \frac{\sigma}{\rho}k^3 - \frac{\mu_0}{\rho}\left(\frac{1+\chi}{2+\chi}\right)\right.$$
$$\left.\times\left[1 - \frac{2\chi}{(2+\chi)^2 \mathrm{e}^{2kh} - \chi^2}\right] M^2 k^2\right\}, \tag{55}$$

depends on h, g_0, σ, χ and the magnetization M. For high filling levels, $kh \gg 1$, the critical magnetization M_C^R and the critical wave number k_C^R of the

Rosensweig instability read

$$\left(M_C^R\right)^2 = \frac{2}{\mu_0}\left(\frac{2+\chi}{1+\chi}\right)\sqrt{\rho g_0 \sigma}, \quad k_C^R = \sqrt{\rho g_0/\sigma}\,. \tag{56}$$

There is a second characteristic value of the magnetization $\widetilde{M} < M_C^R$ above which the dispersion curve is no longer monotonic. A simple calculation gives: $\widetilde{M} = \sqrt[4]{3/4}\, M_C^R$, *i.e.*, $\widetilde{M} \approx 0.93\, M_C^R$. For $M > \widetilde{M}$ the dispersion relation $\omega_0^2(k)$ shows a minimum so that there is a range of k-values in which the group velocity is negative; this is referred to as anomalous dispersion. Thus, for M approaching M_C^R there is a band of driving frequencies in which surface waves with three different k-values can be generated.

For real fluids the viscosity of the fluid must be taken into account. Using the expression of Kumar and Tuckerman [73] the dispersion relation subject to appropriate boundary conditions reads now

$$-\omega^2 + \omega^2 X(k,\omega) + \omega_0^2(k) = 0\,. \tag{57}$$

The magnetization enters *via* the inviscid dispersion relation $\omega_0(k)$ (55) so that

$$-\omega^2 + \omega^2 X(k,\omega) = \frac{\nu^2}{q\coth(qh) - k\coth(kh)}$$
$$\times \left\{q\left[4k^4 + (k^2+q^2)^2\right]\coth(qh) - k\left[4k^2g^2 + (k^2+q^2)^2\right]\tanh(kh) - \right\}. \tag{58}$$

$$\left\{\cdot\frac{4k^2q(k^2+q^2)}{\cosh(kh)\sinh(qh)}\right\}, \tag{59}$$

with $q = \sqrt{k^2 + \mathrm{i}\omega/\nu}$, where ν denotes the kinematic viscosity. In order to determine the critical parameters a_C and k_C of the Faraday instability for a given field value H and a given driving frequency f, we used a special version of the algorithm proposed by Kumar and Tuckerman. The surface deformation is described by a Floquet-ansatz

$$\zeta \propto \mathrm{e}^{\mathrm{i}\beta\Omega t} \sum_{n=-\infty}^{\infty} \zeta_n \mathrm{e}^{2\mathrm{i}n\Omega}\,. \tag{60}$$

Here $\beta = 1$ ($\beta = 0$) corresponds to a subharmonic (harmonic) response of the surface deformation respectively. The Floquet-ansatz leads to tridiagonal system of equations

$$0 = \left[-\omega^2 + \omega^2 X(\omega,k) + \omega_0^2(k)\right]\zeta_n + \frac{ak\tanh(kh)}{2}(\zeta_{n+1} + \zeta_{n-1})\,, \tag{61}$$

with $\omega = (2n+\beta)\Omega$. Eqs. (61) can be numerically approximated by an appropriate cutoff. The solvability condition yields the neutral stability curves

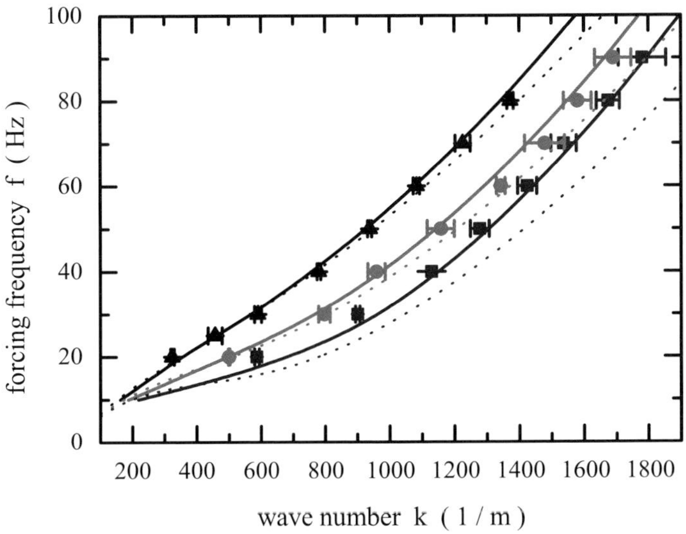

Fig. 24. Forcing frequency f *versus* wave number k of the standing wave pattern at slightly supercritical acceleration amplitudes. The solid lines show critical values resulting from Eqs. (57, 58, 60, 61) with the parameters $\rho = 1110\,\mathrm{kg/m^3}$, $\chi = 0.8$, $\sigma = 0.02\,\mathrm{N/m}$ and $\eta = 45.77 \times 10^{-3}\,\mathrm{Pa\,s}$. The dashed lines are calculated for $\eta = 0$. The magnetization was $M = 0$, $0.7\,M_C^R$, and $0.87\,M_C^R$ from top to bottom, respectively.

$a^{SH}(k)$ and $a^{H}(k)$ for the subharmonic (SH) and harmonic (H) surface response, respectively. A minimization of these curves with respect to the wave number k determines the critical acceleration a_C^{SH} (a_C^{H}) and the critical wave number k_C^{SH} (k_C^{H}).

5.5 Experimental and numerical results

Here we first present results of experiments performed at a fixed filling level of $h = 5\,\mathrm{mm}$. The surface tension σ of the ferrofluid was not known, therefore we first performed measurements at $M = 0$, $M = 0.7\,M_C^R$ and $M = 0.87\,M_C^R$. The data of these measurements are presented in Fig. 24. As demonstrated in [73, 74] the Faraday instability provides a versatile method to measure the surface tension. With our investigations we measured the critical acceleration a_C as function of the driving frequency. Furthermore, the critical wave number k_C was determined from video images recorded close to the threshold of the Faraday instability.

Our experimental results are well described by numerical calculations using Eqs. (57–61) and the following parameters: $\eta = 45.77 \times 10^{-3}\,\mathrm{Pa\,s}$, $\chi = 0.8$ and $\sigma = 0.02\,\mathrm{N/m}$. So the surface tension $\sigma = 0.02\,\mathrm{N/m}$ is a direct outcome of our experiments. In order to study the anomalous dispersion branch

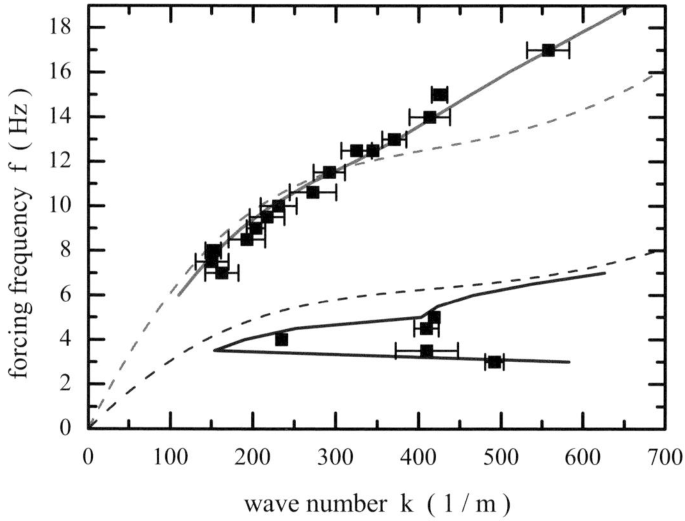

Fig. 25. Forcing frequency f *versus* wave number k as in Fig. 24, here, however, for $M = 0.9\,M_C^R$. For $3\,\text{Hz} \leq f \leq 5.5\,\text{Hz}$ the surface response is harmonic in time, while for $f \geq 6\,\text{Hz}$ it is subharmonic. Full lines show critical values as in Fig. 24; dashed curves refer to the idealized case $\eta = 0$.

the external field strength was chosen to be shortly below the critical value. The experiments have been performed in the following way: with M and f set to the desired values, the acceleration of the shaker has been slowly increased up to the critical value of the Faraday instability where the initially flat surface transforms into a standing wave pattern. In order to map the surface deformation we had to increase the acceleration further to about 3–5% above a_c.

In Fig. 25 we present the experimental results obtained at $M = 0.9\,M_R^C$. For forcing frequencies $3\,\text{Hz} \leq f \leq 5.5\,\text{Hz}$ the system responds with a frequency equal to the forcing frequency (harmonic response), while for $f \geq 6\,\text{Hz}$ the response frequency is half the driving frequency (subharmonic response). In contrast to the subharmonic dispersion curve, that shows normal behavior, the harmonic one in the lower part of Fig. 25 is more complex which is well described by the numerical results. We stress that in non-magnetizable fluids temporal harmonic surface waves are realizable only either at very extreme experimental conditions, *i.e.*, low excitation frequencies and very low filling levels (usually below $h = 1\,\text{mm}$) [75, 76] or by using non-Newtonian liquids [77].

In a second experiment M was set 1.3% below the critical Rosensweig value M_C^R (see Fig. 26). Both, the temporal harmonic and the temporal subharmonic dispersion relation now show anomalous behavior for forcing frequencies $f \leq 10\,\text{Hz}$. The experimental findings are well described with the numerical results obtained with the same parameters for η, σ and χ as in

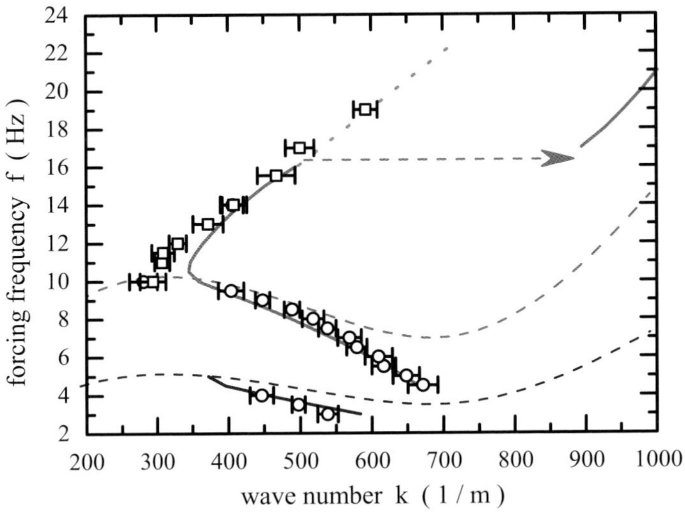

Fig. 26. Forcing frequency f *versus* wave number k as in Fig. 24 for $M = 0.987\, M_C^R$. Open circles denote experimental data on the anomalous dispersion branch, while open squares refer to the normal dispersion branch. The dotted line is a guide for the eye, following the subharmonic-dispersion relation. The arrow is explained in the text. Full lines show critical values as in Fig. 24; dashed curves refer to the idealized case $\eta = 0$.

the experiments at $M = 0$ and $M < \widetilde{M}$. For forcing frequencies $f \geq 16\,\mathrm{Hz}$ the numerical results predict a jump in wave number from $k \approx 500\,\mathrm{m}^{-1}$ to $k \approx 900\,\mathrm{m}^{-1}$ as indicated by the arrow in Fig. 26. However, this jump does not appear in the experiment. The data points rather follow the continuation of the low-frequency dispersion branch (as indicated by the dotted line in Fig. 26 as a guide for the eye). This finding is up to now not fully clear, although the topology of the neutral stability curves could explain this result: for frequencies $f \geq 16\,\mathrm{Hz}$ and $\varepsilon_A \equiv a/a_C - 1 = 0.05$ the band of possible wave numbers k is about one order of magnitude larger than for driving frequencies within the range of the anomalous dispersion relation. Since the experiments have to be performed at finite supercritical ε_A their exists the possibility for the system to select a wave number different from k_C.

6. Convection in colloidal suspensions

Convection in a ferrofluid layer heated from below (Rayleigh–Bénard convection) and exposed to a homogeneous external magnetic field in vertical direction has been considered for a long time [78]. But only later [79] it was understood that the binary nature of the ferrofluid as a mixture of carrier liquid and magnetic particles is important to understand the convection properties.

Ferrofluids, as well as colloidal mixtures in general, differ greatly from molecular mixtures like ethanol-water normally studied in the framework of Rayleigh–Bénard convection. Firstly, the large particle diameter in the nanometer range leads to a very slow concentration diffusion in colloidal mixtures. Secondly, colloids are known to show a very strong Soret effect, *i.e.*, a very strong driving of concentration currents by thermal gradients [80]. This feature has also been explained by the large particle diameter [81]. As a consequence, the solutal diffusion coefficient and the Soret coefficient in ferrofluids and other colloids differ by orders of magnitude from those found in common molecular mixtures. We have investigated this parameter region numerically [82] which has hardly been studied before [83, 84].

6.1 The system and numerical details

The system we have considered is a fluid layer confined between two infinitely extended, rigid, and impermeable parallel plates perpendicular to the direction of gravity. The plates are kept at constant temperatures, the lower plate being the warmer one, allowing for buoyancy-driven instabilities. In the conductive ground state without fluid motion, the temperature field thus shows a linear profile in the vertical direction. The fluid is a binary mixture with a nonvanishing Soret effect which couples the temperature into the concentration field dynamics. In particular, in the ground state the Soret effect will establish a linear profile in the concentration field too. The concentration couples back into the dynamics of the other fields *via* the concentration dependence of the density and thus of the buoyancy force.

For the system of equations governing the system we refer to [85, 86]. Four dimensionless parameters appear in this system. The Prandtl number σ is the ratio of momentum to thermal diffusivity. Typical values for liquids are of the order of 10, and as long as $\sigma \gg 1$, the system is not very sensitive to the exact value. We used $\sigma = 10$ for all our calculations. The Rayleigh number R is the control parameter, the dimensionless temperature difference between lower and upper plate. Below we will mainly use the reduced Rayleigh number $r = R/R_c^0$, $R_c^0 = 1707.76$ being the critical Rayleigh number for the pure fluid. The Lewis number L is the ratio of solutal to thermal diffusivity. For molecular mixtures a typical value is $L = 0.01$. Colloids on the other hand can easily reach values of $\mathcal{O}(10^{-4})$ and smaller. The separation ratio ψ is the ratio of concentration- and temperature-induced density gradients in the ground state. ψ expresses the strength and direction of the Soret effect. In ethanol-water one finds values between -0.5 and 0.5, depending on mean temperature and concentration. In colloids on the other hand values of $\mathcal{O}(10)$ and larger have been found [80]. Below, we will briefly refer to mixtures with $\sigma = \mathcal{O}(10)$, $L = \mathcal{O}(10^{-2})$, and $|\psi| = \mathcal{O}(0.1)$ as liquid molecular mixtures, and to mixtures with $\sigma = \mathcal{O}(10)$, $L = \mathcal{O}(10^{-4})$ or less, and $|\psi| > 1$ as colloidal mixtures.

The patterns we have investigated are those that are the most important ones in liquid molecular mixtures. For negative ψ we have investigated stationary and traveling convection roll patterns (traveling waves, TWs). For positive ψ stationary rolls, squares, and crossrolls were investigated. Squares can be thought as superpositions of two perpendicular roll sets of equal amplitude. Crossrolls also consist of perpendicular roll sets, but of unequal amplitude, such that the full square symmetry is not fulfilled. Two different types of crossrolls exist, a stationary one, and an oscillatory variant where the amplitudes are time-dependent.

We used the Galerkin method to compute the convection patterns. Within this method spatially periodic fields are expanded into a complete set of orthonormal functions (modes) using an ansatz $\sim \cos(lkx)\cos(mky)$ in lateral direction. The basic equations are then projected onto these functions, transforming the system of partial differential equations into an ordinary system of first order in time, with the time-dependent amplitudes as new unknowns. The system is in principle infinite but truncated to allow for a numerical treatment.

6.2 Linear stability, convection at small amplitudes

When $\psi = 0$, concentration variations will only appear as transients, as without Soret effect the concentration field will always equilibrate. In that case, the ground state becomes unstable at the critical temperature difference $r = 1$ of the pure fluid. When the Soret effect is positive, $\psi > 0$, the lighter component will go into the direction of higher temperatures, further destabilizing the ground state, and thus convection sets in at smaller R than R_c^0. For a negative Soret effect ($\psi < 0$) the opposite is the case. The critical R is larger than R_c^0, and except for very small $|\psi|$ that are not of interest here the first bifurcation out of the ground state is backwards leading to unstable small amplitude convection. Furthermore, the critical Rayleigh number quickly diverges when $|\psi|$ grows. Arbitrary small amplitude convection thus does not exist at all for colloids, and we can ignore negative ψ here.

For $\psi > 0$ the first instability is always stationary. For the reduced Rayleigh number of the threshold $r_{stab}(k)$ the power law dependence

$$r_{stab}(k, L, \psi) = h(k)\frac{L}{\psi} \tag{62}$$

on L and ψ holds for a broad range of ψ and L including the relevant parameters for colloids. This behavior can be seen in Fig. 27 where r_{stab} is plotted as function of L and ψ for $k = 0$, the critical wave number for typical colloids, and $k_c^0 = 3.116$, the critical wave number for pure fluids. This behavior can also be confirmed by examining the semianalytical results for the ground state stability found in [87]. The convective threshold in colloids thus lies very low compared to molecular mixtures.

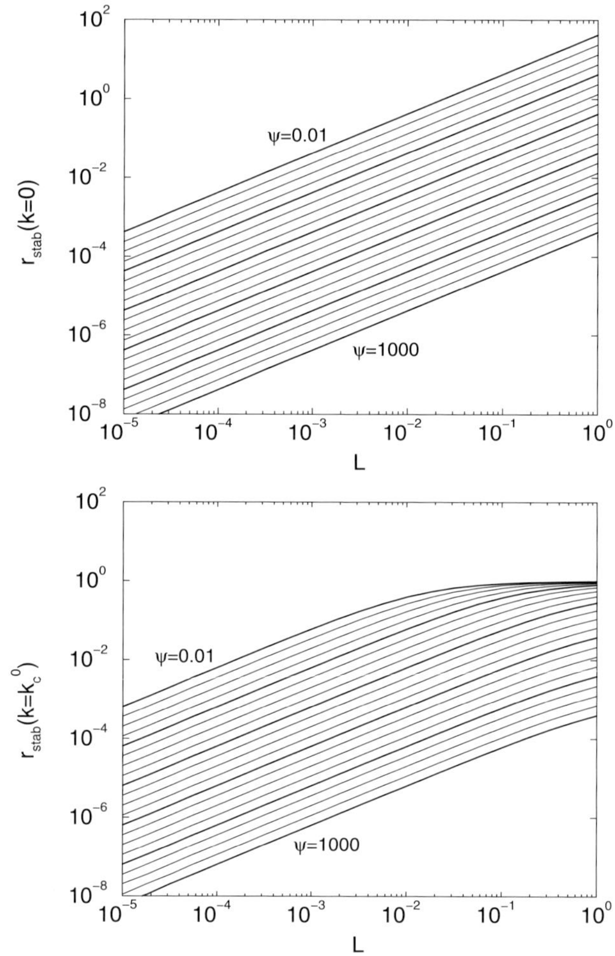

Fig. 27. Reduced stability thresholds of the conductive state, $r_{\text{stab}}(k, \psi, L)$, against non oscillatory convection with wave numbers $k = 0$ (top) and $k = k_c^0$ (bottom) *versus* L. Between thick lines ψ changes by a factor of 10.

That the threshold decreases with growing ψ can be explained by the destabilization due to the growing density gradient. In particular, r_{stab} is for $\psi > 0$ always smaller than for pure fluids. The dependence on L can be qualitatively explained by the fact that a larger L leads to faster equilibration of concentration perturbations and thus to increased stability.

A power law does also hold for the initial slope S, that can be defined as

$$S = \frac{\partial N}{\partial r} \qquad \text{at} \qquad r = r_{\text{stab}} \,, \tag{63}$$

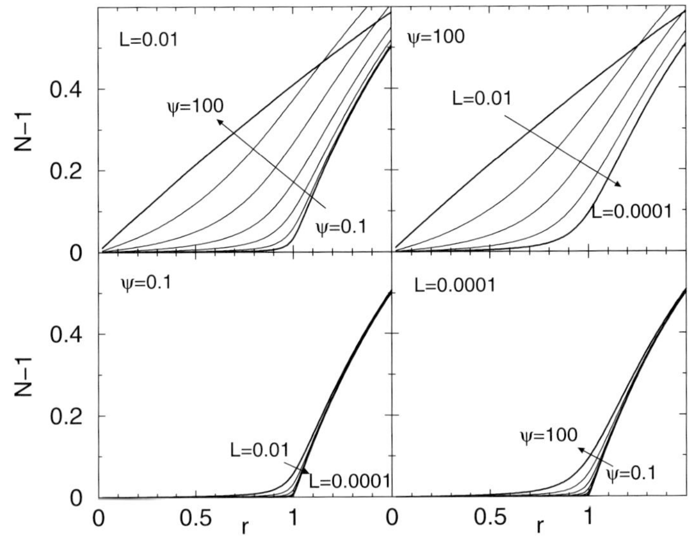

Fig. 28. Nusselt number *versus* r for $\sigma = 10$, $k = k_c^0$, and several (L, ψ)-combinations. The parameter that is varied changes by a factor of $\sqrt{10}$ between the curves in each plot.

where N is the Nusselt number, the ratio of total to conductive heat transport through the system. Here one finds $S \sim L\psi$ for typical colloidal as well as for typical molecular mixtures. The proportionality factor depends also on k and σ. The dependence on σ is very weak as long as $\sigma \gg 1$. The type of pattern does also play a role, but for the two relevant patterns that bifurcate out of the ground state, namely rolls and squares, the difference in the slope is practically constant for colloids and moreover only about 2%.

6.3 Bifurcation scenario for positive ψ

6.3.1 *Bifurcation curves*

Away from the bifurcation point two characteristic convection regimes exist in liquid molecular mixtures that can also be found in colloids, the Soret and Rayleigh regime. These two regimes are visible in the bifurcation diagrams, like in Fig. 28. There we have plotted the Nusselt number N against the reduced Rayleigh number r for roll patterns at $k = k_c^0$. Square patterns also exist for all parameter combinations shown but the difference in the Nusselt number between these two patterns is very small, not only near the onset as discussed above but in the whole r-interval displayed. Except for the highest ψ and L, a sharp bend in the curves separates the Soret region $r < 1$ and the Rayleigh region $r > 1$, especially when both L and ψ are small. Here, the Nusselt number is quantitatively very similar to that of a pure fluid. In particular, $N \approx 1$ in the Soret region, where in a pure fluid the conductive state is still stable and therefore the total heat transport equals the conductive heat transport ($N \equiv 1$).

The explanation for this behavior is as follows: For small ψ the solutal contribution to the density gradient in the ground state is negligible. Nevertheless, for small L the onset of convection lies significantly lower than for pure fluids because the very slow concentration diffusion allows perturbations in the concentration field to grow before they are diffused away. On the other hand, the slow diffusion makes advective mixing very efficient such that only very small convection amplitudes are needed to reach a stationary state for $r < 1$, especially since the concentration gradients are small in the first place. At $r > 1$, however, the density gradient becomes strong enough to allow perturbations to grow even in absence of a concentration gradient. The convection amplitude now quickly grows, assuming values typical for pure fluids, while the weak concentration gradients vanish when the mixture becomes homogenized due to advective mixing.

For large ψ and L, on the other hand, the solutally induced density gradient dominates and solutal diffusion is faster such that the concentration variations are not easily mixed away. The transition between Soret regime and Rayleigh regime is thus much smoother.

The influence of the large ψ and the small L of colloids compensate each other to a certain degree. A detailed inspection has shown that the bifurcation curves depend mainly on the product $L\psi$. Although no strict power law holds as it does near to the onset, changing L and ψ while keeping the product constant changes the curves only very slightly. Colloids therefore show very similar curves $N(r)$ as liquid molecular mixtures with the same $L\psi$, at least when compared on scales $r = \mathcal{O}(1)$ where the onset of convection that scales like L/ψ, not $L\psi$, is very close to zero in both cases.

6.3.2 Pattern selection

It is known that for small L and large ψ square patterns are the first stable form of convection above the onset. The square patterns lose their stability in the Rayleigh region against oscillatory crossrolls. Therein the amplitudes of the two constituting perpendicular roll sets oscillate in counterphase sweeping through a square state of equal amplitude, such that first one and than the other roll set becomes the dominant one for half a period. Beyond this bifurcation point there is another one on the square branch, where the stationary crossrolls bifurcate. They gain stability after the oscillatory solution vanishes in an entrainment process [88]. The amplitude of the dominant roll set then grows with growing r while the other one shrinks until the crossrolls end up at the roll branch and the rolls now gain stability.

We have calculated the Rayleigh numbers of three bifurcation points for a wide range of ψ and L, namely the bifurcation from squares to oscillatory crossrolls, from squares to stationary crossrolls, and from rolls to stationary

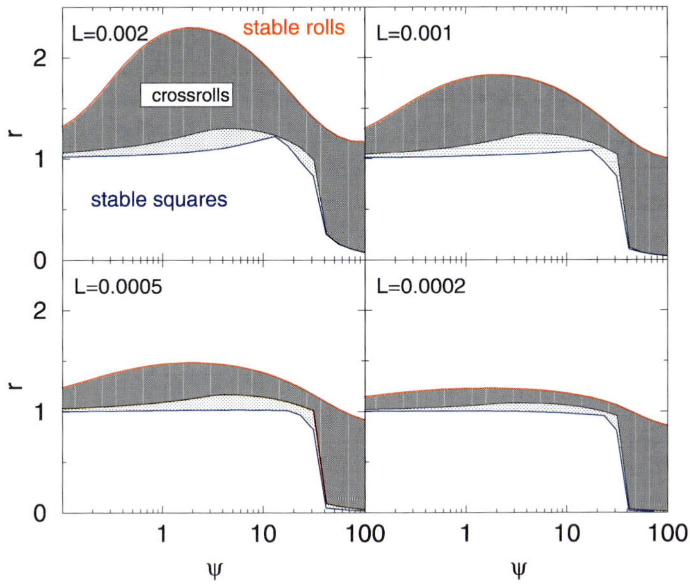

Fig. 29. r–ψ-phase diagrams for squares, oscillatory crossrolls, stationary crossrolls, and rolls for $k = k_c^0$, $\sigma = 10$, and four different L. Rolls are stable in the upper region, squares in the lower. Crossrolls exist in the gray regions.

crossrolls. The resulting phase diagrams for $k = k_c^0$ and $\sigma = 10$ are shown in Fig. 29. For up to $\psi \approx 40$ we always found the pattern sequence of squares, oscillatory crossrolls, stationary crossrolls, and rolls. Crossrolls can be found in the gray areas. Note however that the boundary between the light and dark gray area marks the bifurcation point from squares to stationary crossrolls. The latter exist in the whole dark gray area but will probably still be unstable towards oscillatory crossrolls in the lower part. The smaller L the smaller the crossroll region becomes: For $L = 0.0002$ squares are stable in practically the whole Soret region, rolls in the Rayleigh region with a very small transition region above $r = 1$.

For $\psi > 40$ the behavior changes abruptly. Squares are still stable directly above onset but lose their stability very quickly. Crossrolls dominate almost the whole Soret region instead. In the Rayleigh region rolls remain stable.

6.4 Bifurcation scenario for negative ψ

In a wide parameter region typical for molecular mixtures at negative ψ the first instability of the ground state will be an oscillatory one, leading to a pattern of traveling waves (TWs) [89] and related structures. Appearing at a heating rate r_{osc} the TWs bifurcate backwards, gain stability at a saddle node, and finally end up at another bifurcation point that connects their branch to the station-

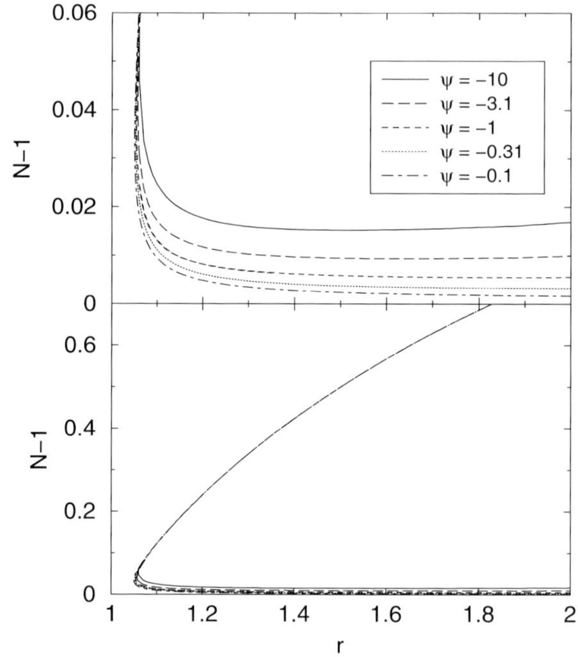

Fig. 30. Bifurcation diagram of Nusselt number *versus* r of stationary rolls for $\sigma = 10$, $k = k_c^0$, and different negative ψ such that $|L\psi| = 10^{-4}$. Top plot is a magnified part of the bottom plot.

ary rolls branch at the heating rate r^*. The rolls that have been unstable before gain stability here. For $\psi < -1$ the stabilizing effect of the concentration gradient overcompensates the destabilizing effect of the thermal gradient and neither a stationary nor an oscillatory threshold exists at finite r. Although the ground state is always stable convection at larger amplitudes is in principle still possible however, even for colloid parameters $\psi \ll -1$, as long as the advection is strong enough to sufficiently equilibrate the concentration field such that its stabilizing effect vanishes. Again, the mixing can be expected to be more effective when L is small.

Figure 30 shows the stationary roll branches for a fixed product $L\psi = -10^{-4}$. Although both parameters span over two orders of magnitude the Nusselt number is almost independent of the product $L\psi$ on the upper branch, demonstrating again the compensating effect of increasing $|\psi|$ while making L smaller at the same time. Only the lower branch shows some variations as can be seen better in the magnified top plot.

So convection in the form of stationary rolls is possible at moderate r as Fig. 30 demonstrates, but we found that these structures are not stable. The reason lies in the divergence of the lower stability boundary r^*. It is known that r^*

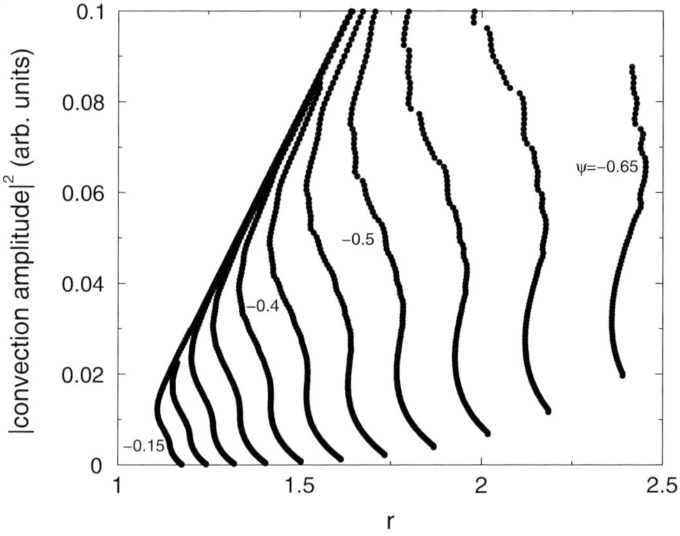

Fig. 31. Bifurcation diagrams of squared TW convection amplitudes *versus* r for $\sigma = 10$, $k = k_c^0$, $L\psi = -10^{-4}$, and several values of ψ between -0.15 and -0.65.

diverges already for moderate L when $|\psi|$ is increased. For $L = 0.01$ for example this is the case at $\psi \approx -0.45$. In this case, reducing L in parallel does not compensate the effect. On the contrary: r^* diverges even earlier for fixed $L\psi = -10^{-4}$ [82].

Not only the bifurcation point between TWs and rolls is shifted towards higher heating rates, the whole TW branch moves to the right when $|\psi|$ is increased, and here too the reduction of L does not compensate this behavior. This can be seen in the last plot Fig. 31. Already at $\psi < -0.65$, still not unusual for molecular mixtures, no TWs can be found for $r < 2$ when $L\psi = -10^{-4}$.

To conclude: We have numerically studied the bifurcation properties of Rayleigh–Bénard convection in a parameter region far away from what is accessible with commonly studied molecular mixtures but easily accessible with colloidal mixtures. Both, for positive as well as for negative ψ the bifurcation curves for rolls do not change much going from molecular mixtures to colloids, as long as the product $L\psi$ is kept constant.

For colloids with positive ψ we found the pattern sequence of squares, oscillatory crossrolls, stationary crossrolls, and rolls that are known already from the molecular liquid case. For not too large ψ the r interval where crossrolls are found becomes very small in the limit $L \to 0$. The region $r < 1$ and $r > 1$ are dominated by squares and rolls respectively. For ψ larger than about 40, however, squares exist as stable form of convection only immediately above onset. The region $r < 1$ is dominated by crossrolls instead.

For $\psi < 0$, we found only the ground state to be stable for typical colloidal parameters. Of the two relevant structures for molecular fluids only the roll branch still exists for moderate r, but is unstable. TWs don't exist at all for not too large r.

Acknowledgement

This work was supported by the Deutsche Forschungsgemeinschaft (SFB 277) and by INTAS (03-51-6064). We gratefully acknowledge the contributions of B. Fischer, R. Hempelmann, A. V. Kityk, K. Knorr, C. E. Krill III, S. May, F. Meyer, B. Müller, H. W. Müller, H. Natter, C. Wagner, S. Wiegand as coauthors of different parts of the work reviewed here and discussions with them.

References

1. B. Huke and M. Lücke, Rep. Prog. Phys. **67** (2004) 1731.
2. B. Huke and M. Lücke, Phys. Rev. E **62** (2000) 6875.
3. B. Huke and M. Lücke, Magnetohydrodynamics **37** (2001) 222.
4. B. Huke and M. Lücke, J. Magn. Magn. Mater. **252** (2002) 132.
5. B. Huke and M. Lücke, Phys. Rev. E **67** (2003) 051 403.
6. V. I. Kalikmanov, Physica A **183** (1992) 25.
7. Y. A. Buyevich and A. O. Ivanov, Physica A **190** (1992) 276.
8. P. Debye, Phys. Z. **13** (1912) 97.
9. K. I. Morozov and A. V. Lebedev, J. Magn. Magn. Mater. **85** (1990) 51.
10. A. F. Pshenichnikov, V. V. Mekhonoshin, and A. V. Lebedev, J. Magn. Magn. Mater. **161** (1996) 94.
11. G. A. van Ewijk, G. J. Vroege, and A. P. Philipse, J. Phys.: Condens. Matter **14** (2002) 4915.
12. P. T. Cummings and L. Blum, J. Chem. Phys. **85** (1986) 6658.
13. T. Sato, T. Iijima, M. Seki, and N. Inagaki, J. Magn. Magn. Mater. **65** (1987) 252.
14. A. O. Ivanov and O. B. Kuznetsova, Phys. Rev. E **64** (2001) 041 405.
15. R. E. Rosensweig, *Ferrohydrodynamics*. Cambridge University Press, Cambridge (1985).
16. J. Embs, H. W. Müller, M. Lücke, and K. Knorr, Magnetohydrodynamics **36** (2000) 387.
17. J. Embs, H. W. Müller, C. E. Krill, F. Meyer, H. Natter, B. Müller, S. Wiegand, M. Lücke, R. Hempelmann, and K. Knorr, Magnetohydrodynamics **37** (2001) 222.
18. J. P. Embs, H. W. Müller, C. E. Krill, F. Meyer, H. Natter, B. Müller, S. Wiegand, M. Lücke, R. Hemplemann, and K. Knorr, Z. Phys. Chem. **220** (2006) 153.
19. B. Huke and M. Lücke, Phys. Rev. E **62** (2000) 6875.
20. J. D. Jackson, *Klassische Elektrodynamik*. de Gruyter, Berlin (1982).
21. T. Weser and K. Stierstadt, Z. Phys. B **59** (1985) 257.
22. H. P. Klug and L. E. Alexander, *X-Ray Diffraction Procedures for Polycrystalline and Amorphous Materials*. Wiley, New York (1975).
23. B. E. Warren and B. L. Averbach, J. Appl. Phys. **23** (1952) 497.
24. B. E. Warren, *X-Ray Diffraction*. Dover Publications Inc., Dover, New York (1990).
25. C. E. Krill and R. Birringer, Philos. Mag. A **77** (1998) 621.

26. H. Natter, M. Schmelzer, M. S. Löffler, C. E. Krill, A. Fitch, and R. Hempelmann, J. Phys. Chem. B **104** (2000) 2467.
27. R. Haberkorn, C. E. Krill, and R. Birringer, *Handbook of Nanostructured Materials and Nanotechnology, Vol.2: Spectroscopy and Theory*. Academic Press, London (2000) Chap. 3.
28. S. J. Provencher, Comp. Phys. Commun. **27** (1982) 213.
29. J. P. McTague, J. Chem. Phys. **51** (1969) 133.
30. W. F. Hall and S. N. Busenberg, J. Chem. Phys. **51** (1969) 137.
31. M. I. Shliomis, Sov. Phys. JETP **34** (1972) 1291.
32. J. Embs, H. W. Müller, C. Wagner, K. Knorr, and M. Lücke, Phys. Rev. E **61** (2000) R2196.
33. K. Henjes, J. Magn. Magn. Mater. **117** (1992) L311.
34. T. Weser and K. Stierstadt, Z. Phys. B **59** (1985) 253.
35. M. Holderied, L. Schwab, and K. Stierstadt, Z. Phys. B **70** (1988) 431.
36. O. Ambacher, S. Odenbach, and K. Stierstadt, J. Phys. I France **86** (1992) 29.
37. S. Odenbach and H. Gilly, J. Magn. Magn. Mater. **152** (1996) 123.
38. The specifications for the ferrofluid APG933 are as follows: $\mu_0 M_s = 20\,\mathrm{mT}$, $\eta = 500\,\mathrm{mPa\,s}$, density $\rho = 1.09\,\mathrm{g/cm^3}$, $\Phi_m = 3.3\%$, susceptibility $\chi = 0.9$, mean diameter of the particle's magnetic core 10 nm, thickness of the polymeric coating 2 nm.
39. J. T. Waldron, Y. P. Kalmykov, and W. T. Coffey, Phys. Rev. E **49** (1994) 3976.
40. One arrives at relation (12) also from the phenomenological low-frequency equations given in Ref. [90]. Also the relaxation equation $\partial_t \boldsymbol{M} = -\frac{1}{\tau_B}(\boldsymbol{M} - \boldsymbol{M}_{eq}) + \frac{1}{2}(\nabla \times \boldsymbol{v}) \times \boldsymbol{M} + \frac{\mu_0}{6\eta\Phi_h}(\boldsymbol{M} \times \boldsymbol{H}) \times \boldsymbol{M}$ suggested by Shliomis [31] reduces under the approximations done in the text to Eq. (12), however, with $\tau_\perp$ being replaced by $\tau_\perp^S = \tau_B/(1 + \frac{1}{2}\alpha_0\mathcal{L}_0)$. We prefer to fit our data with the field dependence of $\tau_\perp$ [39, 42] that has been derived without a semi-microscopic estimate of the Onsager coefficient, which couples linear and spin-angular momentum [31].
41. M. I. Shliomis, Chem. Eng. Commun. **67** (1988) 275.
42. M. A. Martsenyuk, Y. L. Raikher, and M. I. Shliomis, Sov. Phys. JETP **38** (1974) 413.
43. A. Leschhorn and M. Lücke, Z. Phys. Chem. **220** (2006) 89.
44. E. Blums, A. Cebers, and M. M. Maiorov, *Magnetic Fluids*. Walter de Gruyter, Berlin (1997).
45. S. Odenbach, Lect. Notes Phys. m**71** (2002).
46. S. Odenbach (Ed.), Lect. Notes Phys. **594** (2002).
47. J. P. Embs, S. May, C. Wagner, A. V. Kityk, A. Leschhorn, and M. Lücke, Phys. Rev. E **73** (2006) 036 302.
48. A. Leschhorn and M. Lücke, Z. Phys. Chem. **220** (2006) 219.
49. B. U. Felderhof and H. J. Kroh, J. Chem. Phys. **110** (1999) 7403.
50. B. U. Felderhof, Phys. Rev. E **62** (2000) 3848.
51. M. I. Shliomis, Phys. Rev. E **64** (2001) 060 501.
52. H. W. Müller and M. Liu, Phys. Rev. E **64** (2001) 061 405.
53. A. Leschhorn and M. Lücke, J. Phys.: Condens. Matter **18** (2006) 2633.
54. Minimizing

$$\sigma = \sum_{i=1}^{K} [M^{eq}(H_k) - M_k]^2 \tag{64}$$

with respect to w_i, in which $\{H_k, M_k\}_{k=1}^{K}$ indicate the experimental data, leads to an ill-posed problem, resulting in large positive and negative magnetic weights w_i. To avoid these unphysical results one introduces an additional quantity $\rho = \sum_{i=1}^{N} w_i^2$

and one now minimizes

$$\sigma' = \sigma + \rho\widetilde{\alpha}\,, \tag{65}$$

with respect to w_i where $\widetilde{\alpha}$ refers to the so-called regularization parameter ($\widetilde{\alpha} = 0$ leads to the initial ill-posed problem). Using a small but finite positive $\widetilde{\alpha}$ allows for the computation of the distribution of the magnetic weights w_i as function of the particle diameters d_i.

55. B. U. Felderhof, Phys. Rev. E **64** (2001) 063 502.
56. M. I. Shliomis, Phys. Rev. E **64** (2001) 063 501.
57. B. U. Felderhof, Phys. Rev. E **64** (2001) 021 508.
58. T. K. McNab, R. A. Fox, and J. F. Boyle, J. Appl. Phys. **39** (1968) 5703.
59. P. C. Fannin, P. A. Preov, and S. W. Charles, J. Phys. D **32** (1999) 1583.
60. P. C. Fannin, J. Magn. Magn. Mater. **136** (1994) 49.
61. J. Embs, C. Wagner, K. Knorr, and M. Lücke, Europhys. Lett. **78** (2007) 44 003.
62. M. D. Cowley and R. E. Rosensweig, J. Fluid Mech. **30** (1967) 671.
63. B. Abou, G. Néron de Surgy, and J. E. Wesfreid, Phys. Rev. E **7** (1997) 1159.
64. J. Weilepp and H. R. Brand, J. Phys. II (Paris) **6** (1996) 419.
65. D. Salin, Europhys. Lett. **21** (1993) 667.
66. A. Lange, Magnetohydrodynamics **39** (2003) 65.
67. T. Mahr, A. Groismann, and I. Rehberg, J. Magn. Magn. Mater. **159** (1996) L45.
68. H. W. Müller, Phys. Rev. E **58** (1998) 6199.
69. F. Pétrélis, É. Falcon, and S. Fauve, Eur. Phys. J. B **15** (2000) 3.
70. T. Mahr and I. Rehberg, Phys. Rev. Lett. **81** (1998) 89.
71. T. Mahr and I. Rehberg, Europhys. Lett. **43** (1998) 23.
72. J. Browaeys, J.-C. Bacri, C. Flament, S. Neveu, and R. Perzynski, Eur. Phys. J. B **9** (1999) 335.
73. K. Kumar and L. S. Tuckerman, J. Fluid Mech. **279** (1994) 49.
74. P. Huber, V. P. Soprunyuk, J. P. Embs, C. Wagner, M. Deutsch, and S. Kumar, Phys. Rev. Lett. **94** (2005) 184 504.
75. C. Wagner, H. W. Müller, and K. Knorr, Phys. Rev. E **68** (2003) 066 204.
76. H. W. Müller, H. Wittmer, C. Wagner, J. Albers, and K. Knorr, Phys. Rev. Lett. **78** (1997) 2357.
77. C. Wagner, H. W. Müller, and K. Knorr, Phys. Rev. Lett. **83** (1999) 308.
78. S. Chandrasekhar, *Hydrodynamic and Hydromagnetic Stability*. Dover Publications Inc., Dover, New York (1981).
79. P. J. Stiles, M. Kagan, and P. J. Blennerhassett, J. Magn. Magn. Mater. **168** (1994) 545.
80. E. Blums, S. Odenbach, A. Mezulis, and M. Maiorov, Phys. Fluids. **10** (1998) 2155.
81. K. I. Morozov, Lect. Notes Phys. **584** (2002) 38.
82. B. Huke, H. Pleiner, and M. Lücke, Phys. Rev. E **75** (2007) 036 203.
83. A. Ryskin, H.-W. Müller, and H. Pleiner, Phys. Rev. E **67** (2003) 046 302.
84. A. Ryskin and H. Pleiner, Phys. Rev. E **71** (2005) 056 303.
85. J. K. Platten and J. C. Legros, *Convection in Liquids*. Springer, New York (1984).
86. M. Lücke W. Hort, and S. J. Linz, Phys. Rev. A **45** (1992) 3737.
87. St. Hollinger and M. Lücke, Phys. Rev. E **52** (1995) 642.
88. Ch. Jung, B. Huke, and M. Lücke, Phys. Rev. Lett. **81** (1998) 3651.
89. R. W. Walden, P. Kolodner, A. Passner, and C. M. Surko, Phys. Rev. Lett. **55** (1985) 496.
90. V. G. Bashtovoy, B. M. Berkovsky, and A. Vislovich, *Introduction to Thermomechanics od Magnetic Fluids*. Hemisphere Publishing Corporation, Washington (1988).

Combinatorial Fabrication of Thin Film-Libraries and Evaluation of their Piezoelectricity by Ultrasonic Piezo-Mode Imaging

By Daniela Rende[1], Kerstin Schwarz[2,*], Ute Rabe[2], Wilhelm F. Maier[1,**], and Walter Arnold[2,***]

[1] Universität des Saarlandes, Lehrstuhl für Technische Chemie, Geb. C4 2, D-66123 Saarbrücken, Germany
[2] Fraunhofer Institut für Zerstörungsfreie Prüfverfahren, Geb. C 3. 1. Universität, D-66123 Saarbrücken, Germany

Combinatorial Chemistry / High-Throughput / AFM / Ultrasonic Piezo-Mode

Piezoelectric materials are used in an ever increasing number of technical products. The need for new and improved piezoelectric materials is growing rapidly. Conventional methods are too slow to explore the potential parameter spaces. A combinatorial workflow has been developed to produce mixed oxide thin film libraries and screen them for piezoelectric properties by atomic force microscopy in the ultrasonic piezo-mode. Validation and new findings are reported. Determination of piezoelectric properties in an automated fashion was only possible within the same class of materials.

1. Introduction

1.1 Combinatorial chemistry

Considering the complexity of parameters on which the properties of materials depend, it is very difficult to predict which chemical composition will show the favored activity. As a result, the development of new functional materials in

* *Present address*: AG der Dillinger Hüttenwerke, Werkstr. 1, D-66763 Dillingen/Saar, Germany
** Corresponding author. E-mail: w.f.maier@mx.uni-saarland.de
*** Corresponding author. E-mail: arnold@izfp.fraunhofer.de

Z. Phys. Chem. **222** (2008) 587–600

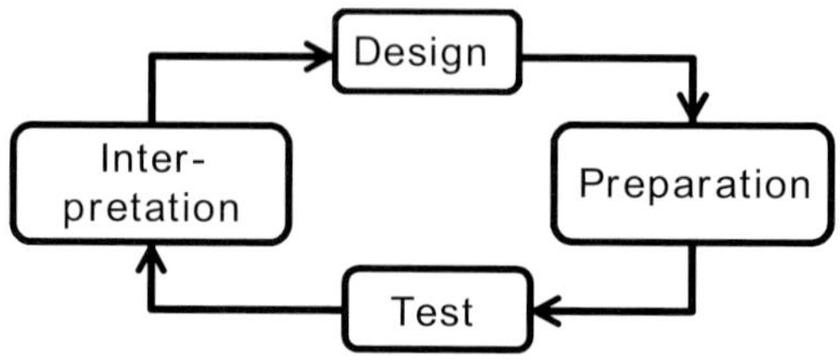

Fig. 1. Schematic drawing of a combinatorial workflow.

the usual way by carrying out large numbers of single experiments is very arduous. Additionally, performing single experiments on a large scale is very time- and cost consuming. Since the 1980's combinatorial strategies for drug discovery [1, 2] are well established in the pharmaceutical industry. More recently, the principles of combinatorial chemistry have been applied to the development of materials and have lead to an increasing efficiency by searching for new materials, such as catalysts [3–7], fluorescent dyes [8], liquid crystals [9], polymers [10], or intumescent coatings [11].

In combinatorial chemistry a large number of samples of interest is prepared in parallel by combinatorial (combinations of components or parameters) or high-throughput (numerical variation of components) variation. The samples are positioned in arrays on plates, usually are called as libraries. These libraries are screened for the desired property by appropriate rapid or parallel analysis. The discovery mode (search for new compositions or formulations) is often referred to as *primary screening*, while the optimization mode (fine tuning of compositions and recipes) are referred to as *secondary screening*. In *primary screening* measurement accuracy is compromised to maximize sample throughput, while in *secondary screening* throughput is reduced for the benefit of realistic conditions and accuracy of measurements. *Secondary screening* is often used to optimize lead structures discovered in *primary screening* studies.

This process of discovery and optimization leads faster to the goal and saves costs. Persidis showed an economic study illustrating the cost savings associated with the use of a combinatorial experiment in comparison to the conventional approach [12].

Figure 1 shows the components of a typical combinatorial workflow. Such a workflow is a repetitive process including the design of the experiments, the synthesis of the samples, the testing of the samples and the interpretation of the data. Due to the competitive advantages the combinatorial chemistry enters also in the production and characterization of thin films. The use of combinatorial or high- throughput methodologies concerning all kinds of coatings is increasing rapidly.

A primary experimental challenge of a high-throughput workflow is to miniaturize and parallelize the sample preparation and characterization. Therefore it is necessary to develop efficient synthesis and rapid measurements technologies on small scales.

1.2 Examples for the search of new coatings using combinatorial approach

Schrof and coworkers described an in-house built combinatorial hardware for organic coatings that consists of a dosing device, a mixing stage and the coating apparatus [13, 14]. Depending on the viscosity of the formulation, three different kinds of dosing devices are available. For mixing, 16 parallel mechanical stirrers can be used. The mixing conditions and the temperature can be adjusted to the formulation. The desired films are generated by an automated 8-fold doctor-blading process. For the evaluation of the mechanical coating properties, such as hardness and elasticity, automated micro-indentation experiments are carried out, and for determining the gloss and the transparency of the films imaging techniques are utilized.

McFarland and coworkers synthesized a thin film library by automated serial electrochemical deposition [15]. The films were tested to find improved materials for photo-electrochemical hydrogen production. The high-throughput photoelectric screening included the measurements of flat-band potential, bias-dependent photocurrent, and action spectra.

Chisholm *et al.* have reported a combinatorial workflow at General Electric Company (GE) for protective clear coatings for plastic substrates [16–18]. The preparation of the 48 liquid coatings is done with a commercially available handling robot. The liquid was applied into the wells of a template composite. To spread the liquid coatings within these wells and to evaporate the solvents, a custom built apparatus was used [19]. The curing of the coatings occurred under UV radiation or at higher temperatures. To measure the optical clarity of clear coatings, a high-throughput screening technique based on a fiber optic probe was used. The probe was automatically positioned over the individual coating, the spectra of the back-scattered light were measured and analyzed by a computer. This procedure took 3 min. The measurement results were compared with conventional hazemeter measurements and the values correlated very well.

Bach *et al.* [20–22] and Wicks *et al.* [23] developed at Bayer Corporation an application of high-throughput experimentations for coatings. They studied the cure kinetics of polymer systems to find new catalysts for the deblocking of isocyanates in poly-isocyanate acrylate thermoset systems. For the film testing they used a known amount of a non-reactive dye in the compositions. Then the compositions were cured under the required conditions. Afterwards they extracted the dye from the cured films by using an organic solvent in a certain time interval. By this method, the dye concentration is related to the degree of cross-linking.

High-throughput techniques were also established, leading to the development of new dielectric ($Zr_{0.2}Sr_{0.2}Ti_{0.6}O_2$) [24] and magnetic (Co-doped TiO_2) [25] materials.

Koinuma and co-workers fabricated organic thin films by a combinatorial molecular beam epitaxy (MBE) system [26]. The presented method is well suited to quickly optimize the fabrication of organic films and to analyze the mechanism of organic film growth.

Bendersky and Takeuchi presented a combinatorial pulsed laser deposition (PLD) system to prepare composition spread libraries of oxides [27]. This system was equipped with a flexible, two-dimensional, automated, overlapping-shutter system and a rotatable heater. They studied composition spreads of the systems $BaTiO_3$–$SrTiO_3$, ZnO–MgO, and $LaMnO_3$–$CaMnO_3$. The libraries were studied by electron energy-dispersive spectroscopy (EDS), by scanning X-ray micro-diffractometry, and by transmission electron microscopy (TEM).

Ludwig presented the combinatorial fabrication and determination of magnetic multilayer films [28]. The $Fe_{50}Co_{50}/Co_{80}B_{20}$ multilayer films are deposited by an automated magnetron-sputtering process on a wafer using a lift-off mask and a special sputtering routine. After dicing off the substrate the magnetic properties of the films were analyzed.

Wessler *et al.* presented a combinatorial approach to prepare and analyze thin film libraries to find more efficient microwave dielectrics [29]. The thin films are produced by chemical solution deposition, using a special photolithographically pre-structured substrate. The characterization of the films is performed by X-ray diffraction measurements and measurements with scanning probe microwave microscopy.

Cremer and Neuschütz described a combinatorial workflow to optimize metastable multi-component hard coatings [30]. Composition spreads of Ti-Al-Hf-N and Ti-Al-Si-N have been deposited by reactive magnetron sputtering. Subsequently the coatings were characterized by XRD in a fully automated way. In addition, they determined the oxidation resistance of the coatings. The results illustrated the effectiveness of the combinatorial approach.

1.3 Scanning probe microscopy as a high-throughput screening method

Scanning probe microscopy (SPM) is used as an automated characterization method in some combinatorial approaches [31, 32]. In the meantime various automated SPM measurement systems are commercially available, several companies offer such applications to test materials in a fully automated way.

He and coworkers described a combinatorial workflow to determine the molecular chain structure and intrinsic mechanical properties of high-density polyethylene (HDPE) by scanning probe microscopy [33]. The HDPE films were obtained by the melt-draw method. First, the nanostructure and the friction properties of the polymer films were investigated using friction force microscopy (FFM) and afterwards the elastic modulus was measured by nanoindentation.

A systematic investigation into the surface properties of siloxane rubber/ carbon black (CB) nanocomposites has been performed by Schubert and coworkers, using an automated atomic force microscope [34]. They studied the influence of CB concentration and the curing rate of the siloxane rubber matrix on roughness and conductivity.

Schubert *et al.* used this automated technique to characterize the photoembossed relief structures of polymer films [35]. A library of 72 films was prepared by dispensing a given amount of a sample onto a chemically patterned substrate consisting of hydrophilic areas separated by fluorinated hydrophobic barriers.

Sormana and Meredith determined the mechanical properties of combinatorial segmented polyurethaneurea gradient libraries [36]. They carried out high-throughput AFM and infrared Fourier-transform spectroscopy.

1.4 Ultrasonic piezo-mode and atomic force microscopy

Imaging of piezoelectric materials on a nano-scale, using the technique called piezo-mode AFM was first presented in 1992 by Güthner and Dransfeld [37]. The piezo-mode technique is based on the detection of the local electromechanical vibration of ferroelectric samples due to their piezoelectricity. A sinusoidal voltage is applied directly to a conductive AFM cantilever while scanning the surface of a piezoelectric material. This causes the electric field between the tip and the ground to excite locally a ferroelectric film into vibration due to the inverse piezoelectric effect. The amplitude and phase of the surface displacement is sensed by the tip and transferred into the cantilever vibrations. The ac frequency is higher than the cut-off frequency of the feedback loop of the AFM, but lower than the frequency of first flexural mode of the cantilever.

Many groups discussed the different forces that act on the cantilever in piezo-mode AFM, *e.g.* electrostatic and electromechanical forces [38–41].

Eng *et al.* used the piezo-mode AFM to reconstruct the distribution of the polarization in the surface of $BaTiO_3$ ceramics by three-dimensional measurements [42].

Muñoz-Saldaña *et al.* investigated the domain configuration on lead zirconate titanate $Pb(Zr,Ti)O_3$ (PZT) ceramics by using piezoelectric force microscopy in combination with topography measurements [43].

Durkan *et al.* performed quantitative localized measurements of surface polarization charge density on ferroelectric thin films [44]. Changes in the symmetry of the hysteresis behavior of piezoelectric films were used to investigate differences in the materials [45].

The piezoelectric properties of the thin-film PZT samples were investigated with the ultrasonic piezo-mode technique [46]. In this technique (Fig. 2),

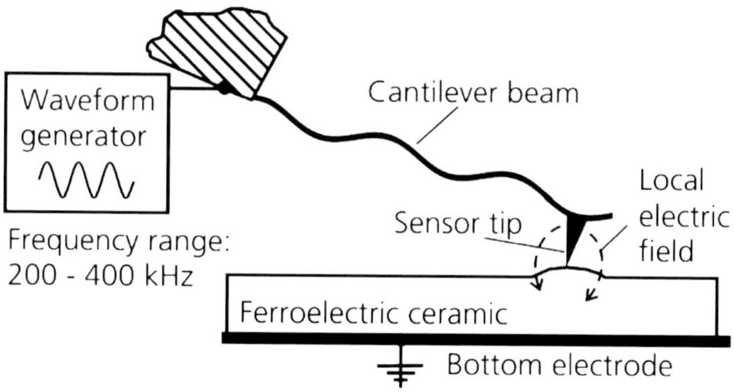

Fig. 2. Principle of the ultrasonic piezo-mode. A sinusoidal voltage is applied to the conducting cantilever. When the tip is in contact with the sample, the ac voltage excites locally the piezoelectric film at the tip position into vibrations due to the inverse piezoelectric effect. This in turn excites the cantilever into flexural vibrations which are detected by the photo-diode of the AFM.

the ac frequency of excitation is close to a contact resonance entailing an enhanced signal through resonance amplification by the Q-value of the cantilever vibration, typically about 100. The frequency of the first contact resonance was chosen, which is located between the frequencies of the first and the second flexural modes of the clamped-free cantilever. All samples were carefully grounded to avoid electrostatic charging.

The output signal of the photo-diode of the AFM representing the cantilever vibrations is fed to a heterodyne down-converter, which in addition receives a reference signal at the excitation frequency from the frequency generator. It shifts the signal to a fixed 20 kHz intermediate frequency. For imaging, a fixed excitation frequency is selected and the down-converted signal is evaluated by a lock-in amplifier whose output is fed into an auxiliary channel of the AFM as imaging quantity while the sample surface is scanned. A topography image of the surface is acquired simultaneously with the piezo-mode image.

2. Experimental section

2.1 Development of the combinatorial workflow

The first milestone was the development of a high-throughput method to find new piezoelectric thin films by using a non-destructive characterization tool [47]. For the automated preparation of the thin film libraries the sol–gel technique was selected, because the automation of the film preparation with a handling robot can easily be realized. In addition, the easy accessibility of a lot of elements and the good mixing of single elements on an atomic level speak for the sol–gel method. The well-investigated lead zirconate titanate $Pb(Zr,Ti)O_3$ (PZT) was chosen as starting and reference material to develop and test the high-throughput workflow. A PZT sol–gel-synthesis from Yi *et al.* [48] was improved to produce PZT sols in a combinatorial way by using a pipetting robot.

First, several validation experiments had been carried out before the high-throughput experiments were started. The optimization of the recipe leads to precursor solutions which are stable over the time of the library preparation. Furthermore, the produced sols can be broadly varied in composition and doping additives. For the production of the different sols and the application onto the substrate, a pipetting robot (Perkin Elmer MultiPROBE II) was used. Several sols were premixed in vials, which were then used for deposition on the library plate. As a substrate for the thin film libraries, a specially pre-structured 6" silicon wafer with a Pt/Ir coating [29] was used. The wafer supplied a 9×9 array of circular grooves of distinct depth. The individual thin-film samples are generated by depositing the sols onto the circular grooves.

The design of the library was accomplished by using special in-house developed software [49]. This software calculated the individual recipe based on the desired parameter variation and generated an optimized pipetting list, which was transferred as a software command directly to the liquid handling robot.

Seventeen undoped sols $Pb(Zr,Ti)O_3$ (PZT) of different compositions were produced by varying the ratio of lead to zirconium to titanium as piezoelectric reference materials. Thereby only such compositions were chosen which map PZT-phases associated with piezoelectric properties. Furthermore, 33 sols with a constant ratio of Pb : Zr : Ti ($Pb_{1.1}Zr_{0.58}Ti_{0.42}$), but doped with different elements were prepared. The composition $Pb_{1.1}Zr_{0.58}Ti_{0.42}$ was selected for doping, since it belongs to the well known piezoelectric phases in the PZT-phase diagram [46].

For the production of the sols precursor solutions of lead acetate, $Pb(CH_3CO_2)_2 \cdot H_2O$, zirconium propoxide, $Zr(C_3H_7O)_4$, and titanium isopropoxide, $Ti[(CH_3)_2CHO]_4$ with a concentration of 0.5 M were used. The precursors were dissolved in hot acetic acid, CH_3COOH. The reagents were pipetted by the robot into 2-ml sampling flasks. The sols with the composition $Pb_{1.1}Zr_{0.58}Ti_{0.42}$ were doped with 6 mol % of one of the doping elements which are shown in Tables 1 and 2.

Afterwards the sols were shaken for 5 min. Then the robot pipetted 1 μl sol onto a circular sector of the library plate. After the first sequence of depositions the coated wafer was calcined at 500 °C for 5 min. This procedure was repeated for a second time. Finally the library was slowly heated to 500 °C and kept at this temperature for 2 h.

For the analysis of the prepared films the AFM ultrasonic piezo-mode technique as a non-destructive characterization tool was used. The atomic force microscope Dimension 3000 tool "programmed move" (Dimension 3000, Soft-

Table 1. Doping solutions 0.25 M, solvent CH_3COOH.

Compound	manufacturer
$Ba(CH_3COO)_2$	Fluka
$K(CH_3COO)$	unknown
$Li(CH_3COO) \cdot 2H_2O$	Aldrich
$Mg(CH_3COO)_2 \cdot 4H_2O$	Merck
$Ni(CH_3COO)_2 \cdot 4H_2O$	Aldrich
$Na(CH_3COO) \cdot 3H_2O$	Janssen
$Sr(CH_3COO)_3$	Aldrich

Table 2. Doping solutions 0.25 M, solvent CH_3OH.

Compound	manufacturer
$Bi(NO_3)_3 \cdot 5H_2O$	Fluka
$Cr(NO_3)_3 \cdot 9H_2O$	Aldrich
$Ce(NO_3)_3 \cdot 6H_2O$	Merck
$Co(NO_3)_2 \cdot 6H_2O$	Fluka
$Ca(NO_3)_2 \cdot 4H_2O$	Merck
$Dy(NO_3)_3 \cdot 5H_2O$	Aldrich
$Er(NO_3)_3 \cdot 5H_2O$	Aldrich
$Eu(NO_3)_3 \cdot 6H_2O$	Strem Chemicals
$Fe(NO_3)_3 \cdot 9H_2O$	Merck
$Ga(NO_3)_3 \cdot H_2O$	Aldrich
$Gd(NO_3)_3 \cdot 6H_2O$	Aldrich
$La(NO_3)_3 \cdot 6H_2O$	Merck
$Nd(NO_3)_3 \cdot 6H_2O$	Aldrich
$Pr(NO_3)_3 \cdot 6H_2O$	Heraeus
$RbNO_3$	Fluka
$Sm(NO_3)_3 \cdot 6H_2O$	Aldrich
$AgNO_3$	ABCR
TeO_2	B Plank
$V(acac)_3$	unknown
$VO(acac)_2$	Acros
$Y(NO_3)_3 \cdot 6H_2O$	ABCR
$Yb(NO_3)_3 \cdot x\, H_2O$	ChemPur

ware Version 5.12rb, Digital Instruments, Santa Barbara, CA) was employed allowing to take AFM images automatically of each of the 50 samples of the library. Before starting the program, all settings for the positioning of the cantilever and for imaging such as image size, scan velocity, frequency of excitation for the ultrasonic piezo-mode *etc.* are defined by the user. The piezo-mode images of all samples are carried out automatically in sequence.

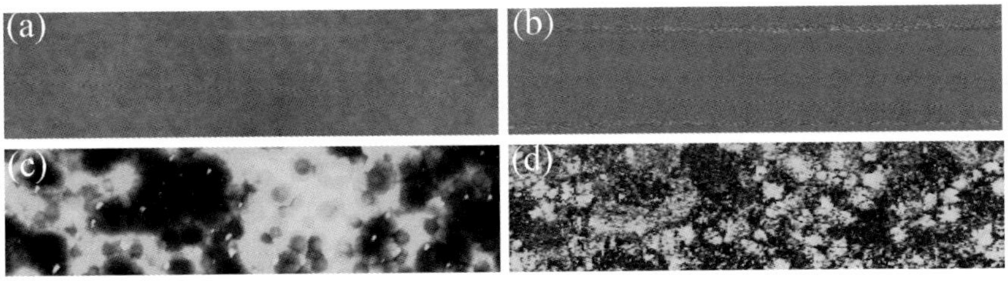

Fig. 3. (a) and (c) topography images of two different samples doped with Bi and Sr, respectively. The image size is $10 \times 2.5\,\mu m^2$. The grey scale covers 40 nm difference in height. Images (b) and (d) are the corresponding piezo-mode images of the same surface areas as shown in (a) and (c). The amplitude range of the ultrasonic piezo-mode images is (b) 2 V and (d) 1 V.

By starting the program the cantilever moves to the first sample, goes into contact and scans the predefined area. After having finished an entire image, the topography and the piezo-mode images measured in parallel are saved, and the cantilever moves to the next sample, repeating the measuring procedure. For the measurements presented here AFM-cantilevers from nanosensors, NanoWorld AG, Neuchatel, Switzerland (type EFM, coated with Pt-Ir layer) were used. The first contact resonance frequency of this type of cantilevers was found to lie typically between 200 and 400 kHz for ceramic samples.

Figure 3a and c show examples of topography images of two different samples and Fig. 3b and d show the corresponding ultrasonic piezo-mode images. In Fig. 3b there is no contrast in the ultrasonic piezo-mode image. Figure 3d shows dark and bright areas caused by high and low local vibration amplitudes, respectively. This variation, in contrast to the ultrasonic piezo-mode images refers to ferroelectric domains. By comparing the topography images with the ultrasonic piezo-mode images, a strong variation in the contrast can be seen.

All undoped 17 reference samples showed ferroelectric domains. Furthermore, samples with the following doping elements showed also a contrast in the piezo-mode images: Ba, Sr, Ce, Li, Na, K, Co, Cu, La, Rb, Ag, and Te. This contrast indicates the presence of piezoelectricity.

To check the results of the piezo-mode images impedance measurements were employed. Schröter *et al.* developed a fully automated modular unit for electric characterization [50]. This unit consists of a measurement head, an x-y-z table and an impedance analyzer. The first test to measure the existing library did not succeed, because tiny dust inclusions induced short circuits. To do these sensitive electric measurements the library had to be produced again under clean room conditions in order to avoid such dust inclusions. Several samples of the newly prepared library could be examined by impedance spectroscopy yielding ε. For the PZT coating doped with strontium an ε-value of 194.36 was obtained which is typical for perovskites. These results corroborate the ultrasonic AFM piezo-mode measurements.

Table 3. Precursor solutions 0.2 M

Compound	manufacturer
$Al(C_5H_7O_2)_3$	Aldrich
$Ba(CH_3CO_2)_2$	Fluka
$Bi(CH_3CO_2)_3$	Sigma-Aldrich
$Co(CH_3CO_2)_2 \cdot 4H_2O$	Lancaster
$Fe(C_5H_7O_2)_3$	Acros
$La(C_5H_7O_2)_3 \cdot x\, H_2O$	Aldrich
$La(NO)_3 \cdot x\, H_2O$	Merck
$Mg(CH_3CO_2)_2 \cdot 4H_2O$	Aldrich
$Mg(NO_3)_2$	Merck
$Mn(CH_3CO_2)_3 \cdot 2H_2O$	Merck
$Nd(CH_3CO_2)_3 \cdot x\, H_2O$	Sigma-Aldrich
$Ni(CH_3COO)_2 \cdot 4H_2O$	Aldrich
$Pb(CH_3CO_2)_2 \cdot Pb(OH)_2$	Adrich
$Sb(CH_3CO_2)_3$	Aldrich
$Sb(OC_3H_7)_3$	Alfa Aesar
$Sn(OC_3H_7)_4$	Alfa Aesar
$Sr(CH_3CO_2)_2$	Aldrich
$Ti(OC_4H_9)_4$	Merck
$Y(NO_3)_3 \cdot 6H_2O$	ABCR
$Y(OC_3H_7)_3$	Lancaster
$Zn(NO_3)_2 \cdot 6H_2O$	unknown
$Yb(NO_3)_3 \cdot 5H_2O$	Chempur
$V(OC_3H_7)_3$	unknown
$VO(C_5H_7O_2)_2$	Acros
$Mo(OC_3H_7)_5$	Alfa Aesar
$Mo(C_5H_7O_2)_5$	unknown
W-ethylacetate[1]	self-made
$W(OC_3H_7)_6$	Alfa Aesar
$Ca(CH_3CO_2)_2 \cdot H_2O$	Sigma-Aldrich
$Ca(NO_3)_2 \cdot 4H_2O$	Merck

[1] J. P. Cronin, D. J. Tarico, A. Agrawal, R. L. Zhang, United States Patent Number 5,277,986 (1994).

2.2 Searching for new piezoelectric materials

After the high-throughput workflow was developed and successfully tested, the search for new piezoelectric materials could be started. To create highly diverse coating libraries, a sol–gel-synthesis had to be found, which is tolerant for a large pool of elements. An acidic sol–gel-synthesis developed by Jéhanno *et al.* [51] fulfilled this requirement.

We started with a binary (50 : 50) mixed oxide composition spread. Thirty different elements were chosen for this composition spread. These elements are shown in Table 3. The different precursors were dissolved in a 1 : 1-mixture

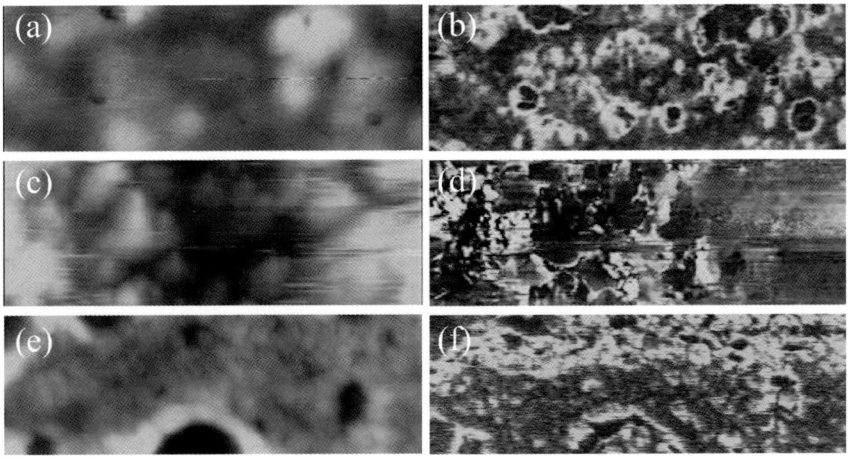

Fig. 4. Topography images of three different samples: (a) $Co_{50}Fe_{50}$, (c) $Bi_{50}Zn_{50}$, and (e) $Al_{50}Ba_{50}$. The image size is $2 \times 0.5\,\mu m^2$. The grey scale covers (a) 30 nm, (b) 80 nm, and (c) 60 nm in height. The ultrasonic piezo-mode images (b), (d), and (f) are taken at the same surface areas as shown in (a), (c), and (e). The amplitude range of the ultrasonic piezo-mode images is (b) 7 V, (d) 7 V, and (f) 4 V.

of propionic acid and isopropanol. These mixtures were stirred and heated at 80 °C for 30 min.

The 50 : 50 sols were premixed in the wells of microtiter plates followed by the application of the sols onto the areas of the wafer. The library was heated on a hot plate at 50 °C for 5 min. This procedure of coating and heating was repeated 5 times for each wafer. Finally, the libraries were calcined at 500 °C for 1 h. Four libraries with 325 coatings were produced. The films on these libraries could not be measured in an automated way, because the employed AFM instrument did not allow the automated measurement of the contact-resonance frequency.

The undoped and doped PZT films of the test system consisted of samples differing only slightly in the ratio of the compounds or in doping. As all test samples were based on the same matrix material the contact-resonance frequencies of the different coatings were very similar. The same frequency could be set for the measurement of the whole test library. In contrast, all films on the new libraries were of different materials having also different contact resonance frequencies varying from 302 kHz to 422 kHz. Therefore all coatings were investigated one by one.

The next step was the preparation of composition spread libraries of the materials, which show piezoelectric properties and the study of these libraries by AFM ultrasonic piezo-mode. Until now, the following mixed oxide samples with the following compositions showed a contrast in the ultrasonic piezo-mode images: $Co_{50}Fe_{50}$, $Bi_{50}V_{50}$, and $Bi_{50}Zn_{50}$.

Figure 4a, c, and e show examples of topography images of three different samples. Figure 4b, d, and f show the corresponding ultrasonic piezo-mode images. The variation in contrast, which can be seen in the ultrasonic piezo-mode images, is again caused by the different local vibration amplitudes. This indicates ferroelectric domains. Some topography effects are reflected in the piezo-mode image if the roughness of the coating is too high, but a clear difference between the surface-topography and the contrast in the piezo-mode images can still be seen. The biggest disadvantage of the AFM piezo-mode method as high-throughput technique is the lack of quantification of the measurement results. This is caused by the fact that the amplitude of the cantilever signal depends on many parameters rendering presently a comparison from film to film difficult [47], so that a validation of measured data needs further efforts. A calibration of the measurement method is required allowing to rank the data of the piezo-mode measurements.

3. Conclusion

A combinatorial workflow to produce mixed oxide thin-film libraries and to determine their piezoelectric properties has been developed. It was shown that the AFM ultrasonic piezo-mode technique allows the determination of piezoelectric properties in the thin films. Several problems concerning the automation of the AFM ultrasonic piezo-mode technique as screening method and the calibration of this technique have still to be solved.

Other methods like the impedance spectroscopy have other disadvantages such as the high demand on the quality of the films. The ultrasonic piezo-mode images of the films are also influenced by their topography, but to a lesser extent than electric measurements.

Acknowledgement

This work was supported by grants from the German Science Foundation through the "SFB 277" as well as the "Graduate College 232 "Materials for efficient energy use". The authors thank B. Wessler from Siemens AG for the generous supply of structured wafers and C. Schroeter from Siemens AG for the impedance spectroscopy measurements.

References

1. S. Borman, Chem. Eng. News **77** (1999) 33.
2. E. M. Gordon, R. W. Barrett, W. J. Dower, S. P. A. Fodor, and M. A. Gallop, J. Med. Chem. **37** (1994) 1385.
3. R. Schlögl, Chemie Heute (1999/2000) 26.
4. W. F. Maier, Chemie Heute (2000/2001) 114.
5. W. F. Maier, Angew. Chem. **111** (1999) 1294.

6. J. M. Newsam, *Design of Catalysts and Catalysts Libraries.* In *Combinatorial Catalysis and High Throughput Catalyst Design Testing.* E. G. Derounane (Ed.), Kluwer Academic, Dordrecht (1999).
7. M. Orschel, J. Klein, H.-W. Schmidt, and W. F. Maier, Angew. Chem. **111** (1999) 2961.
8. M.-S. Schiedel, C. A. Briehn, and P. Bäuerle, Angew. Chem. **113** (2001) 4813.
9. O. Deeg, P. Kirsch, D. Pauluth, and P. Bäuerle, Chem. Commun. (2002) 2762.
10. D. F. Acevedo, M. C. Miras, and C. A. Barbero, J. Combi. Chem. **7** (2005) 513.
11. M. Jimenez, S. Duquesne, and S. Bourbigot, Fr. Ind. & Eng. Chem. Res. **45** (2006) 7475.
12. A. Persidis, *Economics of Combinatorial Chemistry and Combinatorial Technologies.* In *Combinatorial Chemistry and Technology.* S. Miertus and G. Fassina (Eds.), Marcel Dekker, Inc., New York (1999).
13. W. Schrof, S. Lehmann, J. Hadeler, G. Oetter, G. Dralle-Voss, E. Beck, W. Paulus, and S. Bentz, Athen's Conf. Coat.: Sci. Technol., Proc., 27th, Athen (2001).
14. R. Iden, W. Schrof, J. Hadeler, and S. Lehmann, Macromol. Rapid Commun. **24** (2003) 63.
15. T. F. Jaramillo, S.-H. Baeck, A. Kleinman-Shwarsctein, K.-S. Choi. G. D. Stucky, and E. McFarland, J. Comb. Chem. **7** (2005) 264.
16. B. J. Chisholm, R. A. Potyrailo, J. N. Cawse, R. E. Shaffer, M. Brennan, C. Molaison, D. Whisenhunt, B. Flanagan, D. Olson, J. Akhave, D. Saunders, A. Mehrabi, and M. Licon, Prog. Org. Coat. **45** (2002) 313.
17. B. J. Chisholm, R. A. Potyrailo, J. N. Cawse, M. Brennan, and C. Molaison, Athens Conf. Coat.: Sci. Technol., Proc. 26th, (2000) 23.
18. B. J. Chisholm, R. A. Potyrailo, J. N. Cawse, R. E Shaffer, M. Brennan, C. Molaison, D. Whisenhunt, and D. Olson, Proc. Int. Waterborne, High Solids, Powder Coat. Symp., (2002) 151.
19. J. N. Cawse, D. Olson, B. J. Chisholm, M. Brennan, T. Sun, W. Flanagan, D. Olson, J. Akhave, A. Mehrabi, and D. Saunders, Prog. Org. Coat. **47** (2003) 112.
20. H. Bach, C. A. Gambino, L. K. Gindin, R. M. Konitnsey, P. D. Lunney, and K. G. Spitler, Polym. Prepr. (Am. Chem. Soc., Div. Polym. Chem.) **44** (2003) 70.
21. H. Bach and C. Guertler, PCT Paten Appl. WO 02/42745 A2 20020530 (2002).
22. H. Bach, C. Gurtler, and S. Nowak, Eur. Coat. J. **3** (2004) 22.
23. D. A. Wicks and H. Bach, Proc. Int. Waterborne, High-Solids, and Powder Coatings Symposium, 29th (2002), 1.
24. R. B. Van Dover, L. F. Schneemeyer, and R. M. Fleming, Nature **392** (1998) 162.
25. Y. Matsumoto, M. Murakami, T. Shono, T. Hasegawa, T. Fukumura, M. Kawasaki, P. Ahmet, T. Chikyow, S.-Y. Koshihara, and H. Koinuma, Science **291** (2001) 854.
26. K. Itaka, M. Yamashiro, J. Yamaguchi, S. Yaginuma, M. Haemori, and H. Koinuma, Appl. Surf. Sci. **252** (2006) 2562.
27. L. A. Bendersky and I. Takeuchi, Mocromol. Rapid Commun. **25** (2004) 675.
28. A. Ludwig, Appl. Surf. Sci. **223** (2004) 78.
29. B. Wessler, V. Jéhanno, W. Rossner, and W. F. Maier, Appl. Surf. Sci. **223** (2004) 30.
30. R. Cremer and D. Neuschütz, Surf. Coat. Techn. **146–147** (2001) 229.
31. A. P. Smith, J. F. Douglas, J. C. Meredith, E. J. Amis, and A. Karim, J. Polym. Sci., Part B: Polym. Phys. **39** (2001) 2141.
32. J.-B. D. Green and G. U. Lee, Langmuir **16** (2000) 4009.
33. B. Du, J. Zhang, Q. Zhang, D. Yang, T. He, and O. K. C. Tsui, Macromolecules **33** (2000) 7521.
34. R. Neffati, A. Alexeev, S. Saunin, J. C. M. Brokken-Zijp, D. Wouters, S. Schmatloch, U. S. Schubert, and J. Loos, Macromol. Rapid Commun. **24** (2003) 113.
35. N. A. Adams, B.-J. De Gans, D. Kozodaev, C. Sánchez, C. W. M. Bastiaansen, D. J. Broer, and U. S. Schubert, J. Comb. Chem. **8** (2006) 184.

36. J.-L. Sormana and J. C. Meredith, Macromol. Rapid Commun. **24** (2003) 118.
37. P. Güthner and K. Dransfeld, Appl. Phys. Lett. **61** (1992) 1137.
38. S. V. Kalinin and S. A. Bonnell, Phys. Rev. B **65** (2002) 125408.
39. A. Gruverman, O. Auciello, and H. Tokumoto, Integr. Ferroelectr. **19** (1998) 49.
40. K. Franke and M. Weihnacht, Ferroelectr. Lett. **19** (1995) 25.
41. K. Franke, Ferroelectr. Lett. **19** (1995) 35.
42. L. M. Eng, H.-J. Güntherodt, G. A. Schneider, U. Köpke, and J. Muñoz-Saldaña, Appl. Phys. Lett. **74** (1999) 233.
43. J. Muñoz-Saldaña, M. J. Hoffmann, and G. A. Schneider, J. Mater. Res. **18** (2003) 1777.
44. C. Durkan, D. P. Chu, P. Migliorato, and M. E. Welland, Appl. Phys. Lett. **76** (2000) 366.
45. S. Dunn and R. W. Whatmore, J. Europ. Ceram. Soc. **22** (2002) 825.
46. U. Rabe, M. Kopycinska, S. Hirsekorn, J. Muñoz Saldaña, G. A. Schneider, and W. Arnold, J. Phys. D: Appl. Phys. **35** (2002) 2621.
47. D. Rende, K. Schwarz, U. Rabe, W. F. Maier, and W. Arnold, Progr. Solid State Chem. **35** (2007) 361.
48. G. Yi, Z. Wu, and M. Sayer, J. Appl. Phys. **64** (1988) 2717.
49. J. Scheidtmann, J. Saalfrank, and W. F. Maier, Stud. Surf. Sci. Catal. **145** (2003) 13.
50. C. Schroeter, B. Wessler, A. Schoenecker, U. Keitel, and L. M. Eng, J. Appl. Phys. **100** (2006) 114114.
51. V. Jéhanno, B. Wessler, W. Rossner, G. Frenzer, and W. F. Maier, *Preparation of dielectric thin-film libraries by sol–gel Techniques*. In *Combinatorial and High-Throughput Discovery and Optimization of Catalysts and Materials*. R. A. Potyrailo and W. F. Maier (Eds.), CRC, Boca Raton, New York, Oxon (2007).

Keywords

Zeitschrift für Physikalische Chemie

International journal of research in physical chemistry and chemical physics

Founded 1887 by W. Ostwald and J. H. van't Hoff, continued by K. F. Bonhoeffer

Professor Dr. H. Baumgärtel	Institut für Physikalische u. Theoretische Chemie, Freie Universität Takustraße 3, D-14195 Berlin
Professor Dr. U. K. Deiters	Institut für Physikalische Chemie, Universität Luxemburger Straße 116, D-50939 Köln
Professorin Dr. K. Kohse-Höinghaus	Physikalische Chemie I, Universität Universitätsstraße 25, D-33615 Bielefeld
Professor Dr. D. M. Kolb	Abteilung für Elektrochemie, Universität Ulm Albert-Einstein-Allee 47, D-89081 Ulm
Professor Dr. J. Troe	Institut für Physikalische Chemie der Universität Tammannstraße 6, D-37077 Göttingen
Professor Dr. K.-M. Weitzel	Institut für Physikalische Chemie der Universität Postfach, D-35032 Marburg
Professor Dr. H.-J. Werner	Universität Stuttgart, Institut für Theoretische Chemie Pfaffenwaldring 55, D-70569 Stuttgart
Professor Dr. R. Winter	Physikalische Chemie I, Universität Otto-Hahn-Straße 6, D-44227 Dortmund
Professor Dr. J. Wintterlin	Department Chemie, Ludwig-Maximilians-Universität Butenandtstraße 11, Haus E, D-81377 München

Contributions should, preferably, be sent to the Editor-in-Chief, or to a member of the Editorial Board who is familiar with the research reported.

See "Instructions for Authors" 3rd cover page !

Z. Phys. Chem.

www: z-phys-chem.de